Tutorials in
Metamaterials

Tutorials in Metamaterials

Edited by
Mikhail A. Noginov
Viktor A. Podolskiy

CRC Press
Taylor & Francis Group
Boca Raton London New York

CRC Press is an imprint of the
Taylor & Francis Group, an **informa** business

CRC Press
Taylor & Francis Group
6000 Broken Sound Parkway NW, Suite 300
Boca Raton, FL 33487-2742

© 2012 by Taylor & Francis Group, LLC
CRC Press is an imprint of Taylor & Francis Group, an Informa business

No claim to original U.S. Government works

Printed in the United States of America on acid-free paper
Version Date: 2011919

International Standard Book Number: 978-1-4200-9218-9 (Hardback)

Visit the Taylor & Francis Web site at
http://www.taylorandfrancis.com

and the CRC Press Web site at
http://www.crcpress.com

Contents

9 Spatial Dispersion and Effective Constitutive Parameters of
   Electromagnetic Metamaterials
   Chris Fietz and Gennady Shvets

Preface

Metamaterials—composite media with unusual optical properties—have revolutionized the landscape of optical science and engineering over the past decades. Metamaterials have transformed science-fiction-like concepts of superresolution imaging and optical cloaking to the realm of science laboratories, and further promise to transform these into the realm of our everyday life.

The new era of optical metamaterials calls for the development of experimental and theoretical methods capable of analyzing optical behavior on the multitude of scales, starting from the nanometer scale of individual inhomogeneity, and moving on to the micrometer of the metamaterial, and to an even larger scale of the metamaterials-based device. Future progress in the areas of photonics, plasmonics, and metamaterials critically depends on our ability to answer this call.

This book is a collection of self-contained tutorials describing metamaterial photonics aimed at upper undergraduates and graduate students, as well as experts in physics and engineering who are willing to familiarize themselves with the state of the art in the metamaterials research. It starts with the most general reviews and progresses to more specialized topics.

The book is organized as follows:

In Chapter 1, N. M. Litchinitser, I. R. Gabitov, A. I. Maimistov, and V. M. Shalaev review linear and nonlinear properties of photonic metamaterials and their potential applications, which include negative index metamaterials and magnetic metamaterials as well as transformation optics.

In Chapter 2, A. Boltasseva discusses advantages, drawbacks, and challenges of a broad spectrum of fabrication techniques enabling photonic metamaterials, ranging from electron-beam lithography, focused ion beam milling, and nanoimprint lithography to direct laser writing.

In Chapter 3, A. Alù and N. Engheta describe the recent achievements in microwave metamaterial research, emphasizing their unusual guiding, scattering, and emission properties.

In Chapter 4, D. Tan, K. Ikeda, and Y. Fainman describe the novel applications of dielectric metamaterials for light guiding, steering, and refraction.

In Chapter 5, M. A. Noginov describes the efforts to compensate and eliminate one of the main limitations of metamaterials, optical loss, by introducing optical gain into the metamaterial matrix.

In Chapter 6, V. A. Podolskiy describes the response of uniaxial media, emphasizing the unusual properties of strongly anisotropic, hyperbolic, metamaterials for subwavelength light confinement, guiding, and imaging.

Chapter 7, by L. Alexeyev and E. Narimanov, focuses on another unique topic of hyperbolic media: its ability to strongly modify the radiative decay of nearby atoms and molecules.

In Chapter 8, S. Linden and M. Wegener describe light propagation and refraction in chiral and bianisotropic metamaterials. Several motifs leading to chiral and bianisotropic response are discussed, and the relationship between the geometry of building blocks and effective response of the metamaterial is outlined.

Finally, in Chapter 9, C. Fietz and G. Shvets address the fundamental issues of spatial dispersion in metamaterials, and use the developed formalism to extract effective properties of metamaterials from experimental measurements.

About the Editors

Prof. Mikhail A. Noginov graduated from Moscow Institute for Physics and Technology with a Master of Science degree in electronics and automatics in 1985. In 1990 he received a Ph.D. degree in physical–mathematical sciences from the General Physics Institute of the USSR Academy of Sciences, Moscow.

Dr. Noginov has been affiliated with the General Physics Institute of the USSR Academy of Sciences as junior staff research scientist, then staff research scientist (1985–1991); Massachusetts Institute of Technology, Cambridge; Center for Materials Science and Engineering as a postdoctoral research associate (1991–1993); Alabama A&M University, Huntsville, as assistant research professor, then associate research professor (1993–1997); and the Department of Physics, Center for Materials Research, Norfolk State University (NSU), Norfolk, Virginia, as research associate professor, then assistant professor, and associate professor; currently, he is a professor (1997 to the present). In 2010, Dr. Noginov was named NSU Eminent Scholar 2010–2011.

Dr. Noginov has published two books, five book chapters, over 100 papers in peer reviewed journals, and over 100 publications in proceedings of professional societies and conference technical digests (many of them invited). He is a member of Sigma Xi, OSA, SPIE, and APS, and has served as a chair and a committee member of several conferences of SPIE and OSA. He regularly serves on NSF panels and reviews papers for many professional journals. Since 2003, Dr. Noginov has been a faculty advisor of the OSA student chapter at NSU. His research interests include metamaterials, nanoplasmonics, random lasers, solid-state laser materials, and nonlinear optics.

Prof. Viktor Podolskiy earned his B.S. degree in applied mathematics and physics from Moscow Institute for Physics and Technology in 1998, followed by an M.S. in computer science and a Ph.D. in physics from New Mexico State University in 2001 and 2002, respectively. Upon completion of the Ph.D. program, Dr. Podolskiy joined the Electrical Engineering Department, Princeton University, as postdoctoral

research associate. Between September 2004 and December 2009, Dr. Podolskiy worked in the physics department of Oregon State University as assistant professor, and later as associate professor. Since January 2010, Dr. Podolskiy has been an associate professor in the Department of Physics and Applied Physics, and a member of Photonics Center at the University of Massachusetts–Lowell.

Dr. Podolskiy's research is focused on theory and modeling of optical properties of nano- and microstructured composites, metamaterials, and plasmonic systems. He has presented over 25 invited talks, and coauthored over 50 peer-reviewed publications, and over 60 conference proceedings, and holds three U.S. patents.

Contributors

Leonid Alekseyev
Department of Electrical Engineering
Princeton University
Princeton, New Jersey

Andrea Alù
Department of Electrical and Computer
    Engineering
University of Texas at Austin
Austin, Texas

Alexandra Boltasseva
Purdue University
Birck Nanotechnology Center
West Lafayette, Indiana

Nader Engheta
Department of Electrical and Systems
    Engineering
School of Engineering and Applied
    Science
University of Pennsylvania
Philadelphia, Pennsylvania

Yeshaiahu Fainman
Ultrafast Nanoscale Optics Group
Department of Electrical and Computer
    Engineering
University of California, San Diego
La Jolla, California

Chris Fietz
Department of Physics
University of Texas at Austin
Austin, Texas

Ildar Gabitov
Department of Mathematics
The University of Arizona
Tucson, Arizona

Kazuhiro Ikeda
Ultrafast Nanoscale Optics Group
Department of Electrical and Computer
    Engineering
University of California, San Diego
La Jolla, California

Stefan Linden
Karlsruhe Institute of Technology (KIT)
Institute of Applied Physics
Karlsruhe, Germany

Natalia Litchinitser
Department of Electrical Engineering
University at Buffalo
The State University of New York
Buffalo, New York

Andrei Maimistov
Moscow State Engineering Physics
   Institute
Department of Solid State Physics
Moscow, Russia

Evgenii Narimanov
Purdue University
Birck Nanotechnology Center
West Lafayette, Indiana

Mikhail Noginov
Department of Physics
Center for Materials Research
Norfolk State University
Norfolk, Virginia

Viktor Podolskiy
Department of Physics and Applied
   Physics
University of Massachusetts at Lowell
Lowell, Massachusetts

Vladimir Shalaev
Purdue University
Birck Nanotechnology Center
West Lafayette, Indiana

Gennady Shvets
Department of Physics
University of Texas at Austin
Austin, Texas

Dawn Tan
Ultrafast Nanoscale Optics Group
Department of Electrical and Computer
   Engineering
University of California, San Diego
La Jolla, California

Martin Wegener
Karlsruhe Institute of Technology (KIT)
Institute of Applied Physics
Karlsruhe, Germany

1

Linear and Nonlinear Metamaterials and Transformation Optics

Natalia M. Litchinitser
Department of Electrical Engineering, The State University of New York at Buffalo, Buffalo, New York

Ildar R. Gabitov
Department of Mathematics, University of Arizona, Tucson, Arizona

Andrei I. Maimistov
Department of Solid State Physics, Moscow Engineering Physics Institute, Moscow, Russia

Vladimir M. Shalaev
School of Electrical and Computer Engineering and Birck Nanotechnology Center, Purdue University, West Lafayette, Indiana

1.1 Introduction

Albert Einstein said once: "The important thing is not to stop questioning." Indeed, it is the scientific curiosity of a Russian physicist, Victor Veselago, in 1968 that led to the emergence of an entirely new area of modern optics: the optics of metamaterials [1–3]. In optics, interactions between light waves and materials are usually characterized by two parameters—dielectric permittivity ε and magnetic permeability μ—that explicitly enter Maxwell's equations, or by their product, the index of refraction, defined as $n = \pm\sqrt{\varepsilon\mu}$. In common transparent optical materials, the index of refraction and dielectric permittivity are positive numbers that are greater or equal to 1 and $\mu \approx 1$. The refractive index can be modified to some degree by altering the chemical composition of the material or using electrical, thermal, or nonlinear effects. Nevertheless, the refractive index is typically greater than one (air) and less than four (silicon). Veselago investigated in detail the question of whether or not ε and μ can simultaneously take negative values, leading to a negative index of refraction. It is noteworthy that the discussions of backward waves and negative index of refraction date back to the beginning of the twentieth century (see, e.g., [4]); nowadays, we can generalize Veselago's question and ask whether one could engineer ε and μ (or n) to take any value beyond the limits imposed by nature. *Meta* in Greek means "beyond," so the ultimate goal of metamaterials research is to create materials with properties and functionalities that have not been found in nature.

Metamaterials technology offers unique opportunities for "engineering" refractive indices that were not previously accessible and for gaining control over the spatial refractive index distribution. Negative index of refraction, antiparallel phase and energy velocities, magnetism in optics, backward phase-matched nonlinear wave interactions, and cloaking are just few peculiar, and perhaps counterintuitive at first glance, phenomena enabled by these artificial structures. Such unprecedented freedom for designing the materials' properties is made possible by the metamaterials' constituent components: meta-atoms.

Optical meta-atoms are engineered structures that have characteristic dimensions less than the wavelength of light. Their electric and magnetic properties can be carefully designed and tuned by changing the geometry, size, and other characteristics of meta-atoms. As a result, entirely new kinds of optical materials that have not been found in nature emerged, including negative-index metamaterials (NIMs), so-called near-zero index materials, as well as magnetic positive-index materials (PIMs) with nearly any ε and μ. The first optical NIMs, demonstrated in 2005 at 1500 nm, are shown in Figure 1.1a,b [5]. Since then, significant progress has been made in producing NIMs in the visible wavelength range. Recently, a material with a negative index at a wavelength as short as 580 nm has been reported [6]. Although many unusual physical properties can already be demonstrated in two-dimensional metamaterials, many practical applications require bulk structures. Recently, considerable progress has been made in the area of fabrication of three-dimensional metamaterials at optical frequencies [7–10]. Several examples of fabricated structures are shown in Figures 1.1c–g.

In this chapter, we review the linear and nonlinear properties of photonic metamaterials and their potential applications. Section 1.2 focuses on unique properties of magnetic and negative-index materials and their experimental realization at optical frequencies. In Section 1.3, we discuss recent theoretical and experimental progress in nonlinear

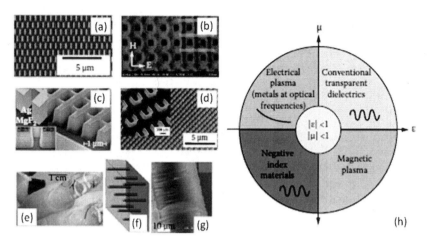

FIGURE 1.1 Metamaterials. (a) One of the first optical NIMs consisting of paired gold nano-rods and having a negative refractive index at 1500. (From V. M. Shalaev et al., *Opt. Lett.* 30, 3356, 2005.) (b) Fishnet NIM structure having negative refractive index at 580 nm. (From S. Xiao et al., Yellow-light negative-index metamaterials, arXiv:0907.1870.) (c) Diagram and SEM image of the 21-layer fishnet structure consisting of alternating layers of silver (Ag) and magnesium fluoride (MgF_2). (Adapted from J. Valentine et al., *Nature* 455, 376, 2008. With permission.) (d) Field-emission scanning electron microscopy images of the four-layer SRR structure. (From N. Liu et al., *Nat. Mater.* 7, 31, 2008. With permission.) (e)–(g) Silver-filled membrane, schematic of silver nanowires in an alumina membrane, and SEM picture of the etched side wall of the membrane. (From M. A. Noginov et al., *Appl. Phys. Lett.* 94, 151105, 2009.) (h) Dielectric permittivity ε and magnetic permeability μ diagram.

optics of negative-index and magnetic metamaterials. New and exciting developments in the field of gradient index and transformation optics are reviewed in Section 1.4. Finally, Section 1.5 outlines existing challenges and future directions in the field.

1.2 Magnetic and Negative-Index Metamaterials at Optical Frequencies

Figure 1.1h illustrates various combinations of material parameters (dielectric permittivity and magnetic permeability) and the corresponding index of refraction. In this diagram, most conventional optical materials belong to the region where $\varepsilon > 0$ and $\mu > 0$ corresponding to $n > 0$, where propagating waves are allowed. The second ($\varepsilon < 0$ and $\mu > 0$) and the fourth ($\varepsilon > 0$ and $\mu < 0$) quarters of the diagram correspond to opaque materials that cannot support any propagating waves. Finally, the materials with parameters corresponding to the third quarter of the diagram ($\varepsilon < 0$ and $\mu < 0$) are transparent and allow wave propagation, but their refractive indices are negative. The NIMs and magnetic PIMs ($\mu \neq 1$) in optics became possible only with the emergence of metamaterials. Another important range of parameters corresponds to $|\varepsilon| < 1$, $|\mu| < 1$. Metamaterials with such ε and μ have been shown to be useful for designing cloaking applications and near-zero refractive index waveguides.

As discussed in the Introduction, a majority of naturally existing optical materials are nonmagnetic. However, it has been recognized for some time that magnetism at optical frequencies may lead to new fundamental physics and novel applications.

1.2.1 NIMs Design

One of the most remarkable new classes of materials enabled by bringing magnetism to an optical wavelength range is NIMs. In NIMs, ε and μ are negative in the same range of wavelengths. As a result, the refractive index is also negative. Many of the unique properties of NIMs were predicted in the original paper published by Victor Veselago in 1968 (1). However, the main obstacle to experimental observation of these properties is that NIMs have not been found in nature. Over 30 years later, the first experimental demonstrations of NIMs at microwave frequencies were reported [2,3]. In these experiments, following the approach proposed by Pendry et al. [11], NIMs were built of pairs of subwavelength concentric SRRs providing negative μ and straight wires responsible for negative ε.

In the last four years, several approaches to the realization of optical NIMs structures have been developed by several different groups worldwide. One of the first metamaterials with a negative index of refraction at optical frequencies was demonstrated using pairs of metallic nanorods shown in Figure 1.1a [5]. The origin of a negative refractive index in a composite material built with such paired nanorods can be understood as follows [12–14] (see Figure 1.2). The electric resonances of individual nanorods originate from the excitation of the surface waves on the metal–air interface. While such surface waves, known as surface plasmon polaritons, cannot be excited with the plane wave in a semi-infinite medium, they are excited in the finite-size nanorods. In a paired nanorod configuration, two types of plasmon polariton waves can be supported: symmetric and antisymmetric. The electric field, oriented parallel to the nanorods, induces parallel currents (symmetric plasmon polariton wave) in both nanorods, leading to the excitation of a dipole moment. The magnetic field, oriented perpendicular to the plane of the

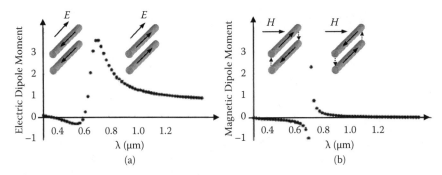

FIGURE 1.2   (a) The electric dipole moment of the coupled nanorod pair with the length of the nanorod $a = 162$ nm, the separation between two nanorods $d = 80$ nm, and the nanorod diameter $b = 32$ nm. (b) The magnetic dipole moment of the system in (a). (Adapted from T. A. Klar et al., *IEEE J. Selected Top. Quantum Electron.* 12, 1106, 2006.)

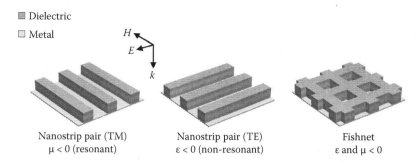

Dielectric
Metal

Nanostrip pair (TM)
$\mu < 0$ (resonant)

Nanostrip pair (TE)
$\varepsilon < 0$ (non-resonant)

Fishnet
ε and $\mu < 0$

FIGURE 1.3 Paired metal nanostrips responsible for resonant magnetic response, nonresonant electric response, and the combined fishnet structure.

nanorods, excites antiparallel currents (antisymmetric plasmon polariton wave) in the pair of nanorods. Combined with the displacement currents between the nanorods, they induce a resonant magnetic dipole moment. The excited moments are codirected with the incident field when the wavelength of the incident light is above the resonance, and they are counter-directed to the incident fields at wavelengths below the resonance. The excitation of such plasmon resonances for both the electric and magnetic field components results in the resonant response of the refractive index. In particular, the refractive index can become negative at wavelengths below resonance.

It is important that the electric and magnetic resonances occur at similar wavelengths, but this is quite difficult to achieve. Therefore, an alternative approach was proposed that combines the resonant magnetic response of one set of parallel metal nanostrips and the nonresonant electric response of another set of nanostrips that run perpendicular to the first set, resulting in a so-called nano-fishnet structure, as shown in Figure 1.3. Finally, it should be mentioned that simultaneously negative ε and μ is a sufficient, but in fact, not a necessary condition for obtaining negative refractive index [15]. Indeed, the following condition $\varepsilon'|\mu| + \mu'|\varepsilon| < 0$ guarantees a negative refractive index assuming that $\varepsilon = \varepsilon' + i\varepsilon''$ and $\mu = \mu' + i\mu''$ where $\varepsilon'(\mu')$ and $\varepsilon''(\mu'')$ are real and imaginary parts of the dielectric permittivity and magnetic permeability, respectively. This condition is valid for passive NIMs and indicates that a negative refractive index can be achieved even when $\varepsilon' < 0$ and $\mu' > 0$ provided that $\mu'' \neq 0$, implying that the material is inherently lossy.

The transmission properties of NIMs are often characterized by the figure of merit (FOM) defined as the ratio of the real and imaginary parts of the refractive index, $F = |n'|/n''$. The larger the FOM, the better the NIMs' transmission properties are. Another factor affecting the overall transmission is impedance mismatching. The highest $FOM \approx 3$ for two-dimensional optical NIMs has been demonstrated at $\lambda = 1.4\ \mu m$ using the self-supporting fishnet structure consisting of rectangular dielectric voids in parallel metal films [16]. In Section 1.3, we discuss several approaches to loss compensation in negative NIMs.

1.2.2 Unique Properties of NIMs

One of the most fundamental properties of NIMs is the antiparallel orientation of the phase velocity and the Poynting vector. This property directly follows from Maxwell's

6

Tutorials in Metamaterials

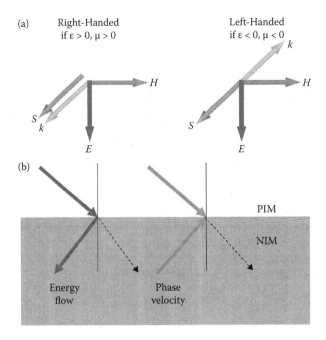

FIGURE 1.4 (a) A right-handed triplet of E, H, and k vectors in PIMs and a left-handed triplet in NIMs. (b) Refraction of light at a PIM–NIM interface.

equations. Let us consider Maxwell's equations that the case of uniform plane waves can be written in the form

$$k \times E = \frac{\omega}{c} \mu H$$

$$k \times H = -\frac{\omega}{c} \varepsilon E.$$

(1.1)

Equations 1.1 suggest that if $\varepsilon > 0$ and $\mu > 0$ in the same frequency range, vectors E, H, and k form a "right-handed" triplet as shown in Figure 1.4a, and the refractive index is positive. As usual, the Poynting vector defined as

$$S = \frac{c}{4\pi} E \times H$$

(1.2)

is parallel to the k-vector. However, if $\varepsilon < 0$ and $\mu < 0$ are in the same frequency range, vectors E, H, and k form a "left-handed" triplet as shown in Figure 1.4b, and the sign of the refractive index is negative. Owing to this distinct feature, NIMs were referred to as left-handed materials by Veselago in his original paper (1). The Poynting vector in NIMs is antiparallel to the k-vector. Therefore, phase velocity defined as $v_p = \frac{\omega}{k}$ is codirected with the energy velocity determined by the Poynting vector in the PIMs and counter-directed with the Poynting vector in the NIMs. The opposite directionality of the phase velocity and the Poynting vector is taken as the most general definition of the NIMs.

The second remarkable property of NIMs (that in fact defined their name) is that these materials have a negative index of refraction. As a result, light refracts "negatively" in contrast to conventional, or "positive," refraction, as shown in Figure 1.4b. It should be emphasized that, in general, a material exhibiting negative refraction is not equivalent to a negative refractive index material. Negative refraction associated with NIMs has been demonstrated at microwave frequencies in a metamaterial wedge and in the visible frequency range at the interface between a bimetal Au-Si3N4-Ag waveguide and a conventional Ag-Si3N4-Ag slot waveguide using plasmons [17]. Also, negative refraction at optical frequencies was demonstrated in photonic crystals (PC) [18]. However, it should be mentioned that the main limitation of PCs for realization of many unusual phenomena associated with a negative index of refraction is that the size of their characteristic features is comparable to the wavelength of light. On the contrary, optical metamaterials with a feature size much smaller than the wavelength of light are predicted to enable many truly remarkable phenomena.

The third fundamental characteristic of NIMs is the inherent frequency dependence of both ε and μ. This property originates from the fact that NIMs are resonant structures, that is, negative ε and μ occur in close proximity to the electric and magnetic resonances. As a result, the refractive index is negative only in a limited range of frequencies. In fact, the same material may act as a NIM in one range of frequencies, and as a conventional positive-index material (PIM) at other frequencies.

1.2.3 Unique Applications of NIMs

In addition to extraordinary fundamental physical properties, NIMs give rise to unique functionalities and device applications. A very unusual property of NIMs is the possibility of imaging using a flat slab of NIM with $n = -1$ surrounded by a conventional medium with $n = 1$. Moreover, under the appropriate conditions, the NIM slab not only refocuses propagating field components, but also reamplifies the evanescent field components that decay exponentially with distance from source (Figure 1.5a) through the excitation of plasmon resonance on the NIM surface. These evanescent field components responsible for the imaging of the high frequency and correspondingly small-scale features of the object cannot be restored by conventional lenses, inevitably limiting its resolution. Thus, at least in an ideal (lossless) case, an imaging system based on a NIM slab, named a "superlens" by Pendry [19], has the potential for significantly improved resolution in the image plane. Unfortunately, a superlensing effect is extremely sensitive to losses in the NIM slab.

A somewhat simpler version of the superlens, a so-called "poor man's superlens," has been proposed for applications that involve distances much smaller than a wavelength [20–22]. In this case, a subwavelength resolution can be realized using single-negative ($\varepsilon < 0$) materials such as metals at the UV or visible frequencies. In particular, it was proposed to use subwavelength sheets of silver to obtain near-field focusing for TM-polarized light. Following this strategy, subwavelength imaging on a scale of a few tens of nanometers has been demonstrated experimentally. Improved ($\lambda/20$) resolution was further demonstrated when silver was replaced by silicon carbide, which provides better performance in terms of losses [23]. However, an inherent limitation of this type of lens is its near-field performance.

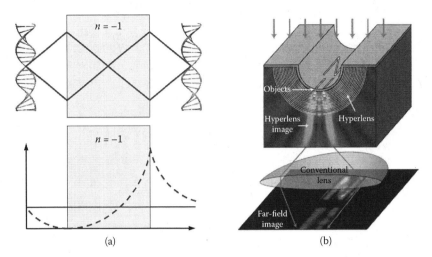

FIGURE 1.5   (a) Focusing of propagating (upper plot) and evanescent (lower plot) waves by NIM superlens. (From J. B. Pendry et al., *Phys. Rev. Lett.* 85, 3966, 2000.) (b) A hyperlens. Z. Liu et al., *science* 315, 1686, 2002.

A promising solution to far-field imaging was proposed independently of each other by Engheta and Narimanov [24,25]. This device is based on recently proposed strongly anisotropic metamaterials that feature opposite signs of the two permittivity tensor components [26–28]. Such metamaterials have been shown to support propagating waves with very large wave numbers (that would evanescently decay in ordinary dielectrics) [26,29]. After reamplifying and refocusing the evanescent field components as Pendry's superlens does, a hyperlens converts those evanescent waves into propagating waves. Once all the components are propagating waves, they can easily be imaged by a conventional lens (microscope) in the far field. A schematic of experimental realization of the hyperlens is shown in Figure 1.5b [30]. Finally, Zhang's group demonstrated another type of far-field superlens by using a surface grating to convert evanescent waves into propagating waves [31].

It is worth noting that the majority of the unique properties of NIMs most efficiently reveal themselves when NIMs are combined with conventional positive-index materials. As will be illustrated in the next section, NIMs and PIMs can be combined either spatially or spectrally.

1.3 Loss Management and Nonlinear Optics in Metamaterials

In addition to unusual linear properties, combinations of PIMs and NIMs fundamentally change nonlinear optical interactions. In particular, the antiparallel wave and Poynting vectors in NIMs enable a novel backward phase-matching mechanism that facilitates new regimes of second-harmonic generation (SHG) [32–37] and optical parametric amplification (OPA) [37–42].

As mentioned in the introduction, both ε and μ are frequency dependent in NIMs. Therefore, the same material can reveal negative-index properties at one wavelength and positive-index properties at another wavelength, forming the basis for fundamentally new regimes of the nonlinear optical interactions. In particular, one of the most fundamental properties of the NIMs–antiparallel wave and Poynting vectors enables new "backward" phase-matching conditions leading to unusual regimes of the SHG and OPA.

1.3.1 Second-Harmonic Generation

The basic idea of backward phase matching can be understood as follows. It is assumed that the metamaterial is a NIM at the fundamental frequency ω_1 and a PIM at the second-harmonic frequency ω_2. If the energy flow of the fundamental frequency travels from left to right, the phase of the wave at the same frequency should move in the opposite direction, that is, from right to left. The phase-matching requirement $k_2 = 2k_1$ can be satisfied if the phase of the second harmonic also travels from right to left. Since the second harmonic propagates in the PIM, its energy flow is co-directed with the phase velocity and, therefore, the energy propagates from right to left as well. On the contrary, in conventional PIM materials, the wave and Poynting vectors propagate in the same direction at both the fundamental and second-harmonic frequency and backward phase matching generally does not occur.

One of the important differences between the SHG in the NIM and PIM cases is reflected in the Manley–Rowe relations given by $|A_1|^2 - |A_2|^2 = C$, where A_1 and A_2 are slowly varying amplitudes of the fundamental and second-harmonic waves, respectively. In the conventional PIM case, the Manley–Rowe relations require that the sum of the squared amplitudes be constant [36,37]; however, in the NIM case, their difference is constant. This unusual form of Manley–Rowe relations in NIMs is a direct result of the fact that the Poynting vectors for the fundamental and the second harmonic are antiparallel, while their wavevectors are parallel. An example of an important implication of this new Manley–Rowe relations is now briefly described.

It is important that the boundary conditions for the fundamental and second-harmonic waves in the NIM case are specified at opposite interfaces of the slab of a finite length L in contrast to the PIM case, where both conditions are specified at the front interface. Owing to such boundary conditions, the conversion at any point within the NIM slab depends on the total thickness of the slab. In the limit of a semi-infinite NIM, both fundamental and second-harmonic waves disappear at infinity. Therefore, a 100% conversion efficiency of the incoming wave at the fundamental frequency to the second harmonic frequency propagating in the opposite direction is expected. As a result, the NIM slab acts as a nonlinear mirror [35,36].

1.3.2 Loss Compensation

An important potential application of backward phase matching realized in NIMs is the compensation of losses. As discussed earlier, the majority of the photonic metamaterials realized to date consist of metal–dielectric nanostructures that have highly controllable magnetic and dielectric responses. The problem, however, is that these structures have losses that are difficult to avoid, especially in the visible range of frequencies. Losses in such structures have various origins and are still not completely quantified. Some known sources of loss in photonic metamaterials stem from surface roughness, quantum size and chemical interface effects, the resonant nature of their magnetic response, or the fundamental loss properties of their constitutive components: metals [43]. Therefore, various approaches to the realization of low-loss metamaterials, including advancements in fabrication techniques [44], development of novel metamaterials designs [45], and introduction of gain in the metamaterials structures (optical amplification) [37–42], are being developed simultaneously.

One of the first conceptual approaches to managing losses in metamaterials has been discussed by Shamonina et al. [46] and Ramakrishna et al. [47]. They found that the effect of losses in bulk negative refractive index metamaterials (NIMs) can be reduced by replacing a slab of bulk NIM with a stack of alternating negative- and positive-index layers. Further improvement has been predicted by making the positive-index layers out of gain materials such as optically pumped semiconductors [48].

Preliminary estimations of the required gain levels and initial numerical simulations of realistic loss-compensated NIMs have been performed by Klar et al. [14], who proposed immersing the NIM comprising of the double silver strips into the gain medium and found that the structure becomes transparent at a gain level of $12 \cdot 10^3$ cm^{-1}. The required amount of gain can be achieved by applying dye molecules (e.g., Rhodamine 6G) or semiconductor quantum dots (e.g., CdSe) on top of the NIM. Following numerous theoretical studies, two recent experimental demonstrations proved the possibility of realization of such loss compensation techniques in practice [49,50]. First, it was shown that 44-nm-diameter nanoparticles with a gold core and dye-doped silica shell enable complete compensation of the loss of localized surface plasmons by gain [49]. Moreover, latest experiments by Shalaev's group showed that placing the gain material between the metal strips significantly reduces the required gain because of the enhanced local fields in that area and indeed leads to complete loss compensation [50].

Recently, a new approach to overcoming losses in metamaterials over the entire negative-index frequency range was proposed [42,51]. The proposed technique is based on a well-known three-wave mixing process that takes place in nonlinear media exhibiting second-order susceptibility (in $\chi^{(2)}$ materials), and induces optical parametric amplification. However, the unique electrodynamics of negative-index metamaterials (NIMs) give rise to a rather unusual regime of OPA, making it a promising candidate for converting lossy NIMs into transparent or even amplifying negative-index media.

Optical parametric amplification refers to a process of amplification of a light signal through mixing with pump light in a nonlinear material. Specifically, the photon flux in the signal wave grows through coherent energy transfer from a higher-frequency intense pump wave. A photon from an incident pump laser is divided into two photons, one of which is a photon at the signal frequency. The strong pump field with angular frequency and wavenumber (ω_3, k_3) and a weak signal with (ω_1, k_1) generate a difference frequency idler with (ω_2, k_2). The OPA process requires both momentum and energy conservation. Momentum is conserved when the phase-matching condition for the wavevectors, $k_3 = k_1 + k_2$, is satisfied, whereas energy conservation requires that $\omega_3 = \omega_1 + \omega_2$.

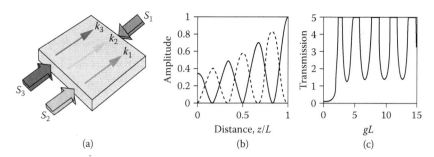

FIGURE 1.6 Optical parametric amplification in metamaterials. (a) Schematic representation of backward OPA in a NIM slab. S_1, S_2, and S_3 are the magnitudes of the Poynting vectors of the signal, idler, and pump beam, respectively, and k_1, k_2, and k_3 are their wavevectors. (b) Signal (solid line) and idler (dashed line) field distributions inside the NIM. All waves propagate in the z-direction, and the signal enters NIM at $z = L$, where L is the slab thickness. (c) Signal amplification in the NIM slab. g is a nonlinear coupling coefficient. (From N. M. Litchinitser and V. M. Shalaev, *Nat. Photonics* 3, 75, 2009.)

Metamaterials represent a unique environment for the realization of a backward OPA. Indeed, one of the most fundamental properties of NIMs is that the wavevector, $\boldsymbol{k}$ (and phase velocity), and the Poynting vector, $\boldsymbol{S}$, are antiparallel. Also, as a metamaterial's properties are frequency dependent, it is possible that the same structure may possess negative-index properties at one frequency but positive-index properties at another. Thus, an OPA with counter-directed energy flows can be realized with all three waves having co-directed wavevectors, as shown in Figure 1.6a. Indeed, if the pump and idler wave propagate in the positive-index regime, and the signal wave frequency belongs to the negative-index regime, the energy flow of the signal wave will be antiparallel to that of the pump and the idler, as shown in Figure 1.6a. This will create an effective feedback mechanism. Such a NIM-based OPA was shown theoretically to exhibit peculiar oscillatory distributions of the fields, shown in Figure 1.6b. As a result, the scheme supports oscillations without a cavity as each spatial point serves as a source for the generated wave in the backward direction.

Another remarkable property of such a NIM-based OPA, illustrated in Figure 1.6c, is that the originally lossy NIM becomes transparent (or amplifying) for a broad range of pump field intensities and slab thicknesses, provided that the absorption (loss) coefficient at the idler frequency is equal to or larger than that at the signal frequency. A significant advantage of the OPA approach as compared to other loss-compensating techniques, such as metamaterials combined with some gain media, is that it is not limited to the narrow wavelength range defined by the laser transitions of the laser gain media (for example, rhodamine 6G or quantum dots). Therefore, it can be made tunable in a wide frequency range.

1.3.3 Backward Waves in Nonlinear PIM-NIM Couplers

Backward waves can also facilitate an effective feedback mechanism in wave-guiding structures. This novel mechanism will be exemplified by a nonlinear coupler with one

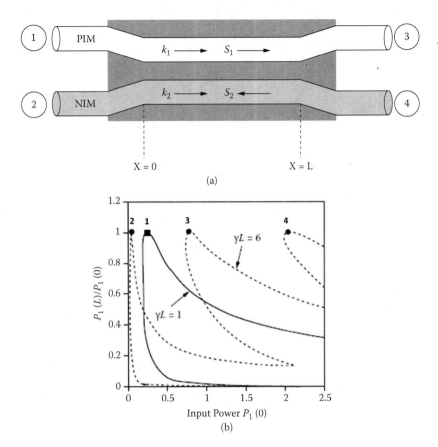

FIGURE 1.7 (a) A schematic of a nonlinear PIM-NIM coupler. Light is initially launched into channel 1 (PIM). A wavevector k and a Poynting vector S are parallel in the PIM channel and anti-parallel in the NIM channel, enabling a new backward-coupling mechanism. (b) Transmission coefficient defined as the ratio of output and input powers $P1(L)/P1(0)$ as a function of input power $P1(0)$ for $\gamma L = 1$ (solid line) and $\gamma L = 6$ (dashed line) when normalized linear coupling coefficient $\kappa L = 6$. Transmission resonances are indicated by the numbers 1, 2, 3, and 4.

channel filled with NIM. It turns out that introducing a NIM in one of the channels dramatically changes both the linear and nonlinear transmission characteristics of such a coupler [52,53]. In contrast to the previous case of SHG and OPA, in a PIM-NIM coupler, PIMs and NIMs are combined spatially as shown in Figure 1.7a. While nonlinear couplers made of conventional PIM materials have attracted significant attention owing to their strong potential for all-optical processing applications, they typically lack an important functionality—optical bistability—if no external feedback mechanism is introduced. In contrast, it was found that the PIM-NIM coupler exhibits bistable transmission, as shown in Figure 1.7b, and supports gap solitons originating from antiparallel phase and energy velocities in the NIM channel and parallel phase and energy velocities in the PIM channel, which results in an "effective" feedback mechanism.

These effects have no analogies in conventional PIM-PIM couplers composed of uniform (homogeneous) waveguides with no feedback mechanism. These optical couplers have a strong potential for all-optical processing applications, including buffering, switching, and power-limiting functionalities.

1.4 Numerical Modeling and Optimization Methods

Due to the complexity of metamaterial structures, the realization of all these remarkable properties and their potential applications would not have been possible without the availability of advanced and specialized numerical modeling tools.

It may appear that since nanostructured metamaterials are electromagnetic structures described by Maxwell's equations, conventional computational methods should apply to their design and analysis. However, detailed studies prove that there are a number of numerical challenges specific to modeling metamaterials owing to (1) the subwavelength nature of metamaterial structures; (2) the essentially dispersive nature of negative-index metamaterials, which often leads to an extremely large number of unit cells that are not tractable by existing computational methods and resources, especially in 3D geometries; (3) singularities and instabilities caused by high-index contrast, the presence of metals, and interfaces between positive and negative refractive index materials. Finally, software developed for modeling conventional optical materials and devices does not take into account any magnetic properties, as it is commonly assumed that $\mu \approx 1$ at optical frequencies [54]. Therefore, numerical analysis quickly became an essential component of metamaterial research. Although the field of optical metamaterials is very new, there are a number of numerical approaches and tools developed for designing, characterizing, and modeling linear and nonlinear wave propagation in metamaterials.

1.4.1 Optimization Algorithms

The typical design of metamaterials starts with the unit cell or "meta-atom," for example, a split-ring resonator, a pair of nanorods, coupled strips, or fishnet structure. The dimensions of these unit cells and the material properties form a hyperspace for the global optimization problem [55]. The optimization problem consists of finding the best point in this hyperspace with respect to a particular figure of merit; for example, in the case of negative-index materials, the figure of merit is often defined as the ratio of the real and imaginary parts of the index of refraction.

Several optimization methods have been developed and successfully applied to designing of metamaterials, including: (1) simulated annealing [56]; (2) genetic algorithm [57]; and (3) particle swarm optimization [58]. These approaches have been successfully utilized for designing novel metamaterials for parallel nanoscale optical sensing and imaging.

The efficient optimization procedures incorporate realistic fabrication constraints, for example, the lowest dimension and accuracy that could be achieved, constraints on material properties that can be employed, and an understanding that the fabrication process itself is known to alter the properties of the elementary materials, which can, for example, result in increased loss compared to bulk material.

1.4.2 Numerical Modeling of Combined Functionalities

Multiscale and multiphysics simulations become increasingly important in the context of metamaterials and their applications. Indeed, although light propagates in optical metamaterials as in "effective" media that can be characterized by effective refractive indices, metamaterials properties are determined by their underlying subwavelength structure. Therefore, at least three important scales may be identified: (1) nano- (or subwavelength) scale simulations, which account for the structure of meta-atoms and enable material properties optimization; (2) device-level simulations, which allow for the study of novel functionalities enabled by the unique physical properties of metamaterials; and (3) system-level simulations that effectively treat each individual metamaterial-based component as a black box so that various functionalities can be combined together for a particular application.

1.4.3 Modeling of Nonlinear Effects in Metamaterials

In the past few years, a number of unusual nonlinear wave interactions were predicted to occur in metamaterials. In a majority of theoretical studies to date, nonlinear metamaterials were considered as a uniform media with prescribed dielectric permittivity and magnetic permeability, and the nanostructured nature of these artificial materials was not taken into account. Local field enhancement and other effects are likely to alter nonlinear interactions of electromagnetic waves with metamaterials. Therefore, availability of efficient and reliable numerical modeling tools that take into account actual nanostructure and near-field effects is essential for developing future applications of metamaterials.

Recently, an efficient method for introducing third-order nonlinearities in optical nanostructured materials, including photonic metamaterials, was developed [59]. Two examples of structures that were designed and studied using this approach are shown in Figure 1.8. The method uses scalar H-field frequency domain formulation; it is shown to produce fast and accurate results without superfluous vector E-field formalism. A standard TM representation using cubic nonlinear susceptibility is problematic due to an intractable implicit equation; the technique described in detail in [59] alleviates this problem.

This approach was implemented in a commercial finite element solver (COMSOL Multiphysics) and validated with 3D full-wave simulations solved for E-field. The 2D H-field formulation was found to exhibit a substantially faster performance and converged over a broader range of nonlinearities compared with the 3D formulation.

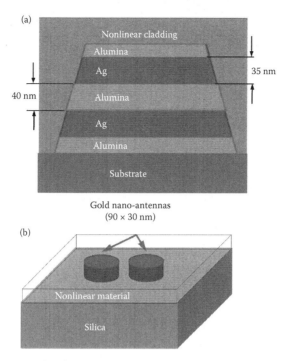

FIGURE 1.8 Two examples of nonlinear nanostructures simulated using the approach described in the text: (a) silver nano-strips enabling magnetic response across visible range, (b) gold nano-antennas providing strong field enhancement.

1.5 Gradient-Index Metamaterials and Transformation Optics

The subject of gradient-index media dates from the 1850s [60]. The examples of artificial gradient-index structures include the Maxwell fisheye lens [61] developed in 1854; the Luneburg lens [62] used in the microwave region of the spectrum and for acoustical imaging; the Wood lens, made in 1905 [63]; and the gradient-index fiber developed for telecommunication applications [64]. Gradient-index media can also be found in nature. Examples of such media include human eyes, common mirages such as a pool of water appearing on a road, or more exotic phenomena such as complex mirages, illustrated in Figure 1.9.

In the past, numerous techniques have been used for manufacturing gradient-index structures in glasses and plastics. Those gradient-index structures were characterized by two important features: the depth of the gradient and the magnitude of the index-of-refraction change Δn. Some typical values of Δn that can be achieved in conventional optics are $\Delta n \sim 0.02–0.05$ [60]. Metamaterial technology offers exceptional flexibility with respect to the profile, depth, and absolute value of the refractive index change, and precise control over its distribution, enabling unprecedented opportunities for light manipulation.

(a)   (b)

FIGURE 1.9 Two classic examples of superior atmospheric mirage: (a) Crocker Land, and (b) Flying Dutchman.

The enormous potential of refractive index engineering in metamaterial structures was recently exemplified by the first theoretical and experimental demonstrations of an invisibility (cloaking) device [65–71]. However, cloaking is just one of numerous prospective applications of these structures. In this section, we discuss several examples of unusual physical phenomena and potential applications of gradient-index metamaterials, as well as a powerful method for the design of these structures: the transformation method.

1.5.1 Graded-Index Transition Metamaterials

It is worth noting that the most unusual properties of NIMs are revealed at the interface of positive and negative-index materials. Particularly, the right-handed triplet formed from the electric and magnetic fields and the wavevector in the positive-index material (PIM) undergoes an abrupt change to form a left-handed triplet in the NIM. A topologically critical phenomenon such as this leads to antiparallel directions of the wave and Poynting vectors in the NIM.

Let us consider wave propagation in a gradient-index structure, as schematically shown in Figure 1.10. Assuming ε and μ are linearly changing from positive to negative values, propagation of the TE-polarized component (E is perpendicular to the plane of propagation) is described by the following wave equation that directly follows from Maxwell's equations in the case of spatially dependent ε and μ:

$$\frac{\partial^2 E_y}{\partial x^2} + \frac{\partial^2 E_y}{\partial z^2} - \frac{1}{\mu}\frac{\partial \mu}{\partial x}\frac{\partial E_y}{\partial x} + \frac{\omega^2}{c^2}\varepsilon\mu E_y = 0. \tag{1.3}$$

Here, E_y is the amplitude of the electric component of the harmonic wave at frequency ω, and c is the speed of light in vacuum. The components of the magnetic field are related to E_y through Maxwell's equations. The corresponding equation for the TM wave can be obtained from Equation 1.3 by replacing E with $-H$, and μ with ε, respectively. It was

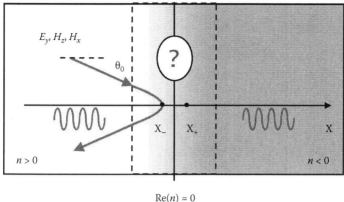

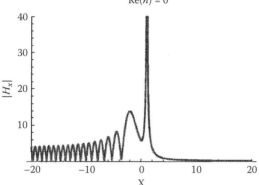

FIGURE 1.10 (a) The real parts of the dielectric permittivity and magnetic susceptibility as functions of a longitudinal coordinate. (b) The absolute value of the normalized magnetic field component as a function of a normalized longitudinal coordinate (solid line) and the boundary between the PIM and the NIM (short-dashed line).

shown analytically [72] that at the point in space where real parts of ε and μ cross zero, the field components E_y, H_x, and H_z are given by

$$E_y = E_0 \exp(i\beta z - i\omega t) + O((1 - x/h)^2),$$

$$H_x = -E_0 \sqrt{\varepsilon_0/\mu_0}\,(1 - x/h)^{-1} \sin\theta \ \exp(i\beta z - i\omega t) + O((1 - x/h)^2), \qquad (1.4)$$

$$H_z = 2i\pi E_0 (h/\lambda) \sin^2\theta \ \ln\sqrt{a}(1 - x/h) \exp(i\beta z - i\omega t) + O((1 - x/h)^2),$$

where β is the propagation constant, h is the half-width of the gradient-index transition layer, θ is the incidence angle, and E_0 is the electric field amplitude at the zero-index point.

These asymptotics (4) predict the resonant field enhancement of the H_x component near the zero-refractive-index point under oblique incidence of the TE wave on such a gradient-index metamaterial layer. Numerical simulations confirmed this prediction. In a simplified way, the origin of the anomalous field enhancement shown in Figure 1.10b can be understood as a spatial analog of the well-known resonance occurring in the spectral domain when, for example, light interacts with a harmonic oscillator. It is noteworthy that because the wavelength of light becomes very large in the vicinity of the zero-index point, the system can effectively be considered static-like. In the case of the TM wave, the thin layer near the zero-index point can be considered a very thin capacitor that accumulates infinitely large electric field energy if the effects of dissipation and spatial dispersion are neglected. Note that such energy accumulation occurs only for obliquely incident waves since the electric field at oblique incidence has a nonzero component in the direction of propagation. Since the electric displacement D must be continuous, the electric field E anomalously increases as the refractive index tends to zero. Likewise, for the TE wave considered herein, the magnetic field has a nonzero component in the direction of propagation, and the magnetic field energy accumulates in the vicinity of the zero-index point in space. Such a thin layer near the zero-index point can be considered to be a short solenoid that stores the magnetic field energy. In this case, H anomalously increases as the refractive index tends to zero. Finally, owing to the singularities of the magnetic field components, the x and z components of the Poynting vector are also singular at the zero-index point. In addition, the z component of the Poynting vector changes sign while passing the point where $\mathrm{Re}(n) = 0$.

1.5.2 Transformation Optics and Cloaking

While originally the entire field of metamaterials was stimulated by the development of NIMs, yet another exciting branch of modern optics developed: transformation optics [65,73–75]. First applied to the development of the cloak, transformation optics is now considered a very general and powerful design tool that offers unparalleled opportunities for controlling light propagation through careful refractive index engineering. Recently, various functionalities and novel device applications enabled by this approach have been proposed, including image-processing operations such as translation, rotation, mirroring, and inversion; light concentrators [75–77]; super-and hyper-lenses [78]; and gradient-index waveguides and bends [79–81].

The basic idea of this method is that in order to guide waves along a certain trajectory, either the space should be deformed, assuming that material properties stay the same, or the material properties should be properly modified. More specifically, under a coordinate transformation, the form of Maxwell's equations should remain invariant while new ε and μ would contain the information regarding the coordinate transformation and the original material parameters. Therefore, in order to realize a particular property/light trajectory, it is possible to use an anisotropic material with prescribed components of permittivity and permeability tensors calculated through a particular coordinate transformation. Several examples of structures that were designed using this approach are shown in Figure 1.11.

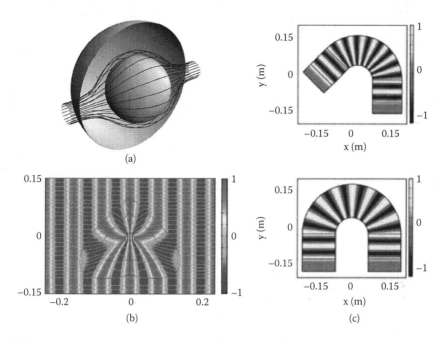

(a)

(b)

(c)

FIGURE 1.11 *See Color Insert.* Example of the structures designed using the transformation optics approach. (a) A cloak. (From J. B. Pendry et al., *Science* 312, 1780, 2006. With permission.) (b) A wave concentrator. (From W. X. Jiang et al., *Appl. Phys. Lett.* 92, 264101, 2008.) (c) A reflectionless waveguide filled with anisotropic and inhomogeneous material. (From W. X. Jiang et al., *Phys. Rev. E* 78, 066607, 2008; M. Rahm et al., *Phys. Rev. Lett.* 100, 063903, 2008; M. Rahm et al., *Opt. Express* 16, 11555, 2008. With permission.)

The general strategy of the transformation approach includes two main steps. In the first step, a coordinate transformation of the space with the desired property is built. In the next step, a set of material properties is calculated that would realize this property of the transformed space in the original space using the following equations:

$$\varepsilon^{i'j'} = \left|\det\left(\Lambda_i^{i'}\right)\right|^{-1} \Lambda_i^{i'} \Lambda_j^{j'} \varepsilon$$

$$\mu^{i'j'} = \left|\det\left(\Lambda_i^{i'}\right)\right|^{-1} \Lambda_i^{i'} \Lambda_j^{j'} \mu \qquad i,j = 1,2,3,$$

(1.5)

where it was assumed that the original space is isotropic, transformations are time invariant, and $\Lambda_\alpha^{\alpha'} = \frac{\partial x^{\alpha'}}{\partial x^\alpha}$ are the elements of the Jacobian transformation matrix.

Let us consider the application of the transformation method for designing a nonmagnetic cylindrical cloak that was proposed and theoretically designed to operate at optical wavelengths [70,71]. An "ideal cloak" is defined as a device that renders objects invisible and that is object-independent, macroscopic, and operates for nonpolarized light and in a wide range of frequencies. Also, it does not reflect, scatter, or absorb any light, introduce any phase shifts, or produce a shadow.

The coordinate transformation that compresses the cylindrical region $0 < r < b$ in space into the shell $a < r' < b$ is given by

$$r' = \frac{b-a}{b}r + a, \quad \theta' = \theta, \quad z' = z. \tag{1.6}$$

The design equations (Equation 1.6) give the following material parameters:

$$\varepsilon_r = \mu_r = \frac{r-a}{r}, \quad \varepsilon_\theta = \mu_\theta = \frac{r}{r-a}, \quad \varepsilon_z = \mu_z = \left(\frac{b}{b-a}\right)^2 \frac{r-a}{r}. \tag{1.7}$$

While magnetism at optical frequencies has been demonstrated previously, it is still considered a challenging task that can only be realized in resonant structures with relatively high loss. However, it was realized that an optical cloak for the TM polarization can be built without any magnetism. In this case, Equations 1.7 are replaced with the following set of reduced parameter equations:

$$\mu_z = 1, \quad \varepsilon_\theta = \left(\frac{b}{b-a}\right)^2,$$

$$\varepsilon_r = \left(\frac{b}{b-a}\right)^2 \left(\frac{r-a}{r}\right)^2. \tag{1.8}$$

A constant (greater than one) azimuthal dielectric permittivity component can easily be achieved in conventional dielectrics. A crucial part of the design is the realization of the required radial distribution of dielectric permittivity varying from zero to one. This can be achieved using subwavelength metal wires aligned along the radii of the annular cloak shell. Other potential implementations of design parameters (Equation 1.8) include chains of metal nanoparticles and thin continuous and semicontinuous strips.

Note that the optical cloak designed using Equation 1.8 is object independent and does not impose any limitations on the size of the object. However, it is not completely reflection-, scattering-, absorption-, phase shift-, and shadow-free, due to the impedance mismatch related to reduced parameters (Equation 1.8) and small but nonnegligible material absorption. Additionally, it is intended to work for TM polarization only. Finally, the device is inherently narrowband. Indeed, since the refractive index of the cloaking shell varies from zero to one (as follows from Equation 1.8), the phase velocity of light inside the shell is greater than the velocity of light in a vacuum. While this condition itself does not contradict any law of physics, it implies that the material parameters must be dispersive.

1.5.3 Carpet Cloaking

Recently, two very different approaches to optical cloaking that solve some of the aforementioned limitations of the cloak designed using Equation 1.8 were proposed and experimentally demonstrated. One of them is a so-called carpet cloak, which is made of a dielectric and operates in a broad range of wavelengths [83–86]. The basic idea behind this approach is illustrated in Figure 1.12a. When the object is placed under a curved reflecting surface with the carpet cloak on top of it, the object appears as if it was the

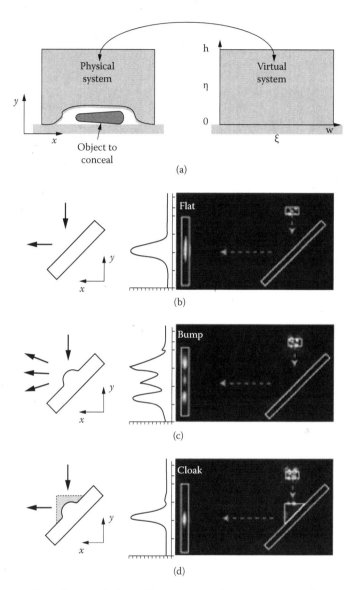

FIGURE 1.12 Optical carpet cloaking. (a) A schematic. (From M. Rahm et al., *Opt. Express* 16, 11555, 2008.) (b) A Gaussian beam reflected from a flat surface. (c) A Gaussian beam reflected from a curved (without cloak) surface. (d) The same curved reflecting surface with cloak. (From J. Valentine et al., *Nat. Mater.* 8, 568, 2009.)

original flat reflecting surface, so it is hidden under a "carpet." This approach avoids both material and geometrical singularities. The cloak region is obtained by varying the effective refractive index in a two-dimensional space. This index profile is designed using quasi-conformal mapping. Figure 1.12 shows the schematics and the experimental results obtained at a wavelength of 1540 nm in the cases of (a) a flat surface, (b) a bump without the cloak, and (c) carpet cloaking [83]. The carpet cloaking approach solves several problems associated with other approaches. The approach can be implemented using nonresonant elements (e.g., conventional dielectric materials) and therefore allows for achieving low-loss and broadband cloaking at optical wavelengths. Indeed, this approach was already demonstrated in a broad range of optical wavelengths extending from 1400 to 1800 nm.

1.5.4 Tapered Waveguide Cloaking

Finally, very recently another extremely simple approach to cloaking was proposed and experimentally demonstrated. In this approach, the metamaterial requiring anisotropic dielectric permittivity and magnetic permeability is emulated by a specially designed tapered waveguide [87]. It was shown that the transformation optics approach allows one to map a planar region of space filled with an inhomogeneous, anisotropic metamaterial into an equivalent region of empty space with curvilinear boundaries (a tapered waveguide).

This approach leads to low-loss, broadband performance in the visible wavelength range. A schematic of the cloak is shown in Figure 1.13. A double convex glass lens

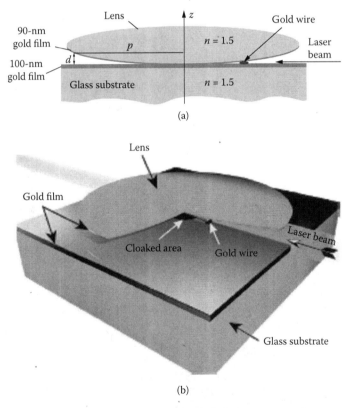

FIGURE 1.13 Tapered waveguide acting as an optical cloak. (From I. I. Smolyaninov et al., *Phys. Rev. Lett.* 102, 213901, 2009.) (a) Cross-sectional sketch for the waveguide experiment. (b) 3D version of the experimental setup.

coated on one side with a gold film is placed with the gold-coated side down on top of a flat gold-coated glass slide. The air gap between these surfaces can be used as an adiabatically changing waveguide. In fact, this geometry is identical to the classic geometry of the Newton rings observation.

It is well known that the modes of the waveguide have cutoff wavelengths. It turns out that for a particular mode in the waveguide of Figure 1.13, the cutoff radius is given by the same expression as that of the radius of the corresponding Newton ring; thus, no photon launched into the waveguide can reach an area within the radius from the point of contact between the two gold-coated surfaces. This approach leads to low-loss, broadband cloaking performance. Importantly, the cloak that has already been realized with this approach was 100 times larger than the wavelength of light, a property that was not demonstrated with any other cloaking techniques to date.

1.6 Summary

To summarize: In less than a decade, the enormous progress in both theoretical and experimental research on metamaterials has been made. Several ideas that would have been considered science fiction only a few years ago have been experimentally demonstrated. New material properties and device functionalities enabled by metamaterials technology have the prospect of contributing to nearly all areas of science and technology, including telecommunications, optical computing, healthcare, national security, and defense.

Acknowledgment

The authors acknowledge the support of the Army Research Office through grants W911NF-09-1-0075, W911NF-09-1-0231, and 50342-PH-MUR.

References

1. V. G. Veselago, The electrodynamics of substances with simultaneously negative values of ε and μ, *Sov. Phys. Usp.* 10, 509, 1968.
2. D. R. Smith, W. J. Padilla, D. C. Vier, S. C. Nemat-Nasser, and S. Schultz, Composite medium with simultaneously negative permeability and permittivity, *Phys. Rev. Lett.* 84, 4184, 2000.
3. R. A. Shelby, D. R. Smith, and S. Schultz, Experimental verification of a negative index of refraction, *Science* 292, 77, 2001.
4. A. Schuster, *An Introduction to the Theory of Optics*, Edward Arnold, London, 1904.
5. V. M. Shalaev, W. Cai, U. K. Chettiar, H. Yuan, A. K. Sarychev, V. P. Drachev, and A. V. Kildishev, Negative index of refraction in optical metamaterials, *Opt. Lett.* 30, 3356, 2005.
6. S. Xiao, U. K. Chettiar, A. V. Kildishev, V. P. Drachev, and V M. Shalaev, Yellow-light negative-index metamaterials, arXiv:0907.1870.
7. J. Valentine, S. Zhang, T. Zentgraf, E. Ulin-Avila, D. A. Genov, G. Bartal, and X. Zhang, Three-dimensional optical metamaterial with a negative refractive index, *Nature* 455, 376, 2008.

8. N. Liu, H. Guo, L. Fu, S. Kaiser, H. Schweizer, and H. Giessen, Three-dimensional photonic metamaterials at optical frequencies, *Nat. Mater.* 7, 31, 2008.

9. M. A. Noginov, Yu. A. Barnakov, G. Zhu, T. Tumkur, H. Li, and E. E. Narimanov, Bulk photonic metamaterial with hyperbolic dispersion, *Appl. Phys. Lett.* 94, 151105, 2009.

10. G. Dolling, M. Wegener, and S. Linden, Realization of a three-functional-layer negative-index photonic metamaterial, *Opt. Lett.* 32, 551, 2007.

11. J. B. Pendry, A. J. Holden, D. J. Robbins, and W. J. Stewart, Magnetism from conductors and enhanced nonlinear phenomena, *IEEE Trans. Microwave Theory Tech.* 47, 2075, 1999.

12. V. A. Podolskiy, A. K. Sarychev, and V. M. Shalaev, Plasmon modes in metal nanowires and left-handed materials, *J. Nonlinear Opt. Phys. Mater.* 11, 65, 2002.

13. V. A. Podolskiy, A. K. Sarychev, and V. M. Shalaev, Plasmon modes and negative refraction in metal nanowire composites, *Opt. Express* 11, 735, 2003.

14. T. A. Klar, A. V. Kildishev, V. P. Drachev, and V. M. Shalaev, Negative-index metamaterials: Going optical, *IEEE J. Selected Top. Quantum Electron.* 12, 1106, 2006.

15. R. A. Depine and A. Lakhtakia, A new condition to identify isotropic dielectric-magnetic materials displaying negative phase velocity, *Microwave Opt. Technol. Lett.* 41, 315, 2004.

16. G. Dolling, C. Enkrich, M. Wegener, C. M. Soukoulis, and S. Linden, Low-loss negative-index metamaterial at telecommunication wavelengths, *Opt. Lett.* 31, 1800, 2006.

17. H. J. Lezec, J. A. Dionne, and H. A. Atwater, Negative refraction at visible frequencies, *Science* 316, 430, 2007.

18. E. Schonbrun, M. Tinker, W. Park, and J.-B. Lee, Negative refraction in a Si-polymer photonic crystal membrane, *IEEE Photon. Technol. Lett.* 17, 1196, 2005.

19. J. B. Pendry, Negative refraction makes a perfect lens, *Phys. Rev. Lett.* 85, 3966, 2000.

20. N. Fang, H. Lee, C. Sun, and X. Zhang, Sub-diffraction-limited optical imaging with a silver superlens, *Science* 308, 534, 2005.

21. R. J. Blaikie, and D. O. S. Melville, Imaging through planar silver lenses in the optical near field, *J. Opt. A: Pure Appl. Opt.* 7, S176, 2005.

22. D. O. S. Melville, and R. J. Blaikie, Super-resolution imaging through a planar silver layer, *Opt. Express* 13, 2127, 2005.

23. T. Taubner, D. Korobkin, Y. Urzhumov, G. Shvets, and R. Hillenbrand, Near-field microscopy through a SiC superlens, *Science* 313, 1595, 2006.

24. Z. Jacob, L. V. Alekseyev, and E. Narimanov, Optical hyperlens: Far-field imaging beyond the diffraction limit, *Opt. Express* 14, 8247, 2006.

25. A. Salandrino and N. Engheta, Far-field subdiffraction optical microscopy using metamaterial crystals: Theory and simulations, *Phys. Rev. B* 74, 075103, 2006.

26. V. A. Podolskiy and E. E. Narimanov, Strongly anisotropic waveguide as a nonmagnetic left-handed system, *Phys. Rev. B* 71, 201101, 2005.

27. V. A. Podolskiy, L. Alekseyev, and E. E. Narimanov, Strongly anisotropic media: The THz perspectives of lefthanded materials, *J. Mod. Opt.* 52, 2343, 2005.

28. R. Wangberg, J. Elser, E. E. Narimanov, and V. A. Podolskiy, Non-magnetic nanocomposites for optical and infrared negative refraction index media, *J. Opt. Soc. Am. B.* 23, 498, 2006.

29. A. A. Govyadinov and V. A. Podolskiy, Meta-material photonic funnels for sub-diffraction light compression and propagation, *Phys. Rev. B* 73, 155108, 2006.

30. Z. Liu, H. Lee, Y. Xiang, C. Sun, X. Zhang, Far-Field optical Hyperlens Magnifying Sub-Diffraction-Limited objects, *Science* 315, 1686, 2007.

31. Z. Liu, S. Durant, H. Lee, Y. Pikus, N. Fang, Y. Xiong, C. Sun, and X. Zhang, Far-field optical superlens, *Nano Lett.* 7, 403, 2007.

32. V. M. Agranovich, Y. R. Shen, R. H. Baughman, and A. A. Zakhidov, Linear and nonlinear wave propagation in negative refraction metamaterials, *Phys. Rev. B* 69, 165112, 2004.

33. N. Mattiucci, G. D'Aguanno, M. J. Bloemer, and M. Scalora, Second harmonic generation form a positive-negative index material heterostructure, *Phys. Rev E* 72, 066612, 2005.

34. G. D'Aguanno, N. Mattiucci, M. J. Bloemer, and M. Scalora, Large enhancement of second harmonic generation near the zero-n gap of a negative index Bragg grating, *Phys. Rev. E* 73, 036603, 2006.

35. V. Shadrivov, A. A. Zharov, and Y. S. Kivshar, Second-harmonic generation in nonlinear left-handed metamaterials, *J. Opt. Soc. Am. B* 23, 529, 2006.

36. A. K. Popov, V. V. Slabko, and V. M. Shalaev, Second harmonic generation in left-handed metamaterials, *Laser Phys. Lett.* 3, 293, 2006.

37. A. K. Popov and V. M. Shalaev, Negative-index metamaterials: Second-harmonic generation, Manley-Rowe relations and parametric amplifications, *Appl. Phys. B* 84, 131, 2006.

38. A. K. Popov and V. M. Shalaev, Compensating losses in negative-index metamaterials by optical parametric amplification, *Opt. Lett.* 31, 2169, 2006.

39. M. V. Gorkunov, I. V. Shadrivov, and Y. S. Kivshar, Enhanced parametric processes in binary metamaterials, *Appl. Phys. Lett.* 88, 071912, 2006.

40. A. B. Kozyrev, H. Kim, and D. W. van der Weide, Parametric amplification in left-handed transmission line media, *Appl. Phys. Lett.* 88, 264101, 2006.

41. A. K. Popov, S. A. Myslivets, T. F. George, and V. M. Shalaev, Four-wave mixing, quantum control and compensating losses in doped negative-index photonic metamaterials, *Opt. Lett.* 32, 3044, 2007.

42. A. K. Popov and S. A. Myslivets, Transformable broad-band transparency and amplification in negative-index films, *Appl. Phys. Lett.* 93, 191117, 2008.

43. V. P. Drachev, U. K. Chettiar, A.V. Kildishev, H.-K. Yuan, W. Cai, and V. M. Shalaev, The Ag dielectric function in plasmonic metamaterials, *Opt. Express* 16, 1186, 2008.

44. A. Boltasseva and V. M. Shalaev, Fabrication of optical negative-index metamaterials: Recent advances and outlook, *Metamaterials* 2, 1, 2008.

45. G. Dolling, M. Wegener, C. M. Soukoulis, and S. Linden, Design-related losses of double-fishnet negative-index photonic metamaterials, *Opt. Express* 15, 11536, 2007

46. E. Shamonina, V. A. Kalinin, K. H. Ringhofer, and L. Solymar, Imaging, compression and Poynting vector streamlines for negative permittivity materials, *Electron. Lett.* 37, 1243, 2001.

47. S. Ramakrishna, J. B. Pendry, M. C. K. Wiltshire, and W. J. Stewart, Imaging the near field, *J. Modern Opt.* 50, 1419, 2003.

48. S. Ramakrishna and J. B. Pendry, Removal of absorption and increase in resolution in a near-field lens via optical gain, *Phys. Rev. B* 67, 201101, 2003.

49. M. A. Noginov, G. Zhu, A. M. Belgrave, R. Bakker, V. M. Shalaev, E. E. Narimanov, S. Stout, E. Herz, T. Suteewong, and U. Wiesner, Demonstration of a spaser-based nanolaser, *Nature* 460, 1110, 2009.

50. S. Xiao, V. P. Drachev, A. V. Kildishev, X. Ni, U. K. Chettiar, H.-K. Yuan, and V. M. Shalaev, Loss-free and active optical negative-index metamaterials, *Nature* 466, 735, 2010.

51. N. M. Litchinitser and V. M. Shalaev, Loss as a route to transparency, *Nat. Photonics* 3, 75, 2009.

52. A. Alu and N. Engheta, in *Negative-Refraction Metamaterials*, Ed. by G. V. Eleftheriades and K. G. Balmain, Wiley, New York, 2005.

53. N. M. Litchinitser, I. R. Gabitov, and A. I. Maimistov, Optical bistability in a nonlinear optical coupler with a negative index channel, *Phys. Rev. Lett.* 99, 113902, 2007.

54. C. Caloz, Introduction to the special issue Numerical modelling of metamaterial properties, Structures and devices, *Int. J. Numerical Modelling: Electronic Networks, Devices and Fields* 19, 83, 2006.

55. A. V. Kildishev, U. K. Chettiar, Z. Liu, V. M. Shalaev, D.-H. Kwon, Z. Bayraktar, and D. H. Werner, Stochastic optimization of low-loss optical negative-index metamaterial, *J. Opt. Soc. Am. B* 24, A34, 2007.

56. D. T. Pham and D. Karaboga, *Intelligent Optimization Techniques*, Springer-Verlag, 2000.

57. R. L. Haupt and S. E. Haupt, *Practical Genetic Algorithms*, Wiley-Interscience, 2004.

58. J. Kennedy, R. Eberhart and Y. Shi. *Swarm Intelligence* Morgan Kaufmann Academic press, San Francisco, 2001.

59. A. V. Kildishev, Y. Sivan, N. M. Litchinitser, and V. M. Shalaev, Frequency-domain modeling of scalar TM wave propagation in optical nanostructures with a third-order nonlinear response, *Opt. Lett.* 34, 3364–3366, 2009.

60. D. T. Moore, Gradient-index optics: A review, *Appl. Opt.* 19, 1035, 1980.

61. J. C. Maxwell, *The Scientific Papers of James Clerk Maxwell*, W. D. Niven, Ed., Dover, New York, 1965.

62. R. K. Luneburg, *Mathematical Theory of Optics*, U. California Press, Berkeley, 1966.

63. R. W. Wood, *Physical Optics*, Macmillan, New York, 1905.

64. E. G. Rawson, D. R. Herriott, and J. McKenna, Analysis of refractive index distributions in cylindrical, graded-index glass rods (GRIN rods) used as image relays, *Appl. Opt.* 9, 753, 1970.

65. J. B. Pendry, D. Schurig, and D. R. Smith, Controlling electromagnetic fields, *Science* 312, 1780, 2006.

66. U. Leonhardt, Optical conformal mapping, *Science* 312, 1777, 2006.

67. U. Leonhardt and T. G. Philbin, General relativity in electrical engineering, *New J. Phys.* 8, 247 doi:10.1088/1367-2630/8/10/247, 2006.

68. D. Schurig, J. J. Mock, B. J. Justice, S. A. Cummer, J. B. Pendry, A. F. Starr, and D. R. Smith, Metamaterial electromagnetic cloak at microwave frequencies, Science Express Manuscript Number 113362, 2006.

69. S. A. Cummer and D. Schurig, D., One path to acoustic cloaking, *New J. Phys.* 9, 45, 2007.

70. W. Cai, U. K. Chettiar, A. V. Kildishev, and V. M. Shalaev, Optical cloaking with metamaterials, *Nat. Photonics* 1, 224, 2007.
71. W. Cai, U. K. Chettiar, A. V. Kildishev, V. M. Shalaev, and G. Milton, Nonmagnetic cloak with minimized scattering, *Appl. Phys. Lett.* 91, 111105, 2007.
72. N. M. Litchinitser, A. I. Maimistov I. R. Gabitov, R. Z. Sagdeev, and V. M. Shalaev, Metamaterials: Electromagnetic enhancement at zero-index transition, *Opt. Lett.* 33, 2350, 2008.
73. L. S. Dolin, On the possibility of comparison of three-dimensional electromagnetic systems with nonuniform anisotropic filling, *Izv. VUZov, Radiofizika* 4, 964, 1961.
74. A. J. Ward and J. B. Pendry, Refraction and geometry in Maxwell's equations, *J. Mod. Opt.* 43, 773, 1996.
75. M. Rahm, D. Schurig, D. A. Roberts, S. A. Cummer, D. R. Smith, and J. B. Pendry, Design of electromagnetic cloaks and concentrators using form-invariant coordinate transformations of Maxwell's equations, *Photonics Nanostruct. Fundam. Appl.* 6, 87, 2008.
76. A. V. Kildishev and V. M. Shalaev, Engineering space for light via transformation optics, *Opt. Lett.* 33, 43, 2008.
77. H. Chen and C. T. Chan, Transformation media that rotate electromagnetic fields, *Appl. Phys. Lett.* 90, 241105, 2007.
78. A. V. Kildishev and E. E. Narimanov, Impedance-matched hyperlens, *Opt. Lett.* 32, 3432, 2007.
79. W. X. Jiang, T. J. Cui, X. Y. Zhou, X. M.Yang, and Q. Cheng, Arbitrary bending of electromagnetic waves using realizable inhomogeneous and anisotropic materials, *Phys. Rev. E* 78, 066607, 2008.
80. M. Rahm, S. A. Cummer, D. Schurig, J. B. Pendry, and D. R. Smith, Optical design of reflectionless complex media by finite embedded coordinate transformations, *Phys. Rev. Lett.* 100, 063903, 2008.
81. M. Rahm, D. A. Roberts, J. B. Pendry, and D. R. Smith, Transformation-optical design of adaptive beam bends and beam expanders, *Opt. Express* 16, 11555, 2008.
82. W. X. Jiang, T. J. Cui, Q. Cheng, J. Y. Chin, X. M. Yang, R. Liu, and D. R. Smith, Design of arbitrarily shaped concentrators based on conformally optical transformation of nonuniform rational B-spline surfaces, *Appl. Phys. Lett.* 92, 264101, 2008.
83. J. Li and J. B. Pendry, Hiding under the carpet: A new strategy for cloaking, *Phys. Rev. Lett.* 101, 203901, 2008.
84. J. Valentine, J. Li, T. Zentgraf, G. Bartal, and X. Zhang, An optical cloak made of dielectrics, *Nat. Mater.* 8, 568, 2009.
85. R. Liu, C. Ji, J. J. Mock, J. Y. Chin, T. J. Cui, and D. R. Smith, Broadband ground-plane cloak, *Science* 323, 366 2009.
86. L. H. Gabrielli, J. Cardenas, C. B. Poitras, and M. Lipson, Silicon nanostructure cloak operating at optical frequencies, *Nat. Photonics* 3, 461, 2009.
87. I. I. Smolyaninov, V. N. Smolyaninova, A. V. Kildishev, and V. M. Shalaev, Anisotropic metamaterials emulated by tapered waveguides: Application to optical cloaking, *Phys. Rev. Lett.* 102, 213901, 2009.

2

Fabrication of Optical Metamaterials

Alexandra
Boltasseva
*School of Electrical and
Computer Engineering
& Birck Nanotechnology
Center, Purdue University,
West Lafayette, USA
DTU Fotonik, Department
of Photonics Engineering,
Technical University
of Denmark, Kongens
Lyngby, Denmark
aeb@purdue.edu*

2.1 Introduction

An optical metamaterial (MM) is an artificially engineered composite with electromagnetic properties that are determined by subwavelength structuring rather than by the constituent materials. Hence, when rationally designed, metamaterials can demonstrate electromagnetic properties that are unattainable with naturally occurring materials. By changing the design of the unit cell or "meta-atom," the optical properties of the metamaterial can be tailored; for example, the values of permeability, μ, and permittivity, ε,

29

can be controlled. For example, one of the properties that cannot be observed in natural materials is a negative refractive index (demonstrated in *negative index metamaterials, or NIMs*) (Veselago 1968; Pendry 2004; Veselago et al. 2006; Veselago and Narimanov 2006; Soukoulis, Kafesaki, and Economou 2006; Shalaev 2007; Soukoulis, Linden, and Wegener 2007). The limitations of the attainable optical responses of naturally occurring materials are imposed by their material properties that are in turn determined by their unit cells—atoms and molecules. By using artificially structured metamaterials, meta-atoms can be designed to exhibit specific electric and magnetic responses in a required frequency range.

2.1.1 Very Brief Historical Excursion

Although Victor Veselago pointed out in 1968 (Veselago 1968) that the combination $\varepsilon < 0$ and $\mu < 0$ leads to a negative refractive index, this idea remained quite unappreciated for many years partially because no naturally occurring NIMs were known. Today's boom in metamaterial-related research, inspired by Sir John Pendry's prediction of the NIM-based superlens in 2000 (Pendry 2000), comes from recent advances in *nanofabrication* techniques that allow different materials to be structured on the nanometer scale. In human-made NIMs for the optical range of frequencies, subwavelength until cells are specifically designed and densely packed into an effective composite material to exhibit the negative effective refractive index.

A first step toward the realization of a material with a negative index of refraction was done by Pendry et al. who suggested using split-ring resonators (SRRs) as meta-atoms to achieve a magnetic permeability different from one (Pendry et al. 1999), which is a needed step for obtaining a negative index. Such magnetic structures were successfully realized in the microwave (Shelby, Smith, and Schultz 2001), THz (Yen et al. 2004; Katsarakis et al. 2005; Zhang et al. 2005b), and near-infrared optical range, 100 THz (Linden et al. 2004) and 200 THz (Enkrich et al. 2005b). For achieving magnetic response in the visible range, another design of paired metal strips was successfully used by the Purdue group to show first a magnetic response in the red part of the spectrum (Yuan et al. 2007) and then across the whole visible range (Cai et al. 2007c).

Combining the negative-permeability structures with the negative-permittivity material (e.g., simple metal wires arranged in a cubic lattice) can result in a negative index material. Such a metamaterial was straightforward to accomplish in the microwave region (Shelby, Smith, and Schultz 2001).

The approaches for moving to shorter wavelengths were initially based on concepts from the microwave regime (such as SRRs) with scaled down unit-cell sizes. The main idea was that the magnetic resonance frequency of the SRR is inversely proportional to its size. Using single SRRs, this approach works up to about 200 THz (Enkrich et al. 2005a; Enkrich et al. 2005b; Klein et al. 2006b). However, this scaling breaks down for higher frequencies for the single SRR case because, for wavelengths shorter than the 200 THz range, the metal starts to strongly deviate from an ideal conductor (Zhou et al. 2005; Klein et al. 2006b). For an ideal metal with infinite carrier density, hence infinite plasma frequency, the carrier velocity and the kinetic energy are zero, even for finite current in a metal coil. For a real metal, velocity and kinetic energy become finite. For a small SRR, nonideal metal behavior leads to a modified scaling law where the frequency approaches a constant and becomes independent of the SRR size (Zhou et al. 2005; Klein et al. 2006b). This scaling limit combined with the fabrication difficulties of making nanometer-scale SRRs along with metal wires led to the development of alternative designs that are more suitable for the THz and optical regimes.

One suitable design for optical NIMs is based on pairs of metal rods (also called "cut-wires") or metal strips, separated by a dielectric spacer. Such structures can provide a magnetic resonance $\mu < 0$ originating from antiparallel currents in the strips. An electric resonance with $\varepsilon < 0$ can result from the excitation of parallel currents (Shalaev et al. 2005) in the same strips. However, normally it is difficult to get the $\varepsilon < 0$ and $\mu < 0$ regions to overlap, so a different design was proposed, the so-called fishnet structure. This structure combines the magnetic-coupled strips (providing $\mu < 0$) with continuous "electric" strips that provide $\varepsilon < 0$ in a broad spectral range. NIMs at optical wavelengths were successfully demonstrated by using coupled cut-wires and fishnet structures (Shalaev et al. 2005; Zhang et al. 2005c; Dolling et al. 2006a; Dolling et al. 2006b; Dolling et al. 2007; Chettiar et al. 2007).

Negative refraction of surface plasmons at visible frequencies was also demonstrated but was confined to a two-dimensional (2D) waveguide (Lezec, Dionne, and Atwater 2007).

2.1.2 Optical Metamaterial Fabrication: Main Challenges

To create a metamaterial operating at optical wavelengths, one should deal with small lattice constants of the "meta-crystal" (less then 300 nm) and tiny feature sizes (about 30 nm) to ensure effective-medium-like behavior. Thus, the fabrication of optical MMs is challenging since we aim at high-precision, high-throughput, and low-cost manufacturing processes. Feature sizes for metamaterials operating in the infrared or visible range can be smaller than the resolution of state-of-the-art photolithography (due to the diffraction limit), thus requiring nanofabrication processes with 100- or sub-100 nm resolution. Due to the limitations of current nanolithography tools, fabricated metamaterials do not really enter the real meta-regime where the unit-cell size is orders of magnitude smaller than the wavelength. However, the feature size is typically small enough compared to the wavelength so that one can describe such material, at least approximately, as a medium with effective permeability μ and permittivity ε (which is in contrast to photonic crystals where the lattice period matches the wavelength).

In addition to subwavelength resolution, careful material choice is required for optical MM fabrication. The choice of the metal is crucial in the optical regime because the overall losses are dominated by losses in the metal components. Thus, their lowest losses at optical frequencies make silver and gold the first-choice metals for optical metamaterials. Since the refractive index is a complex number $n = n' + in''$, where the imaginary part n'' characterizes light extinction (losses), a convenient measure for optical performance of a metamaterial is the figure of merit (FOM), defined as the ratio of the real and imaginary parts of n: $F = |n'| / n''$ (see, e.g., [Zhang et al. 2005a; Shalaev 2007]). Aside from the proper metal choice, loss compensation by introducing optically amplifying materials can be considered for achieving low-loss metamaterials (Klar et al. 2006; Popov and Shalaev 2006a; Popov et al. 2007).

To make use of the novel optical properties of metamaterials, where meta-atoms are designed to efficiently interact with both the electric and magnetic components of light, one has to create a slab of such a material with an interaction length of many light wavelengths. Together with the novel fundamental properties of NIMs, entirely new regimes of nonlinear interactions were predicted for such materials (see, e.g., [Lapine, Gorkunov, and Ringhofer 2003; Zharov, Shadrilov, and Kivshar 2003; Agranovich et al. 2004; Mattiucci et al. 2005; Gorkunov, Shadrilov, and Kivshar 2006; Popov and Shalaev 2006a; Klein et al. 2006a; Popov and Shalaev 2006b; Popov et al. 2007; Litchinitser, Gabitov, and Maimistov 2007; Klein et al. 2007]). For all exciting applications to be within reach, the next issue to address (in addition to losses) is the development of truly three-dimensional (3D) metamaterials in the optical range. The challenging tasks here are to move from planar structures to a 3D slab of layered metamaterial and to develop new isotropic designs.

With the development of low-loss large-scale optical metamaterials, many new device concepts will emerge empowered by optical magnetism and new regimes of linear and nonlinear light–matter interactions. Materials to manipulate an object's degree of visibility, subwavelength imaging systems for sensors and nanolithography tools, and ultra-compact waveguides and resonators for nanophotonics are only few examples of possible metamaterial applications. For example, among the most exciting new applications for 3D low-loss metamaterials are those based on transformation optics (Pendry, Schurig, and Smith 2006; Leonhardt 2006; Kildishev and Shalaev 2008), including a hyperlens that enables subwavelength far-field resolution (Jacob, Alekseyev, and Narimanov 2006; Salandrino and Engheta 2006; Liu et al. 2007; Smolyaninov, Hung, and Davis 2007; Narimanov and Shalaev 2007; Kildishev and Narimanov 2007) and designs for optical cloaking (Pendry, Schurig, and Smith 2006; Schurig et al. 2006; Cai et al. 2007a; Cai et al. 2007b).

In this chapter, the recent progress in fabrication of metal–dielectric nanostructured metamaterials at optical wavelengths is reviewed.

2.1.3 Brief Introduction to Lithographic Patterning

Fabrication of devices with dimensions measured in nanometers requires complicated equipment and a special environment (*cleanroom*). One of the most important methods in micro- and nanofabrication is lithography (see, e.g., [Levinson 2005])—the art and science of writing or printing at the micro- and submicrometer level. Lithography is one of the most common patterning methods capable of structuring material on a fine scale. Typically, structuring on the scale smaller than 10 μm is considered to be *microlithography* (e.g., *photolithography*), and patterning of features smaller than 100 nm—*nanolithography* (e.g., *electron-beam lithography or EBL*). Lithography is a process used in micro- and nanofabrication to selectively remove parts of a thin film or the bulk of a substrate. In photolithography (also called *optical lithography, OL*), a prefabricated *mask* is used as a master from which the final pattern is derived; it uses light to transfer a geometric pattern from a photomask to a light-sensitive chemical photoresist, or simply "*resist*," on the substrate. In EBL, a beam of electrons is scanned across a surface covered with an EBL resist that is sensitive to electrons, thus depositing energy

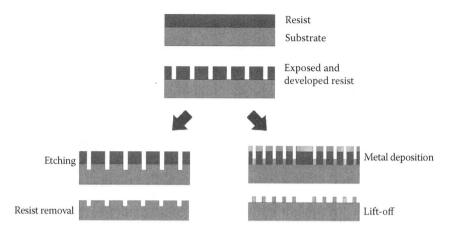

Resist

Substrate

Exposed and
developed resist

Etching

Metal deposition

Resist removal

Lift-off

FIGURE 2.1 Schematic of fabrication steps for making patterns using lithography.

in the desired pattern on the resist film. Resists are the recording and transfer media for lithography: light or electron-beam exposure modifies the resist so that it develops away at exposed regions. When a pattern is defined in the resist, one can transfer it further into the substrate using etching or create structures by depositing some material (e.g., metal) into the developed holes in the resist and lifting off the unneeded resist (Figure 2.1). Both photo- and electron-beam lithography used for optical metamaterial fabrication are described in more details in the following sections.

Various types of lithography are used in different applications depending on the designed pattern, materials, or required resolution. One of the photolithography types is *interference lithography* (IL)—a technique for patterning regular arrays of fine features, without the use of complex optical systems or masks. The basic principle is the use of the interference pattern between two or more coherent light waves to expose the resist. Current research is exploring new operating regimes for photolithography ranging from extreme ultraviolet lithography and x-ray to ion-projection lithography.

2.2 First Experimental Demonstrations: Single Metamaterial Layer

The first proof-of-principle studies of optical metamaterials exhibiting negative index of refraction were performed on 2D structures—a single layer of a metamaterial created on top of a transparent substrate. The first experimental demonstrations of negative refractive index in the optical range were accomplished nearly at the same time for two different metal–dielectric geometries: pairs of metal rods separated by a dielectric layer (Shalaev et al. 2005) and the inverted system of pair of dielectric voids in a metal–dielectric–metal multilayer (Zhang et al. 2005c) (Figure 2.2).

In the first example, an array of pairs of parallel 50-nm-thick gold rods separated by 50 nm of SiO_2 spacer was fabricated using EBL—a standard process where structures were created via EBL patterning, exposure, multimaterial deposition, and the lift-off process (Shalaev et al. 2005; Dolling et al. 2006a; Yuan et al. 2007; Chettiar et al. 2007).

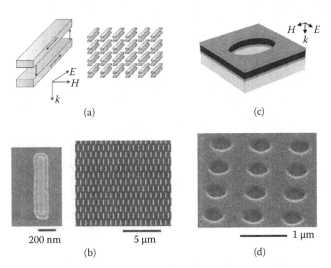

FIGURE 2.2   First experimentally obtained optical negative-index metamaterials: (a) schematic of an array of paired Au nanorods separated by a layer of SiO$_2$ together with (b) field-emission scanning electron microscope (SEM) images of the fabricated array (Au (50 nm)–SiO$_2$ (50 nm)–Au (50 nm) stacks), where a negative refractive index is achieved at telecommunication wavelengths (Shalaev et al. 2005); (c) schematic of a multilayer structure consisting of a dielectric layer between two metal films perforated with a hole on a glass substrate; (d) SEM image of the fabricated structure (Au (30 nm)–Al$_2$O$_3$ (60 nm)–Au (30 nm) stack, 838 nm pitch, 360 nm hole diameter), which exhibits a negative index at about $\lambda \approx 2\,\mu m$. (Reprinted (c,d) with permission from Zhang, S. et al. 2005c. *Physical Review Letters* 95: 137404-4. Copyright (2005) by the American Physical Society.)

The coupled rods-based metamaterial exhibited a negative refractive index of $n' \approx -0.3$ at 1.5 μm (Shalaev et al. 2005). In the other work, IL with a 355 nm UV source was used to define a 2D array of holes in a multilayer structure (60-nm-thick Al$_2$O$_3$ dielectric layer between two 30-nm-thick Au layers), and the structure was shown to exhibit a negative refractive index of −2 around 2 μm (Zhang et al. 2005c). Being the first proof-of-principle experiments, these structures had a high loss coefficient (so that the FOM was small).

2.3 Fabrication of 2D Metamaterials

A 2D metamaterial structure consists of a single functional layer that is normally perpendicular to the incident light (direction of propagation). Such single-layer structures are not suitable for studying complex (bulk) phenomena in metamaterials, and it can be difficult to assign macroscopic properties such as the index of refraction to the monolayer. However, with development of reliable modeling tools and retrieval methods, 2D structures are widely used for proof-of-principle experiments with novel designs and materials. Moreover, the majority of optical metamaterials (Shalaev 2007; Soukoulis, Linden, and Wegener 2007) can be fabricated by lithographic patterning (either electron-beam or photolithography) and evaporation of metal and dielectric films; both of these processes are well-established planar fabrication technologies available in any clean-room. Such planar fabrication approaches are described in the following sections.

2.3.1 Standard Method: EBL

Due to the fact that the required feature sizes for optical metamaterial fabrication are smaller than the resolution of state-of-the-art photolithography, 2D metamaterial layers are normally fabricated using EBL (Shalaev et al. 2005; Dolling et al. 2006a; Dolling et al. 2006b). In EBL, a beam of electrons is used to generate patterns on a surface. Beam widths can be on the order of nanometers, which gives rise to the high nanoscale resolution of the technique. EBL is a serial process wherein the electron beam must be scanned across the surface to be patterned. This technique is quite versatile at the point of initial design and preliminary studies of optical properties of metamaterials since it offers sub-wavelength resolution and almost complete pattern flexibility. In the past 2 years, different metamaterial structures were successfully fabricated by EBL and experimentally investigated by several groups. When compared to the first structures (Shalaev et al. 2005; Zhang et al. 2005c), second-generation NIMs showed improved optical performance via higher figures of merit (i.e., lower losses).

The first substantial improvement of the FOM for a negative refractive index material at the telecommunication wavelength compared to the first samples was achieved in 2006 (Dolling et al. 2006a). The results were obtained for a fishnet structure (Zhang et al. 2005a), which could be viewed as an array of rectangular dielectric voids in parallel metal films (rather than circular voids as in [Zhang et al. 2005c]). A refractive index in the fabricated fishnet structure (lattice constant 600 nm) based on a sandwich of Ag (45 nm)–MgF$_2$ (30 nm)–Ag (45 nm) was reported to reach $n' = -2$ at $\lambda \approx 1.45\ \mu m$.

Moving to shorter wavelengths, two groups reported negative refractive index results in the visible range: the Karlsruhe group ($n' = -0.6$ at 780 nm) (Dolling et al. 2007) and a group at Purdue University ($n' = -0.9$ and $n' = -1.1$ at about 770 nm and 810 nm, respectively) (Chettiar et al. 2007) (Figure 2.3). We note while the demonstrated NIMs at 780 nm and 770 nm were single negative (i.e., even though $n < 0$ was accomplished, μ remained positive at these wavelengths), the Purdue material at 810 nm was double-negative, with both ε and μ simultaneously negative at 810 nm, resulting in a larger (exceeding one) FOM (see, e.g., [Chettiar et al. 2007] for the detailed discussion of the $n' < 0$ requirements: strong [sufficient] condition that $Re(\varepsilon) < 0$ and $Re(\mu) < 0$, or a more general [necessary and sufficient] condition $\varepsilon'\mu'' + \varepsilon''\mu' < 0$). Both designs were based on nano-fishnets made in either Ag–MgF$_2$–Ag (Dolling et al. 2007) or Ag–Al$_2$O$_3$–Ag (Chettiar et al. 2007) multilayer structures with the smallest features below 100 nm. The fabricated structures had good large-scale homogeneity. However, due to sub-100 nm features required for visible-range NIMs (68 nm minimum in-plane feature size at 97 nm thickness of the Ag–MgF$_2$–Ag sandwich [Dolling et al. 2007]), the aspect ratio (height/width) for these NIM stacks can exceed unity. This gives rise to significant fabrication challenges connected to the lift-off procedure, and to increased sidewall roughness (Dolling et al. 2007) (Figure 2.3).

Even though EBL is more frequently used to write relatively small areas, large area all-dielectric planar chiral metamaterials can be fabricated by EBL (Zhang et al. 2006c;

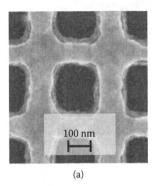 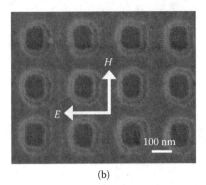

(a)   (b)

FIGURE 2.3 SEM images of nano-fishnets fabricated by electron-beam lithography, metal–dielectric–metal stack deposition, and a lift-off procedure: (a) by the Karlsruhe group ($n' = -0.6$ at 780 nm, Ag (40 nm)–MgF$_2$ (17 nm)–Ag (40 nm) stack, lattice constant 300 nm) (Dolling, G. et al. 2007. *Optics Letters* 32: 53–55) and (b) by the Purdue group ($n' = -0.9$ at 772 nm, Ag (33 nm)–Al$_2$O$_3$ (38 nm)–Ag (33 nm), lattice constant 300 nm) (Chettiar, U. K. et al. 2007. *Optics Letters* 32: 1671–1673). The smallest feature is below 100 nm, side walls are quite rough (Dolling, G. et al. 2007. *Optics Letters* 32: 53–55).

Zhang et al. 2007). In this approach, the use of a charge dispersion layer solves *stitching errors* (misalignment, undesirable shift between two subsequently written patterns caused by sample warping or tilt; such miscalibration between the stage and beam movements results in interruptions at the end of each electron-beam field), which enables the fabrication of good-quality structures without the limitations normally encountered by stitching errors and field alignment.

Main fabrication challenges in making optical metamaterials come from the requirements of small periodicities (near and below 300 nm) and tiny feature sizes (down to several tens of nm). Thus, EBL is still the method of choice for fabricating metamaterials despite low throughput of the serial point-by-point writing and high fabrication costs. Since only small areas (of the order of 100×100 μm) can normally be structured within reasonable time and at reasonable costs, EBL does not offer a solution for the large-scale optical metamaterial fabrication required by applications, where many square centimeters have to be nanopatterned.

2.3.2 Rapid Prototyping: Focused Ion Beam Milling

For rapid prototyping of optical metamaterials, *focused ion beam* (*FIB*) milling technique can be used. In FIB approach, a focused beam of gallium ions is used to sputter atoms from the surface or to implant gallium atoms into the top few nanometers of the surface, making the surface amorphous. Because of the sputtering capability, the FIB method is used as a micromachining tool, to modify or machine materials at the micro- and nanoscale. This technique was successfully employed for fabricating magnetic metamaterials based on SRRs (Enkrich et al. 2005a). Scaling of the SRR structure requires sub-100 nm gap sizes (down to 35 nm) for a 1.5 μm resonance wavelength, thus requiring state-of-the-art nanofabrication tools. For such small features, EBL-based fabrication requires time-consuming tests and careful optimization of writing parameters and processing steps, leading to relatively long overall fabrication times. In contrast, the rapid prototyping of complete structures can be fabricated via FIB writing in times as short as 20 min (Enkrich et al. 2005a). The process is based on FIB writing and corresponds to an inverse process where the FIB removes metal (20 nm of Au) deposited on a glass substrate. So the process is as simple as depositing a uniform layer of metal onto the substrate and then removing it in defined places using FIB, thus immediately creating the designed metal pattern on the substrate. After FIB writing, the structure is ready and no further postprocessing steps are required in contrast to EBL where additional processing steps are performed after the lithographic patterning.

FIB nanofabrication might be of preference for fabricating specific (e.g., SRR-based) designs of metamaterials as in the preceding example. However, for creating an optical NIM, SRRs have to be combined with other metallic structures that can provide negative permittivity (Shalaev 2007). This will give rise to additional processing steps and delicate considerations regarding the choice of process parameters and materials. Moreover, moving SRRs from the telecom to the visible range will bring the process to the limit of size scaling in SRRs (Zhou et al. 2005; Klein et al. 2006b). Due to design and material limitations, FIB technique has certain limitation in its use for fabricating optical metamaterials. Nevertheless, it should be mentioned that for some specific designs and proof-of-principle experiments, for example, including nonlinear materials, FIB is the first choice for rapid prototyping. One example of using FIB for creating a thick (21-layer) metamaterial layer is reported in 3D fabrication section.

2.3.3 Large-Scale Fabrication: Interference Lithography

The only large-scale manufacturing technique used so far by the integrated circuit industry is *optical lithography* (OL). Nowadays, OL continues to be extended and offers new ways of increasing its resolution, for example, by using immersion techniques that meet the industry needs for 45-nm half-pitch nodes (Brueck 2005). One type of OL, namely, IL, is a powerful technique for the fabrication of a wide array of samples for nanotechnology. This fabrication technique is based on the superposition of two or more coherent optical beams forming a standing wave pattern. Being a parallel process, IL provides a low-cost, large-area (up to ~ cm^2) mass-production capability. Moreover, multiple exposures, multiple beams, and mix-and-match synthesis with other lithographic techniques can extend the range of IL applicability (Brueck 2005). IL offers high structural uniformity combined with considerable, but not total, pattern flexibility, while its resolution is now approaching the 20 nm scale (Brueck 2005).

Since optical metamaterial manufacturing is often based on a periodic or quasi-periodic pattern, IL is an excellent candidate for large-area fabrication. Recently, interference lithography was employed for the fabrication of one-dimensional metallic structures (Feth et al. 2006), magnetic metamaterials at 5 μm and 1.2 μm wavelength (Zhang et al. 2005b; Feth et al. 2006), and, as mentioned earlier, NIMs at 2 μm (Fan et al. 2005; Zhang et al. 2005c) (Figure 2.2). Using this technique, the fabrication of square-centimeter-area structures was demonstrated (Fan et al. 2005; Zhang et al. 2005b; Feth et al. 2006) as well as the large-area homogeneity (Feth et al. 2006). For example, a large-scale NIM created

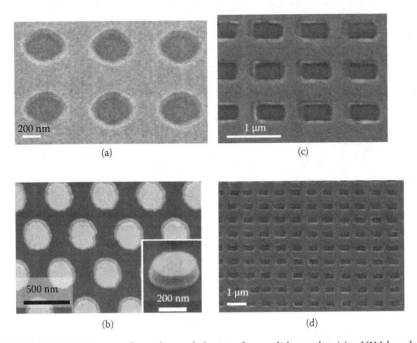

FIGURE 2.4 SEM images of samples made by interference lithography: (a) a NIM based on coupled voids in an Au (30 nm)–Al$_2$O$_3$ (75 nm)–Au (30 nm) multilayer structure (pitch 787 nm, hole sizes 470 and 420 nm) (Zhang, S. et al. 2006a. *Journal of the Optical Society of America B* 23: 434–438) and (b) a hexagonal 2D structure built by Au (20 nm)–MgF$_2$ (60 nm)–Au (20 nm) pillars on a glass substrate (Feth, N. et al. 2006. *Optics Express* 15: 501–507); (c), (d) fishnet structure samples in an Au (30 nm)–Al$_2$O$_3$ (60 nm)–Au (30 nm) multilayer structure (528 nm, 339 nm on long and short sides, respectively) (Ku, Z., and S. R. J. Brueck. 2007. *Optics Express* 15: 4515–4522).

by making elliptical voids in an Au (30 nm)–Al$_2$O$_3$ (75 nm)–Au (30 nm) multilayer stack was found to exhibit $n' \approx -4$ at 1.8 μm (Figure 2.3) (Zhang et al. 2006a). Using different IL-fabricated metamaterial structures, a negative refractive index was obtained over a range of wavelengths: 1.56 μm to 2 μm, 1.64 μm to 2.2 μm, and 1.64 μm to 1.98 μm for NIMs with circles, ellipses, and rectangles (fishnet), respectively (Figure 2.4) (Ku and Brueck 2007).

The results mentioned earlier established IL as a new direction for the design and fabrication of 2D optical MMs and highlighted the advantages of such a large-area patterning technique. The technique is compact, robust, does not require expensive cleanroom equipment and can provide sample areas up to many square centimeters by upscaling the apertures of the optics (Feth et al. 2006).

Given the simplicity and robustness of making a high-quality, single layer of a metamaterial using IL, one can envision further investigations aiming at piling 2D layers to create a 3D structure. Such a transition to 3D fabrication will turn parallel IL process into a step-by-step procedure that would require alignment of subsequent layers. Even though for compact versions of IL, such a transition might result in a time-consuming fabrication process due to multiple alignments, proper technique development would make it possible to optimize and automate the alignment procedure. Thus, IL can be considered as one possible approach for making 3D optical metamaterials.

2.3.4 High-Resolution Large-Scale Fabrication: Nanoimprint Lithography

Another promising direction for the fabrication of production-compatible, large-area, high-quality optical MMs at low processing cost and time is offered by *nanoimprint lithography (NIL)* (Chou, Krauss, and Renstrom 1996; Sotomayor Torres et al. 2003). A next generation lithography candidate, NIL accomplishes pattern transfer by the mechanical deformation of the resist via a stamp rather than a photo- or electroinduced reaction in the resist as in most lithographic methods. Thus, the resolution of the technique is not limited by the wavelength of the light source, and the smallest attainable features are given solely by stamp fabrication. Moreover, NIL provides parallel processing with high throughput. Since metamaterial fabrication requires high patterning resolution, NIL is well suited for large-scale production of optical metamaterials, providing wafer-scale processing using standard cleanroom procedures combined with simplicity and low cost.

Room temperature NIL was also successfully applied to the fabrication of planar, chiral, photonic metamaterials for the study and application of novel polarization effects where both dielectric and metallic metamaterials with feature sizes from micrometric scale down to sub-100 nm were fabricated (Chen et al. 2005).

In 2007, two types of metamaterials with negative refractive index operating at near-IR and mid-IR frequencies, respectively, were fabricated via NIL. The first structure comprised ordered "fishnet" arrays of metal–dielectric–metal stacks that demonstrated negative permittivity and permeability in the same frequency range and hence exhibited a negative refractive index of $n' \approx -1.6$ at a wavelength near 1.7 μm (Wu et al. 2007a) (Figure 2.5). In the mid-IR range, the metamaterial was an ordered array of fourfold symmetric L-shaped resonators (with a minimum feature size of 45 nm) that were shown to exhibit negative permittivity and a magnetic resonance with negative

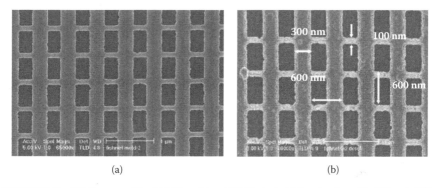

(a) (b)

FIGURE 2.5 SEM images of the (a) NIL mold and (b) fabricated fishnet pattern (Ag (25 nm)–SiOx (35 nm)–Ag (25 nm) NIM stack) (Wu, W. et al. 2007b. *Applied Physics Letters* 90: 063107–3).

permeability near wavelengths of 3.7 μm and 5.25 μm, respectively (Wu et al. 2007b). The smallest achieved feature sizes were about 100 nm and 45 nm for near- and mid-IR NIMs, respectively.

Metallic 2D structures can also be created via simplified NIL method (Chen et al. 2007). The technique is based on direct, hot embossing into metals (e.g., Al) using hard templates such as SiC (Pang et al. 1998). In this approach, metal nanostructures can be obtained by printing directly on metal substrates without any further processing step (when compared to standard lithography and etching/deposition steps), thus simplifying the processing steps and lowering the production cost. Using hard molds, successful pattern transfer directly to Al substrates as well as Au films was achieved by pressing at room temperature (Pang et al. 1998). The main challenge in the technical development of this approach is connected to the fabrication of hard stamps. For the existing designs, this approach cannot be directly applied to the fabrication of optical metamaterials due to specific requirements on geometry and materials.

2.4 Fabrication of 3D Metamaterials

Exploration of optical phenomena in three dimensions as well as realization of metamaterial-based devices clearly requires large-scale 3D metamaterials operating at optical frequencies. The major challenges concern the realistic 3D design and the fabrication of such structures. Different approaches for fabrication of bulk metamaterials are discussed in the following sections.

2.4.1 Stacking Multiple Functional Layers

Stacking up multiple functional layers along the propagation direction constitutes a promising approach for achieving a 3D optical metamaterial. This idea was first verified numerically, when a low-loss optical metamaterial with a thickness much larger than the free-space wavelength in the near-IR region was demonstrated (Zhang et al. 2006b). In the work by Zhang et al., simulations showed that a NIM slab consisting of multiple layers of perforated metal–dielectric stacks (for 100 and 200 layers) would exhibit a small imaginary part of the index over the wavelength range for negative refraction. This established a new approach for thick, low-loss metamaterials at IR and optical frequencies. The theoretically suggested multilayer NIM design (Zhang et al. 2006b) was later modified, and corresponding structures with up to 10 functional layers (up to 21 actual deposited layer) were fabricated (Dolling, Wegener, and Linden 2007; Liu et al. 2008; Valentine et al. 2008).

The silver-based three-functional-layer optical metamaterial (Dolling, Wegener, and Linden 2007) was fabricated by standard EBL, metal and dielectric depositions, and a lift-off procedure with processing steps similar to those for creating a single NIM layer (Dolling et al. 2006a; Dolling et al. 2006b) (Figure 2.6a,b). The measured performance of the fabricated NIMs was found to be close to theory, and the retrieved optical parameters ($n' = -1$ at 1.4 μm) did not change much with the number of functional layers, as expected for an ideal metamaterial. Realization of a three-functional-layer NIM was the

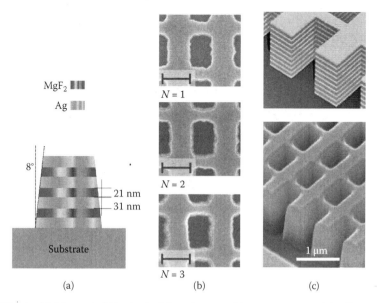

FIGURE 2.6 (a) Schematic (side view) of the 3-functional-layer metamaterial (Dolling, G. et al. 2007. *Optics Letters* 32: 551–553) together with (b) SEM micrographs of fabricated structures with *N* functional layers (400 nm scale bar) (Dolling, G. et al. 2007. *Optics Letters* 32: 551–553); (c) diagram and SEM image of fabricated 21-layer fishnet structure (alternating layers of 30 nm silver and 50 nm magnesium fluoride) (Valentine, J. et al. 2008. *Nature* 455: 07247).

first step toward a 3D photonic metamaterial. However, if one wants to create thicker metamaterial slabs using this approach, the fabrication procedure becomes increasingly difficult. The main challenge is related to the fact that in a standard deposition–lift-off procedure, the total thickness of the deposited layers is limited by the thickness of the patterned e-beam resist. For a successful lift-off procedure, the total deposited thickness should normally be at least 15%–20% less than the thickness of the resist (Figure 2.7a). This value is usually not more than a couple hundred nanometers (for e-beam writing structures with sub-100 nm features). For 2D fabrication via EBL, an aspect ratio exceeding unity poses significant fabrication challenges. Moreover, this fabrication procedure results in nonrectangular side walls, typically with an angle of about 10° with respect to the substrate normal on all sides (Shalaev et al. 2005; Dolling, Wegener, and Linden 2007) (Figure 2.7a). Obviously, this effect becomes particularly large for thick, multilayer structures.

To overcome problems connected to the lift-off procedure and nonrectangular side walls, and to create thicker metamaterial stacks, an alternative approach was suggested (Liu et al. 2008). In this approach, a stack of a 3D optical metamaterial was realized through a layer-by-layer technique similar to that developed for 3D photonic crystal fabrication (Subramania and Lin 2004). In the experiment, a four-layer SRR structure was fabricated. A single SRR layer was fabricated by simple metal evaporation, electron-beam exposure, development, and ion-beam etching of the metal. Since the nonplanar surface of the single SRR layer does not allow simple stacking by a serial layer-by-layer process,

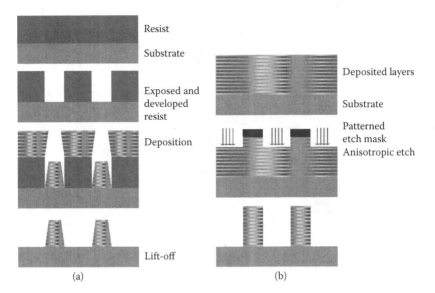

FIGURE 2.7 Schematic of two possible ways of making structures with multiple metal–dielectric layers: (a) the standard deposition–lift-off procedure (Dolling, G. et al. 2007. *Optics Letters* 32: 551–553) that provides trapezoidal final structures and has a total deposited thickness limitation (it has to be at least 15%–20% less than the thickness of the resist for a successful lift-off process) together with (b) the proposed etch-based procedure where a thick planar stack of metal–dielectric layers is deep etched to create a 3D metamaterial slab (Valentine, J. et al. 2008. *Nature* 455: 07247).

the surfaces of the SRR layers were flattened by applying a planarization procedure with dielectric spacers (the roughness of the planarized surface was controlled within 5 nm). The procedures of single-layer fabrication, planarization, and lateral alignment of subsequent layers were repeated several times yielding the four-layer SRR sample (Liu et al. 2008) (Figure 2.8). The developed layer-by-layer techniques can be seen as one of the

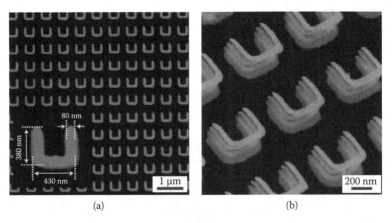

FIGURE 2.8 Field-emission SEM images of the four-layer SRR structure: (a) normal view, (b) enlarged oblique view. (Liu, N. et al. 2008. *Nature Materials* 7: 31–37.)

approaches for the manufacturing of 3D optical metamaterials. However, one has to bear in mind that such a step-by-step method requires additional process development (such as planarization techniques) and careful lateral alignment of different layers that are crucial for successful stacking. Together with EBL fabrication of a single layer, alignment procedures lead to increased fabrication time. Thus, this method is still too costly and has a low throughput for creating large-scale 3D metamaterial slabs for possible applications.

While the 3D metamaterial (Liu et al. 2008) described earlier did not have negative index of refraction, optical metamaterial fabricated later using FIB-based method had negative refractive index with a very high figure of merit of 3.5 (low loss) (Valentine et al. 2008). The 21-layer fishnet metamaterial was fabricated on a multilayer metal–dielectric stack by using focused ion-beam milling (Valentine et al. 2008) (Figure 2.6c). In this work, FIB procedure was optimized for cutting nanometer-sized features with a high aspect ratio providing a 3D structure that was milled on 21 alternating films of silver and magnesium fluoride, resulting in ten functional layers (Figure 2.6c).

In the latter method, replacing complex FIB milling by a process based on deep anisotropic etching may lead to a large-scale, high-throughput fabrication technology of creating 3D multiple-layer metamaterials. In the suggested approach, planar, alternating metal and dielectric layers of desired thicknesses are first fabricated on a substrate. 3D patterning can then be achieved by deep anisotropic etch using an etch mask prepatterned to any required design by lithographic means (Figure 2.7b). However, this approach would require both heavy material and process development including the careful choice of etch-resistant mask materials and anisotropic etch optimization so that both the metal and dielectric layers can be etched.

A 3D optical metamaterial exhibiting negative refraction was also realized in all-semiconductor layered structure (Hoffman et al. 2007). The samples were fabricated by the growth method well established in semiconductor industry: alternating 80-nm-thick layers of $In_{0.53}Ga_{0.47}As$ and $Al_{0.48}In_{0.52}As$ were grown by molecular beam epitaxy on lattice-matched InP substrates. In these samples, doped InGaAs layers were used to provide a plasma resonance of free carriers instead of metal as in previously discussed metamaterials. This design is unique since it relies on an anisotropic dielectric function, instead of overlapping electric and magnetic resonances, which makes it straightforward to design and comparatively low-loss (Hoffman et al. 2007). In addition, this 3D semiconductor metamaterial requires no further fabrication steps beyond the initial growth. Demonstration of a comparatively low-loss, 3D semiconductor metamaterial where negative refraction was observed for all incidence angles across wide wavelength range in the IR region opens up a new research avenue aiming at incorporation of optical metamaterials into semiconductor devices.

2.4.2 Two-Photon Photopolymerization Technology

The *two-photon photopolymerization* (*TPP*) technique has been used extensively in the past few years to realize 3D patterning (Maruo, Nakamura, and Kawata 1997; Kawata et al. 2001). TPP involves polymerization of a material via a nonlinear, multiphoton process that only occurs at the focal point of a tightly focused laser beam, thereby providing

3D control over the location of polymerization. TPP enables the fabrication of complex objects with subdiffraction resolutions (Takada, Sun, and Kawata 2005) since the absorption of light in the material occurs only at the focal region of the laser beam.

While most studies on TPP have focused on the realization of polymeric structures (Kawata et al. 2001; Takada, Sun, and Kawata 2005; Passinger et al. 2007), this technique was recently studied for the fabrication of metallic 3D patterns (Formanek et al. 2006a; Takeyasu, Tanaka, and Kawata 2007). A recent realization of 3D, periodic, metallic structures over large areas was achieved through a TPP technique combined with a microlens array (Formanek et al. 2006b). This technique is a step forward from standard, single-beam laser writing into a polymer matrix, which is time-consuming and thus unlikely to be adopted for large-scale fabrication. The proposed novel method, however, enabled the simultaneously writing of more than 700 polymer structures that were uniform in size. The metallization of the structures was then achieved through the deposition of thin films composed of small silver particles by means of electroless plating (Figure 2.9). A hydrophobic coating on the substrate prevented silver deposition in unwanted areas and allowed the formation of a large number of isolated and highly conducting objects. This metallization is flexible in that it can produce either polymer structures covered with metal or numerous isolated, insulating polymer objects spread over a metallic film, depending on the resin properties and treatment procedures (Formanek et al. 2006a; Formanek et al. 2006b). TPP-based laser writing is now considered to be one of the most promising methods for future manufacturing of large-area, true 3D metamaterials. Offering intrinsic 3D parallel processing capability with acceptable resolution (100 nm [Takada, Sun, and Kawata 2005]), TPP can be successfully combined with selective metal deposition by electroless plating (Formanek et al. 2006a; Formanek et al. 2006b; Takeyasu, Tanaka, and Kawata 2007). While TPP enables arbitrary sculpting below the micrometer scale, electroless plating can provide thin coatings of various metals. Going in this direction, process development and optimization is required in order to include metal–dielectric layers into a polymer template made by TPP.

In addition to electroless deposition, multimaterial deposition (including silver) can be achieved via *chemical vapor deposition* (CVD) (Chi et al. 2007; Rill et al. 2008), which is a chemical process used to deposit high-purity materials. Such fabrication approach can be seen as a 3D analog of planar lithographical pattering (such as EBL) combined with standard evaporation. Recently, high-quality magnetic metamaterials at near-IR frequencies were fabricated using the combination of direct laser writing of polymer templates and silver chemical vapor deposition (Rill et al. 2008) (Figure 2.9). This approach enables rapid prototyping of complex 3D photonic metamaterials. However, metamaterials theory has not yet provided blueprints for 3D metamaterials compatible with this approach, so advances in 3D design and fabrication are yet to come.

Recently, a bianisotropic negative-index metamaterial was fabricated via 3D two-photon direct laser writing and silver shadow evaporation that exhibited a negative real part of the refractive index at around 3.85 μm wavelength (Rill et al. 2009). This fabrication approach uses silver shadow evaporation, and hence cannot be directly extended to an arbitrary number of functional layers to create 3D metamaterial. However, these latest results prove that state-of-the-art 3D direct laser writing can lead to very high quality photonic metamaterials operating at near-IR wavelengths (Rill et al. 2009).

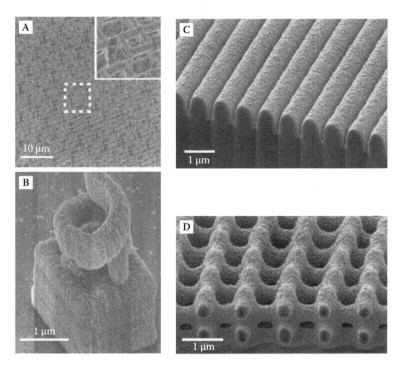

FIGURE 2.9 SEM images of fabricated structures: (a) self-standing empty cubic structures (height ~4.6 µm) connected in pairs; (b) a silver-coated polymer structure composed of a cube (2 µm in size) holding up a spring (inner diameter 700 nm) (Formanek et al. 2006a)—(a) and (b) structures are made by a two-photon induced photopolymerization technique combined with electroless plating (Formanek, F. et al. 2006a. *Applied Physics Letters* 88: 083110–3; Formanek, F. et al. 2006b. *Optics Express* 14: 800–809; Takeyasu, N. et al. 2007. Fabrication of 3D metal/polymer fine structures for 3D plasmonic metamaterials. Proceedings of the OSA meeting "Photonic Metamaterials: From Random to Periodic" (2007); (c) metamaterial corresponding to a planar lattice of elongated SRRs (note visible SiO_2 layer between the polymer (SU-8) template and the silver coating) (Rill, M.S. et al. *Nature Materials* 7: 543–546. 2008); (d) a 3D structure composed of bars (silver coating covers the bars all around) (Rill, M.S. et al. *Nature Materials* 7: 543–546. 2008)—structures (c) and (d) are made by TPP technique combined with silver chemical vapor deposition (Rill, M.S. et al. *Nature Materials* 7: 543–546. 2008).

2.4.3 Fabrication of Complex 3D Structures

Complex 3D metal–dielectric nanostructures can be fabricated today by several techniques. Considerable attention was recently attracted by two methods: *direct electron-beam writing (EBW)* (Griffith et al. 2002) and *focused ion-beam chemical vapor deposition* (*FIB-CVD*) (Morita et al. 2004). These methods offer 3D fabrication that is not possible using traditional layered optical and EBL techniques. The use of EBW was demonstrated for building structures of multiple layers with linewidth resolutions of 80–100 nm using an electron beam to cause direct sintering of 2–10 nm nanoparticles (Griffith et al. 2002), while various free-space Ga- and W-containing nanowirings were successfully fabricated by FIB-CVD, also called *free-space-nanowiring fabrication technology* (Morita et al. 2004). In the EBW process (Griffith et al. 2002), an additive-layer build technique for multiple-material functionality was developed. The processing steps involved spin-coated or drawn-down solutions of thiol-capped nanoparticles (either Ag or Au) onto silicon wafers, glass slides, or polyimide films. The resultant nanoparticle films were patterned by electron beam and showed conductivity for the patterned metals within one order of magnitude of their bulk material properties. Multiple layer fabrication was successfully demonstrated by repeating the processing steps for subsequent layers spun over previously patterned features. Further material and process optimization (e.g., to allow combinations of multiple materials) may lead to EBW adoption for test fabrications of complex metamaterials. While these and similar (Hoffmann 2005) nanofabrication techniques offer unique possibilities for making very complex 3D structures, they all suffer from severe material limitations in terms of what materials can be patterned or deposited. Moreover, such methods are somewhat exotic, complex, and time consuming. Thus they can only be used for making first prototypes or single structures for proof-of-principle studies.

Complex 3D metallic structures can also be fabricated using layer-by-layer repetition of standard planar processes. Recently, 3D tungsten photonic crystals were fabricated using a modified planar silicon MEMS process developed at Sandia National Labs (McCormick et al. 2006). In this approach, a patterned silicon dioxide mold is filled with a 500-nm-thick tungsten film and planarized using a chemical mechanical polishing process. This process is repeated several times, and at the end of the process, the silicon dioxide is released from the substrate, leaving a freely standing thin patterned tungsten film. This method allows the patterning of multilayered (up to 60) samples of large area ($\sim$ cm^2).

Another process for the parallel fabrication of microcomponents is the *LIGA microfabrication technique*. LIGA is an acronym referring to the main steps of the process, that is, deep x-ray lithography, electroforming, and plastic molding (German acronym for Lithographie, Galvanoformung, Abformung [lithography, electroplating, and molding]). These three steps make it possible to mass-produce microcomponents at a low cost. Recently, the LIGA method was used by Sandia National Labs to fabricate 3D photonic lattices (McCormick et al. 2006). In this approach, the 3D lattice is patterned using deep x-ray lithography to create a series of intersecting channels in a polymer. This technique makes it possible to create large area molds that can be filled with gold by electroplating techniques. Large-scale woodpile structures with gold as the matrix material were recently demonstrated (McCormick et al. 2006).

2.4.4 3D Structures by Nanoimprint

As an ultra-high resolution patterning technique simultaneously offering sub-50 nm resolution and sub-10 nm layer alignment capability, NIL can be used in multilayer processes for creating 3D structures. NIL has the potential for very large scale and inexpensive manufacture of structures via multilayer lithography, alignment, and plating techniques. For example, 3D cubic arrays of gold cubes separated by a polymer were recently fabricated (McCormick et al. 2006); such arrays can act as photonic crystals. This rapid-prototyping *contact lithography* approach provides a platform for investigating new structures, materials, and multilayer alignment techniques that are critical for device designs at optical operational wavelengths.

A newly developed nanofabrication technique, namely, *reverse-contact UV nanoimprint lithography* (Kehagias et al. 2007), which is a combination of NIL and contact printing lithography, can also realize 3D polymer structures. In this process, a lift-off resist and a UV cross-linkable polymer are spin-coated successively onto a patterned UV mask-mold. These thin polymer films are then transferred from the mold to the substrate by contact at a suitable temperature and pressure. The whole assembly is then exposed to UV light. After separation of the mold and the substrate, the unexposed polymer areas are dissolved in a developer solution leaving behind the negative features of the original stamp. This method delivers resist pattern transfer without a residual layer, thereby rending unnecessary the etching steps typically needed in the imprint lithography techniques for 3D patterning. This method is reproducible over millimeter-scale surface areas and has already provided encouraging results for fabricating 3D woodpile-like structures for polymer photonic devices (Figure 2.10a) (Kehagias et al. 2007). Combined with multimaterial deposition similar to the TPP case, this approach might also be adapted for future optical MM fabrication.

2.4.5 Self-Assembly and Other Techniques

In searching for ways of creating a true 3D metamaterial, one should also draw attention to the recent advances in the fabrication of 3D *photonic crystals* (PCs) (Yablonovitch 1993; Joannopoulos, Meade, and Winn 1995), artificially designed material systems in which their optical properties largely derive from the system structure rather than from the material itself. Similar to the case of metamaterials, PC fabrication requires different techniques for creating periodic structures built of materials with alternating refractive indices. Recent reviews on the fabrication of 3D PCs tend toward synthesis methods based on *self-assembly* to realize such materials in the optical range. In the self-assembly approach, *opals* have received a strong backing from their ability to be used as

(a)   (b)

FIGURE 2.10 SEM images of 3D structures created using different techniques: (a) a three-layer woodpile-like structure fabricated by the reverse-contact imprinting technique (Kehagias, N. et al. 2007. *Nanotechnology* 18: 175303-1–4), (b) etched side wall of the silver-filled membrane showing loose silver nanowires. (Courtesy of M. Noginov.)

scaffoldings for further templating of other materials (Galisteo et al. 2005). In the fabrication procedure, the formation of the templates (opals) is followed by the subsequent synthesis of guest materials such as semiconductors, metals, and/or insulators and, if desired, additional 2D patterning for the design of new structures. Accurate amounts of silicon, germanium, silica, etc. can be grown in the interior of the opal structures by CVD, allowing the fabrication of multilayer systems of different materials (Galisteo et al. 2005). The use of opals to pattern the growth of other materials (including metals) and subsequent vertical and lateral engineering of the fabricated structures have also been demonstrated (Galisteo et al. 2005). This method, though not directly applicable to metamaterial fabrication at the moment, comprises fabrication steps that are of vital importance for future 3D metamaterials, namely, prepatterned, controllable growth of different materials including metals.

For the realization of 3D, metallic, periodic structures on a large scale, the self-organization of metal-coated colloid particles was successfully employed (Chen et al. 2004). However, similar to opal structures, this method is difficult to employ for metamaterial fabrication requiring the creation of specially designed shapes.

Recently, another self-assembly approach was suggested for the realization of 3D MMs for the microwave and optical frequencies (Logeeswaran et al. 2006). By using a metal–dielectric, stress-actuated, self-assembly method, periodic arrays of metal flap SRRs and hinges were fabricated by lithographic patterning combined with metal deposition, lift-off, and etching procedures. This approach offers large-scale fabrication and can in principle be scalable from microwave (100 GHz) to optical frequencies (300 THz). However, for creating optical metamaterials, the careful choice of materials as well as the close control of geometrical parameters and release etching chemistry is required. Moreover, combined with the fabrication challenges, moving to the optical range will bring up the limit of size scaling in SRR functionality (Zhou et al. 2005).

Recently, it was shown that metamaterials consisting of metal wire arrays exhibit an optical response at frequencies far away from resonances (Elser et al. 2006; Silveirinha, Belov, and Simovski 2007), and electromagnetic waves propagating along the nanowires exhibit negative refraction at a broad frequency band for all angles (Yao et al. 2008). For creating such metamaterial, silver nanowires can be electrochemically deposited into a porous alumina membrane prepared by electrochemical anodization (Yao et al. 2008; Noginov et al. 2009). Such simple fabrication process provides self-standing, 3D optical metamaterial membranes (Figure 2.10b) (Yao et al. 2008; Noginov et al. 2009). In such materials, the dielectric response does not require any resonance; the negative refraction has low loss and occurs in a broad spectral range, for all incident angles, making it an intrinsic optical response of the underlying metamaterials.

2.5 Thin Metal Film Deposition

Aside from the requirements of nanometer-scale resolution and high throughput, highly-controllable *thin-film deposition* methods are needed for the realization of good performance (low loss) optical metamaterials. Whatever approach is chosen as a manufacturing method for the next generation of optical MMs, the possibility of creating thin metal and dielectric films with reduced surface roughness is vital. High *roughness* of the metal film is the main limit to obtaining lower-loss metamaterials since it leads to increased scattering losses in the system and can annihilate the negative-index effect (Yuan et al. 2007).

Depending on the chosen approach, different deposition techniques have to be studied and optimized. For example, in the standard lithography–deposition–lift-off procedure (Shalaev et al. 2005; Dolling et al. 2006a; Yuan et al. 2007; Dolling et al. 2007; Chettiar et al. 2007), the roughness of the e-beam-evaporated metal films can be reduced by the proper choice of the deposition conditions (Yuan et al. 2007). The quality of the deposited metal film depends also on the quality of the dielectric spacer layer. Especially in the case of an unstable metal such as silver (Libardi and Grieneisen 1998; Del Re et al. 2002; Drachev et al. 2007), the roughness of the initial dielectric structure is very important since clusters and lumps on a dielectric surface can, for example, work as seeds for the induced silver restructuring. The conventional way to improve the quality of the metal film is to use a *lower deposition rate* and a dielectric material with *stronger adhesion* and better surface quality. For example, the decrease of the deposition rate from 2 Å/s to 0.5 Å/s resulted in improved surface roughness and, hence, better optical performance of a recent magnetic metamaterial (Yuan et al. 2007). However, the decrease in the deposition rate leads to heating of the structure during deposition that makes the following lift-off process more difficult. Thus, higher complexity, multistep deposition procedures were required (Yuan et al. 2007).

In the case of silver, which is the most used metal for optical MMs (Dolling et al. 2006a; Yuan et al. 2007; Dolling et al. 2007; Chettiar et al. 2007) due to its optical properties, air exposure is also a problem since silver degrades under ambient conditions. The deposition of a dielectric layer on top of the structure (Yuan et al. 2007; Chettiar et al. 2007) can help to prevent silver deterioration. In order to improve the structure further, *annealing* can also be applied (Libardi and Grieneisen 1998).

When considering approaches that are feasible for future 3D MM fabrication, specifically direct laser writing based on TPP or 3D interference lithography yielding complex polymer structures, deposition methods different from standard deposition and sputtering techniques must be developed. Here, metal deposition by *electroless plating* seems to be one of the possible directions. As mentioned earlier, this approach was recently applied to selectively deposit metal on a polymer structure (Formanek et al. 2006a; Formanek et al. 2006b; Takeyasu, Tanaka, and Kawata 2007). Electroless plating has also been shown to be a promising way of achieving controllable deposition of thin metal films with low roughness (average roughness below 2 nm) (Jing et al. 2005). In light of future MM manufacturing where controllable coating of complex 3D structures is required, alternative metal deposition techniques such as *chemical vapor deposition* (Chi et al. 2007; Rill et al. 2008), *polymer-assisted deposition* (Jia et al. 2004), and *nanoparticles-assisted deposition* (Yan, Kang, and Mu 2007) can be explored as well.

2.6 Discussion

The fabrication of optical negative-refractive-index metamaterials is quite challenging, due to the requirements of 100- and sub-100 nm feature sizes of the "meta-atoms" and small periodicities/lattice constants on the order of 300 nm or less. Due to the high-resolution requirement, EBL is still the first choice for fabricating small-area meta-materials (~100 μm × 100 μm) (Shalaev et al. 2005; Dolling et al. 2006a; Dolling et al. 2007; Chettiar et al. 2007). Writing larger areas requires long e-beam writing times and, hence, boosts the operation cost. Thus, this approach is only suitable for proof-of-prin-ciple studies. Similar to EBL, other serial processes, for example, the focused ion-beam milling technique (Enkrich et al. 2005a), are not considered to be feasible for the large-scale metamaterial fabrication required by applications.

One approach to manufacturing high-quality optical MMs on a large scale (~cm² areas) is provided by interference lithography (Fan et al. 2005; Zhang et al. 2005c; Feth et al. 2006). To increase the resolution, IL can be combined with self-assembly tech-niques (Brueck 2005). Moreover, this technique could also be applied to the fabrication of future 3D metamaterials by piling 2D layers into a 3D structure. This step of stacking individual 2D layers made by IL has not been accomplished yet.

Another promising approach to create large-scale, high-quality metamaterials is nanoimprint lithography (Wu et al. 2007a). NIL offers nanoscale resolution; it is a paral-lel process with high throughput and, hence, a good candidate for optical metamaterial fabrication. Since NIL requires a stamp made by other nanofabrication techniques (such as EBL), it is ideal for parallel production of already optimized metamaterials, when the preliminary test structures were patterned via EBL. Thus, NIL can be seen as a large-scale, low-cost process of making EBL-written structures that offers solutions to the intrinsic EBL drawbacks.

The first steps toward the realization of a 3D MM were made by creating multilayer structures (instead of a single functional layer) (Dolling, Wegener, and Linden 2007) and by utilizing a layer-by-layer technique (Liu et al. 2008). Both approaches of making stacked metamaterials still have limitations (similar to challenging lift-off procedure in the first method and alignment requirements in the second). Another possible direction of making multiple-layer MMs is to consider etching of a prefabricated multilayer stack (Valentine et al. 2008). While complex 3D nanostructures can be fabricated today by several techniques (e.g., direct electron-beam writing and focused ion-beam chemical vapor deposition), these methods are too complex and time-consuming to be adapted for large-scale optical MM fabrication.

A fabrication method that is now considered to be one of the most promising approaches for future manufacturing of large-area, true 3D metamaterials is based on two-photon photopolymerization techniques (Formanek et al. 2006a; Takeyasu, Tanaka, and Kawata 2007). While offering subdiffraction resolution (down to 100 nm) (Takada, Sun, and Kawata 2005) due to a nonlinear multiphoton process, this technique possesses intrinsic 3D processing capability. In addition to direct single-beam laser writing of complex structures into a polymer matrix, large-scale 3D polymer structures for future real-life applications can be realized via a 3D, multiple-beam TPP technique (Formanek et al. 2006b; Rill et al. 2008; Rill et al. 2009). When combined with selective metal deposition by electroless plating (Formanek et al. 2006a; Takeyasu, Tanaka, and Kawata 2007), chemical vapor deposition (Rill et al. 2008), or shadow deposition (Rill et al. 2009), TPP can enable 100-nm-scale sculpturing of metal–dielectric structures. However, this approach needs further advances and developments for creating thin, uniform, and smooth layers of different materials on a TPP-fabricated polymer matrix.

3D, multilayered, polymer and metallic structures can also be realized by NIL (Kehagias et al. 2007). This method offers high reproducibility over large areas (millimeter scale) and has been used for fabricating 3D cubic arrays of gold cubes and 3D woodpile-like polymer structures for photonic crystal-based devices (McCormick et al. 2006; Kehagias et al. 2007). Combined with multimaterial deposition, these approaches might also be adapted for future metamaterial fabrication.

To reach real metamaterial applications, several tasks have to be fulfilled: loss reduction, large-scale 3D fabrication, and new isotropic designs. Careful material choice (e.g., new crystalline metals with lower absorption instead of traditional silver and gold) and process optimization (reduced roughness and high uniformity of the materials) can help on the way to creating low-loss optical metamaterials. Another possibility is to introduce a gain material into optical MMs, thus compensating for losses. Even though it is still a long way to truly 3D, isotropic, negative-index metamaterials at optical frequencies, several fabrication approaches do seem to be feasible. With emerging techniques such as nanoimprint, contact lithography, direct laser writing, and possibly new types of self assembly, it seems likely that truly 3D metamaterials with meta-atom sizes much smaller than the wavelength can be created. In the next generation of optical metamaterials, for any chosen manufacturing approach, the careful choice of materials and process optimization will be required in order to obtain high-quality structures. Thus, a discussion of future fabrication tool selection needs to be based on careful considerations of the structural quality achievable with the suggested method and the associated cost.

Acknowledgments

The author would like to acknowledge support from the Danish Research Council for Technology and Production Sciences and SAOT Young Researcher Award grant from Erlangen Graduate School in Advanced Optical Technologies (SAOT), Friedrich-Alexander-Universität Erlangen-Nürnberg.

References

Agranovich, V. M., Y. R. Shen, R. H. Baughman, and A. A. Zakhidov. 2004. Linear and nonlinear wave propagation in negative refraction metamaterials. *Physical Review B* 69: 165112.

Brueck, S. R. J. 2005. Optical and interferometric lithography-nanotechnology enablers. *Proceedings of IEEE* 93: 1704–1721.

Cai, Wenshan, Uday K. Chettiar, Alexander V. Kildishev, G. W. Milton, and Vladimir M. Shalaev. 2007a. Nonmagnetic cloak with minimized scattering. *Applied Physics Letters* 91: 111105.

Cai, Wenshan, Uday K. Chettiar, Alexander V. Kildishev, and Vladimir M. Shalaev. 2007b. Optical cloaking with metamaterials. *Nature Photonics* 1: 224–227.

Cai, Wenshan, Uday K. Chettiar, H. K. Yuan, Vashista C. De Silva, A. V. Kildishev, V. P. Drachev, and V. M. Shalaev. 2007c. Metamagnetics with rainbow colors. *Optics Express* 15 (3333): 3341.

Chen, Y., J. Tao, X. Zhao, Z. Cui, A. S. Schwanecke, and N. I. Zheludev. 2005. Nanoimprint lithography for planar chiral photonic meta-materials. *Microelectronic Engineering* 78–79: 612–617.

Chen, Y., Y. Zhou, G. Pan, and E. Huq. 2007. Nanofabrication of SiC templates for direct hot embossing for metallic photonic structures and metamaterials. *MNE07 Micro- and Nano Engineering Conference Proceedings* (591): 592.

Chen, Z., P. Zhan, Z. Wang, J. Zhang, W. Zhang, N. Ming, C. T. Chan, and P. Sheng. 2004. Two- and three dimensional ordered structures of hollow silver spheres prepared by colloidal crystal templating. *Advanced Materials* 16: 417–422.

Chettiar, U. K., A. V. Kildishev, H. K. Yuan, W. Cai, Shumin Xiao, V. P. Drachev, and V. M. Shalaev. 2007. Dual-band negative index metamaterial: Double negative at 813 nm and single negative at 772 nm. *Optics Letters* 32: 1671–1673.

Chi, Yun, Eddy Lay, Tsung-Yi Chou, Yi-Hwa Song, and Arthur J. Carty. 2007. Deposition of silver thin films using the pyrazolate complex $[Ag(3,5-(CF_3)2C_3HN_2)]_3$. *Chemical Vapor Deposition* 11: 206–212.

Chou, S. Y., P. R. Krauss, and P. J. Renstrom. 1996. Nanoimprint lithography. *Journal of Vacuum Science and Technology B* 14: 4129–4133.

Del Re, M., R. Gouttebaron, J. P. Dauchot, P. Leclère, R. Lazzaroni, M. Wautelet, and M. Hecq. 2002. Growth and morphology of magnetron sputter deposited silver films. *Surface and Coatings Technology* 15–152: 86–90.

Dolling, G., C. Enkrich, M. Wegener, C. M. Soukoulis, and Stefan Linden. 2006a. Low-loss negative-index metamaterial at telecommunication wavelengths. *Optics Letters* 31: 1800–1802.

–––––. 2006b. Simultaneous negative phase and group velocity of light in a metamaterial. *Science* 312: 892–894.

Dolling, G., M. Wegener, and S. Linden. 2007. Realization of a three-functional-layer negative-index photonic metamaterial. *Optics Letters* 32: 551–553.

Dolling, G., M. Wegener, C. M. Soukoulis, and Stefan Linden. 2007. Negative-index metamaterial at 780 nm wavelength. *Optics Letters* 32: 53–55.

Drachev, V. P., U. K. Chettiar, A. V. Kildishev, H. K. Yuan, W. Cai, and V. M. Shalaev. 2007. The Ag dielectric function in plasmonic metamaterials. *Optics Express* 16: 1186–1195.

Elser, J., R. Wangberg, V. A. Podolskiy, and E. E. Narimanov. 2006. Nanowire metamaterials with extreme optical anisotropy. *Applied Physics Letters* 89: 261102-1–3.

Enkrich, C., F. Perez-Williard, D. Gerthsen, J. Zhou, T. Koschny, C. M. Soukoulis, M. Wegener, and S. Linden. 2005a. Focused-ion-beam nanofabrication of near-infrared magnetic metamaterials. *Advanced Materials* 17: 2547–2549.

Enkrich, C., M. Wegener, S. Linden, S. Burger, L. Zschiedrich, F. Schmidt, J. F. Zhou, T. Koschny, and C. M. Soukoulis. 2005b. Magnetic metamaterials at telecommunication and visible frequencies. *Physical Review Letters* 95: 203901-1–4.

Fan, S., S. Zhang, K. J. Malloy, and S. R. J. Brueck. 2005. Large-area, infrared nanophotonic materials fabricated using interferometric lithography. *Journal of Vacuum Science and Technology B* 23: 2700–2704.

Feth, Nils, Christian Enkrich, Martin Wegener, and Stefan Linden. 2006. Large-area magnetic metamaterials via compact interference lithography. *Optics Express* 15: 501–507.

Formanek, F., N. Takeyasu, K. Tanaka, K. Chiyoda, T. Ishihara, and S. Kawata. 2006a. Selective electroless plating to fabricate complex three-dimensional metallic micro/ nanostructures. *Applied Physics Letters* 88: 083110–3.

Formanek, F., N. Takeyasu, T. Tanaka, K. Chiyoda, A. Ishikawa, and S. Kawata. 2006b. Three-dimensional fabrication of metallic nanostructures over large areas by two-photon polymerization. *Optics Express* 14: 800–809.

Galisteo, J. F., F. Garcýa-Santamarýa, D. Golmayo, B. H. Juarez, C. Lopez, and E. Palacios. 2005. Self-assembly approach to optical metamaterials. *Journal of Optics A: Pure and Applied Optics* 7: S244–S254

Gorkunov, M. V., I. V. Shadrilov, and Y. S. Kivshar. 2006. Enhanced parametric processes in binary metamaterials. *Applied Physics Letters* 88: 071912.

Griffith, S., M. Mondol, D. S. Kong, and J. M. Jacobson. 2002. Nanostructure fabrication by direct electron-beam writing of nanoparticles. *Journal of Vacuum Science and Technology B* 20: 2768–2772.

Hoffman, A. J., L. V. Alekseyev, S. S. Howard, K. J. Franz, D. Wasserman, V. A. Podolskiy, E. E. Narimanov, D. L. Sivco, and C. Gmachl. 2007. Negative refraction in semiconductor metamaterials. *Nature Materials* 6: 946–950.

Hoffmann, P. 2005. Comparison of fabrication methods of sub-100 nm nano-optical structures and devices. *Proceedings of SPIE* 5925: 06-1–14.

Jacob, Z., L. V. Alekseyev, and E. E. Narimanov. 2006. Optical hyperlens: Far-field imaging beyond the diffraction limit. *Optics Express* 14: 8247-8257.

Jia, Q. X., T. M. McCleskey, A. K. Burrell, Y. Lin, G. E. Collis, H. Wang, A. D. Q. Li, and S. R. Foltyn. 2004. Polymer-assisted deposition of metal-oxide films. *Nature Materials* 3: 529–532.

Jing, F., H. Tong, L. Kong, and C. Wang. 2005. Electroless gold deposition on silicon(100) wafer based on a seed layer of silver. *Applied Physics A* 80: 597–600.

Joannopoulos, J. D., R. D. Meade, and J. N. Winn. 1995. *Photonic Crystals: Molding the Flow of Light.* Princeton: Princeton University Press.

Katsarakis, N., G. Konstantinidis, A. Kostopoulos, R. S. Penciu, T. F. Gundogdu, M. Kafesaki, E. N. Economou, T. Koschny, and C. M. Soukoulis. 2005. Magnetic response of split-ring resonators in the far-infrared frequency regime. *Optics Letters* 30: 1348–1350.

Kawata, S., H.-B. Sun, T. Tanaka, and K. Takada. 2001. Finer features for functional microdevices. *Nature* 412: 697–698.

Kehagias, N., V. Reboud, G. Chansin, M. Zelsmann, C. Jeppesen, C. Schuster, M. Kubenz, F. Reuther, G. Gruetzner, and C. M. Sotomayor Torres. 2007. Reverse-contact UV nanoimprint lithography for multilayered structure fabrication. *Nanotechnology* 18: 175303-1-4.

Kildishev, A. V., and E. Narimanov. 2007. Impedance-matched hyperlens. *Optics Letters* 32: 3432–3434.

Kildishev, A. V., and Vladimir M. Shalaev. 2008. Engineering space for light via transformation optics. *Optics Letters* 33: 43–45.

Klar, T. A., A. V. Kildishev, V. P. Drachev, and V. M. Shalaev. 2006. Negative-index metamaterials: Going optical. *IEEE Journal of Selected Topics in Quantum Electronics* 12: 1106–1115.

Klein, M. W., C. Enrich, M. Wegener, and S. Linden. 2006a. Second-harmonic generation from magnetic metamaterials. *Science* 313: 502–504.

Klein, M. W., C. Enrich, M. Wegener, C. M. Soukoulis, and S. Linden. 2006b. Single-slit split-ring resonators at optical frequencies: Limits of size scaling. *Optics Letters* 31: 1259–1261,

Klein, M. W., M. Wegener, N. Feth, and S. Linden. 2007. Experiments on second- and third-harmonic generation from magnetic metamaterials. *Optics Express* 15: 5238–5247.

Ku, Z., and S. R. J. Brueck. 2007. Comparison of negative refractive index materials with circular, elliptical and rectangular holes. *Optics Express* 15: 4515–4522.

Lapine, M., M. Gorkunov, and K. H. Ringhofer. 2003. Nonlinearity of a metamaterial arising from diode insertions into resonant conductive elements. *Physical Review E* 67: 065601.

Leonhardt, U. 2006. Optical conformal mapping. *Science* 312: 1777–1780.

Levinson, H. J. 2005. *Principles of Lithography*. SPIE Press Monograph Vol. PM146.

Lezec, H., J. A. Dionne, and Harry A. Atwater. 2007. Negative refraction at visible frequencies. *Science* 316: 430–432.

Libardi, H., and H. P. Grieneisen. 1998. Guided-mode resonance absorption in partly oxidized thin silver films. *Thin Solid Films* 333: 82–87.

Linden, S., C. Enrich, M. Wegener, J. Zhou, T. Koschny, and C. M. Soukoulis. 2004. Magnetic response of metamaterials at 100-terahertz. *Science* 306: 1351–3.

Litchinitser, N. M., I. R. Gabitov, and A. I. Maimistov. 2007. Optical bistability in a nonlinear optical coupler with a negative index channel. *Physical Review Letters* 99: 113902.

Liu, N., H. Guo, L. Fu, S. Kaiser, H. Schweizer, and H. Giessen. 2008. Three-dimensional photonic metamaterials at optical frequencies. *Nature Materials* 7: 31–37.

Liu, Z., H. Lee, Y. Xiong, C. Sun, and X. Zhang. 2007. Far-field optical hyperlens magnifying sub-diffraction-limited objects. *Science* 315: 1686.

Logeeswaran, V. J., M. Saif Islam, M.-L. Chan, D. A. Horsley, W. Wu, and S.-Y. Wang. 2006. Self-assembled microfabrication technology for 3D isotropic negative index materials. *Proceedings of SPIE* 6393: 639305-1–10.

Maruo, S, O. Nakamura, and S. Kawata. 1997. Three-dimensional microfabrication with two-photon-absorbed photopolymerization. *Optics Letters* 22: 132–134.

Mattiucci, M., G. D'Aguanno, M. J. Bloemer, and M. Scalora. 2005. Second harmonic generation from a positive-negative index material heterostructure. *Physical Review E* 72: 066612.

McCormick, F. B., J. C. Fleming, S. Mani, M. R. Tuck, J. D. Williams, C. L. Arrington, S. H. Kravitz, C. Schmidt, G. Subramania, J. C. Verley, A. R. Ellis, I. El-kady, D. W. Peters, M. Watts, W. C. Sweatt, and J. J. Hudgens. 2006. Fabrication and characterization of large-area 3D photonic crystals. IEEE Aerospace Conference Proceedings, IEEE (2006).

Morita, T., K. Nakamatsu, K. Kanda, Y. Haruyama, K. Kondo, T. Kaito, J. Fujita, T. Ichihashi, M. Ishida, Y. Ochiai, T. Tajima, and S. Matsui. 2004. Nanomechanical switch fabrication by focused-ion-beam chemical vapor deposition. *Journal of Vacuum Science and Technology B* 22: 3137–3142.

Narimanov, E., and V. M. Shalaev. 2007. Optics: Beyond diffraction. *Nature* 447: 266–267.

Noginov, M. A., Yu. A. Barnakov, G. Zhu, T. Tumkur, H. Li, and E. E. Narimanov. 2009. Bulk photonic metamaterial with hyperbolic dispersion. *Applied Physics Letters* 94: 151105.

Pang, S. W., T. Tamamura, M. Nakao, A. Ozawa, and H. Masuda. 1998. Direct nano-printing on Al substrate using a SiC mold. *Journal of Vacuum Science & Technology B* 16: 1145–1149.

Passinger, S., S. M. Saifullah, C. Reinhardt, R. V. Subramanian, B. Chichkov, and M. E. Welland. 2007. Direct 3D patterning of TiO_2 using femtosecond laser pulses. *Advanced Materials* 19: 1218–1221.

Pendry, J. B. 2000. Negative refraction makes a perfect lens. *Physical Review Letters* 85: 3966–3969.

-----. 2004. Negative refraction. *Contemporary Physics* 45: 191–202.

Pendry, J. B., A. J. Holden, D. J. Robbins, and W. J. Stewart. 1999. Magnetism from conductors and enhanced nonlinear phenomena. *IEEE Transactions on Microwave Theory and Techniques* 47: 2075–2084.

Pendry, J. B., D. Schurig, and D. R. Smith. 2006. Controlling electromagnetic fields. *Science* 312: 1780–1782.

Popov, A. K., S. A. Myslivets, T. F. George, and Vladimir M. Shalaev. 2007. Four-wave mixing, quantum control and compensating losses in doped negative-index photonic metamaterials. *Optics Letters* 32: 3044–3046.

Popov, A. K., and V. M. Shalaev. 2006a. Compensating losses in negative-index metamaterials by optical parametric amplification. *Optics Letters* 31: 2169–2171.

-----. 2006b. Negative-index metamaterials: Second-harmonic generation, Manley-Rowe relations and parametric amplifications. *Applied Physics B* 84: 131.

Rill, M. S., C. E. Kriegler, M. Thiel, G. von Freymann, S. Linden, and M. Wegener. 2009. Negative-index bianisotropic photonic metamaterial fabricated by direct laser writing and silver shadow evaporation. *Optics Letters* 34: 19–21.

Rill, M. S., C. Plet, M. Thiel, I. Staude, G. von Freymann, S. Inden, and M. Wegener. 2008. Photonic metamaterials by direct laser writing and silver chemical vapour deposition. *Nature Materials* 7: 543–546.

Salandrino, A., and N. Engheta. 2006. Far-field subdiffraction optical microscopy using metamaterial crystals: Theory and simulations. *Physical Review B* 74: 075103.

Schurig, D., J. J. Mock, B. J. Justice, S. A. Cummer, and J. B. Pendry. 2006. Metamaterial electromagnetic cloak at microwave frequencies. *Science* 314: 977–980.

Shalaev, V. M., W. Cai, U. K. Chettiar, H. K. Yuan, A. K. Sarychev, V. P. Drachev, and A. V. Kildishev. 2005. Negative index of refraction in optical metamaterials. *Optics Letters* 30: 3356–3358.

Shalaev, V. M. 2007. Optical negative-index metamaterials. *Nature Photonics* 1: 41–48.

Shelby, R. A., D. R. Smith, and S. Schultz. 2001. Experimental verification of a negative index of refraction. *Science* 292: 77–9.

Silveirinha, M. G., P. A. Belov, and C. R. Simovski. 2007. Subwavelength imaging at infrared frequencies using an array of metallic nanorods. *Physical Review B* 75: 035108-1–12.

Smolyaninov, I. I., Y-J Hung, and C. C. Davis. 2007. Magnifying superlens in the visible frequency range. *Science* 315: 1699–1701.

Sotomayor Torres, C. M., S. Zankovych, J. Seekamp, A. P. Kam, C. Clavijo Ceden˜o, T. Hoffmann, J. Ahopelto, F. Reuther, K. Pfeiffer, G. Bleidiessel, G. Gruetzner, M. V. Maximov, and B. Heidari. 2003. Nanoimprint lithography: An alternative nanofabrication approach. *Materials Science and Engineering: C* 23: 23–31.

Soukoulis, C. M., M. Kafesaki, and E. N. Economou. 2006. Negative-index materials: New fronties in optics. *Advanced Materials* (18): 1941–1952.

Soukoulis, C. M., S. Linden, and M. Wegener. 2007. Negative refractive index at optical wavelengths. *Science* 315: 47–49.

Subramania, G., and S. Y. Lin. 2004. Fabrication of three-dimensional photonic crystal with alignment based on electron beam lithography. *Applied Physics Letters* 85: 5037–9.

Takada, K., H.-B. Sun, and S. Kawata. 2005. Improved spatial resolution and surface roughness in photopolymerization-based laser nanowriting. *Applied Physics Letters* 86: 071122-1–3.

Takeyasu, N., T. Tanaka, and S. Kawata. 2007. Fabrication of 3D metal/polymer fine structures for 3D plasmonic metamaterials. Proceedings of the OSA meeting "Photonic Metamaterials: From Random to Periodic."

Valentine, J., S. Zhang, T. Zentgraf, E. Ulin-Avila, D. A. Genov, G. Bartal, and X. Zhang. 2008. Three-dimensional optical metamaterial with a negative refractive index. *Nature* 455: 07247.

Veselago, V. G. 1968. The electrodynamics of substances with simultaneously negative values of ε and μ. *Soviet Physics Uspekhi* 10: 509–514.

Veselago, V. G., L. Braginsky, V. Shklover, and Ch. Hafner. 2006. Negative refractive index materials. *Journal of Computational and Theoretical Nanoscience* 3: 189–218.

Veselago, V. G., and E. Narimanov. 2006. The left hand of brightness: Past, present and future of negative index materials. *Nature Materials* 5: 759–762.

Wu, W., E. Kim, E. Ponizovskayaly, Z. Liu, Z. Yu, N. Fang, Y. R. Shen, A. M. Bratkovsky, Tong.W., C. Sun, X. Zhang, S.-Y. Wang, and R. S. Williams. 2007a. Optical metamaterials at near and mid-IR range fabricated by nanoimprint lithography. *Applied Physics A* 87: 147–150.

Wu, W., Z. Yu, S.-Y. Wang, R. B. Williams, Y. Liu, C. Sun, X. Zhang, E. Kim, R. Shen, and N. Fang. 2007b. Midinfrared metamaterials fabricated by nanoimprint lithography. *Applied Physics Letters* 90: 063107–3.

Yablonovitch, E. 1993. Photonic bang-gap structures. *Journal of the Optical Society of America B* 10: 283–295.

Yan, Y., S.-Z. Kang, and J. Mu. 2007. Preparation of high quality Ag film from Ag nanoparticles. *Applied Surface Science* 253: 4677–4679.

Yao, J., Z. Liu, Y. Liu, Y. Wang, C. Sun, G. Bartal, A. M. Stacy, and X. Zhang. 2008. Optical negative refraction in bulk metamaterials of nanowires. *Science* 321: 930.

Yen, T. J., W. J. Padilla, N. Fang, D. C. Vier, D. R. Smith, J. B. Pendry, D. N. Basov, and X. Zhang. 2004. Terahertz magnetic response from artificial materials. *Science* 303: 1494–6.

Yuan, H. K., U. K. Chettiar, W. Cai, A. V. Kildishev, A. Boltasseva, V. P. Drachev, and V. M. Shalaev. 2007. A negative permeability material at red light. *Optics Express* 15: 1076–1083.

Zhang, S., W. Fan, K. J. Malloy, S. R. J. Brueck, N. C. Panoiu, and R. M. Osgood. 2005a. Near-infrared double negative metamaterials. *Optics Express* 13: 4922–4930.

-----. 2006a. Demonstration of metal–dielectric negative-index metamaterials with improved performance at optical frequencies. *Journal of the Optical Society of America B* 23: 434–438.

Zhang, S., W. Fan, B. K. Minhas, A. Frauenglass, K. J. Malloy, and S. R. J. Brueck. 2005b. Midinfrared resonant magnetic nanostructures exhibiting a negative permeability. *Physical Review Letters* 94: 037402-1–4.

Zhang, S., W. Fan, N. C. Panoiu, K. J. Malloy, R. M. Osgood, and S. R. J. Brueck. 2005c. Experimental demonstration of near-infrared negative-index metamaterials. *Physical Review Letters* 95: 137404-4.

-----. 2006b. Optical negative-index bulk metamaterials consisting of 2D perforated metal-dielectric stacks. *Optics Express* 14: 6778–6786.

Zhang, W., A. Potts, D. M. Bagnall, and B. R. Davidson. 2006c. Large area all-dielectric planar chiral metamaterials by electron beam lithography. *Journal of Vacuum Science and Technology B* 24: 1455.

-----. 2007. High-resolution electron beam lithography for the fabrication of high-density dielectric metamaterials. *Thin Solid Films* 515: 3714–3717.

Zharov, A. A., I. V. Shadrilov, and Y. S. Kivshar. 2003. Nonlinear properties of left-handed metamaterials. *Physical Review Letters* 91: 037401.

Zhou, J., T. Koschny, M. Kafesaki, E. N. Economou, J. B. Pendry, and C. M. Soukoulis. 2005. Saturation of the magnetic response of split-ring resonators at optical frequencies. *Physical Review Letters* 95: 223902-1-4.

3

Microwave Metamaterials: Selected Features and Sample Applications

Andrea Alù[1,2] and
Nader Engheta[2,*]

[1]*Department of
Electrical and Computer
Engineering, University
of Texas at Austin
1 University Station
C0803, Austin, Texas.*
[2]*Department of Electrical
and Systems Engineering,
University of Pennsylvania
200 South 33rd St., ESE
203 Moore, Philadelphia,
Pennsylvania*

3.1 Introduction

In this chapter, we provide an overview of some of the current developments in the field of metamaterials at microwave frequencies and their exciting potential applications, which may provide a breakthrough in overcoming some of the main limitations of naturally available microwave materials in various setups and technology of interest.

The seminal work of Pendry and his group (see, e.g., References 1 and 2) has envisioned the possibility of lowering the value of real parts of the effective constitutive parameters of artificial materials at microwave frequencies down to negative values, opening up various interesting possibilities for microwave engineers. Inspired by these works, and combining different ideas to synthesize negative permittivity and permeability, Smith, Schultz, and their coworkers [3] realized for the first time in 2000 a composite medium

---

* To whom correspondence should be addressed. E-mail: engheta@ee.upenn.edu

exhibiting a negative index of refraction at microwaves, resulting from the negative real parts of its effective permittivity and permeability in the same frequency range. Since the realization of such negative-refraction metamaterials, the interest in the fundamental aspects and the potential applications of these exotic metamaterials at microwaves has considerably grown. This interest has also recently translated to higher frequency ranges, with several interesting methodologies for the realization of metamaterials at IR (infrared) and optical frequencies.

The surprising features of materials with both real parts of permittivity and permeability being negative or double-negative (DNG) metamaterials had been pointed out almost 40 years ago by Veselago [4], but since then no sample of such material had been found in nature, or synthesized artificially until the work of Smith, Schultz, and their coworkers [3]. That is why the synthesis of DNG materials at microwaves has paved the way for various interesting applications [5–9], some of which we will review in the following. The different properties of this class of metamaterials have been studied by several groups worldwide, with many ideas and suggestions for their potential applications in different setups at microwave frequencies (see, e.g., [5–66]). In our group, we have been very interested in this new venue of materials with anomalous properties, and we have presented a wide range of novel features pertaining to these materials, proposing various applications of interest at microwaves [37–66]. One of the principal ideas that has inspired our earlier works is based on the possibility of having ultrathin, subwavelength cavity resonators in which a layer of the DNG medium can compensate the phase delay caused by a layer of conventional material (i.e., a "double-positive (DPS)" material) [37]. The antiparallel nature of phase and energy velocities in a DNG material [4] indeed ensures the possibility of resonant modes in electrically thin parallel-plate structures containing such DNG–DPS bilayers, which provides a first interesting application of metamaterials to microwave cavities and waveguides. As a matter of fact, this feature has been outlined and proposed in several possible designs in microwave cavities (see, e.g., References 37 and 38). This property may also pave the way to other exciting ideas, such as the possibility of squeezing guided modes below the diffraction limit in parallel-plate waveguides containing a pair of DNG and DPS slabs [39]. Also related to this effect is the possibility of squeezing the dimensions of other microwave components typically associated with a resonant dimension, similar to different classes of antennas and other radiation devices [51–56].

Microwave applications may also benefit from the class of "single-negative" (SNG) metamaterials, in which only one of the material parameters, not both, has a negative real value. These SNG media also exhibit anomalous and interesting properties when they are paired in a conjugate (i.e., complementary) manner, in close analogy with the superlensing properties of a DNG metamaterial in free-space [13]. SNG materials include the ε-negative (ENG) media, in which the real part of permittivity is negative but the real permeability is positive, and the μ-negative (MNG) media, in which the real part of permeability is negative but the real permittivity is positive. The idea of constructing an effective DNG medium by having thin layers of SNG media has been explored in [34] and further discussed and analyzed in [41]. We have analyzed in detail the wave reflection from and transmission through a pair of juxtaposed ENG and MNG slabs, revealing interesting properties such as resonances, transparency, anomalous tunneling, and zero reflection [41]. Such lossless pairs of planar layers may exhibit an "interface resonance" phenomena, leading to total transparency, even if each slab standing alone would be totally opaque. This concept has paved the way to several microwave potential applications for the paired SNG conjugate layers, for antennas, and open radiation problems, as we review in the following, and it has been the precursor to one of the most appealing applications of metamaterials in recent years: invisibility and cloaking. In some recent works [42–45], we have indeed predicted the possibility of using low positive or negative permittivity metamaterials as shells to reduce the scattering from a given object, providing a drastic reduction of visibility and scattering-cancellation-based cloaking.

In some of our recent works, we have also discussed and analyzed the anomalous properties of ε-near-zero (ENZ) metamaterials, which may lead to tailoring the phase pattern of the electromagnetic wave [50] and various supercoupling phenomena [48]. Some of the ENZ applications will be reviewed in this chapter.

More in general, in this chapter we also review some of our results on waveguides, antennas, and radiation applications involving metamaterials at microwave frequencies. We discuss how their unconventional features may be utilized to overcome several physical limitations present in conventional microwave setups that employ standard materials. The applications of these concepts span several areas of microwave technology, from antennas to waveguides to microwave circuits and radars.

3.2 Metamaterial Technology at Microwaves

Metamaterial technology has seen major development at microwave frequencies, for which the realization of resonant inclusions with the size of a fraction of the wavelength of operation is possible. In particular, the field of metamaterials has grown considerably after the realization of ENG microwave metamaterials in the form of dipole-loaded metamaterials [2] and their MNG counterparts formed with split-ring resonators (SRRs) [1].

Advances in the realization of these devices have produced very compact inclusions, which may be of the order of one hundredth of the wavelength of excitation for magnetic resonators [67]. With current technology, resonant electric inclusions cannot be made as small, and therefore, it turns out that MNG metamaterials are relatively "easier" to design and construct at microwave frequencies. (At higher frequencies, e.g., IR and visible domains, the situation is actually reversed, since ENG materials are naturally available at optical frequencies, but the magnetic response of natural materials is very hard to push beyond a certain frequency of about tens of gigahertz.)

The realization of metamaterials with inclusions has the advantage of mimicking the way in which natural materials are "built": artificial inclusions operate and interact with the electromagnetic wave similarly to the way in which atoms and molecules do in natural materials. However, in order for small subwavelength inclusions to strongly interact with the wave, strong resonances are required, which usually are related to narrow bandwidth of operation and strong sensitivity to losses. An example is provided by inductively loaded dipoles [68], which indeed may shrink the typical size of the electric inclusions to a fraction of the wavelength, but at the expense of narrower bandwidth of operation and less robustness. An interesting review of recent work on metamaterial modeling and design, particularly focused at microwave frequencies, may be found in References [69–70].

With the advent of metamaterials, alternative techniques have been suggested for realizing effective constitutive parameters with anomalous properties. Of particular importance, the possibility to tailor the effective permittivity and permeability "experienced" by the electromagnetic wave by exploiting the natural dispersion of guided modes is of special interest for several microwave applications of metamaterials. This technique was first envisioned by Rotman in the early 1960s [71], and it was later introduced by the groups of Marques [72,73] and Hrabar [74] in their metamaterial experiments within closed waveguides at microwave frequencies. In a rectangular waveguide, for instance, the propagation of the dominant TE_{10} mode may be "effectively" regarded and mimic the propagation of a transverse electromagnetic (*TEM*) wave inside a homogeneous "material" with constitutive parameters:

$$\varepsilon_{eff} = \varepsilon_0 \left[\varepsilon_r - c^2/(4 f^2 w^2) \right]$$
$$\mu_{eff} = \mu_0,$$

(3.1)

where ε_0, μ_0 are the free-space constitutive parameters, ε_r is the relative permittivity of the dielectric filling the waveguide, $c = (\varepsilon_0 \mu_0)^{-1/2}$ is the velocity of light in free space, w is the waveguide width, and an $e^{-i2\pi ft}$ time convention is assumed. In other words, as far as the propagation constants are concerned, even with using natural materials such as metals and standard dielectrics with positive values of permittivity, it is still possible to effectively achieve low or negative "effective" permittivity values by varying the width w of a suitably designed rectangular waveguide. By duality, it is possible to tailor the effective permeability "experienced" by a TM electromagnetic wave traveling in the same waveguide. This technique has been successfully employed to realize DNG metamaterials by simply loading a rectangular waveguide below cutoff with SRRs [72–74]. It has been also recently employed to verify the anomalous properties of zero-permittivity metamaterials, as we review in one of the following sections.

In the following, we review one of the most exciting applications of metamaterials at microwaves, highlighting how their exotic electromagnetic properties may come into play for overcoming some conventional physical limitations.

3.3 Waveguides

In this section, we provide an overview of our research on the anomalous features of parallel-plate waveguides partially or totally filled with layers of metamaterials [39]. The geometry of interest is depicted in Figure 3.1, where the region between two perfectly electric conducting (PEC) plates with distance $d = d_1 + d_2$ is filled by a pair of parallel metamaterial layers. The two slabs are characterized by their thicknesses d_1 and d_2, and their constitutive parameters ε_1, μ_1, and ε_2, μ_2.

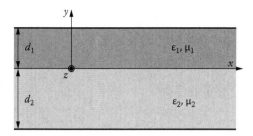

FIGURE 3.1 Geometry of a parallel-plate waveguide filled with a metamaterial bilayer. (From A. Alù, and N. Engheta, in *Guided modes in a Waveguide Filled with a Pair of Single Negative (SNG), Double Negative (DNG), and/or Double Positive (DPS) Layers IEEE Transactions on Microwave Theory and Techniques*, vol. MTT-52, no. 1, pp. 199–210, January 2004.) (With permission.)

We have shown in [39] how the guided modes supported by this simple waveguide satisfy the following dispersion relations:

$$\frac{\mu_1}{k_{t1}^{\text{TE}}}\tan\left(k_{t1}^{\text{TE}}d_1\right) = -\frac{\mu_2}{k_{t2}^{\text{TE}}}\tan\left(k_{t2}^{\text{TE}}d_2\right) \tag{3.2}$$

$$\frac{\varepsilon_1}{k_{t1}^{\text{TM}}}\cot\left(k_{t1}^{\text{TM}}d_1\right) = -\frac{\varepsilon_2}{k_{t2}^{\text{TM}}}\cot\left(k_{t2}^{\text{TM}}d_2\right). \tag{3.3}$$

These dispersion relations yield quite interesting properties if one lets the electromagnetic constitutive parameters to hold opposite signs (in the limit of negligible losses). For instance, we have shown that when the bilayer is formed by juxtaposed ENG and MNG materials, then the solution to these equations in terms of the thickness of the second slab as a function of the other geometrical parameters may yield a unique solution, independent of the total size of the structure. For the two polarizations, the solutions would read [39]:

$$d_2^{\text{TE}} = \frac{\tanh^{-1}\left[\frac{|\mu_1|\sqrt{|k_2|^2+\beta_{\text{TE}}^2}}{|\mu_2|\sqrt{|k_1|^2+\beta_{\text{TE}}^2}}\tanh\left(\sqrt{|k_1|^2+\beta_{\text{TE}}^2}\,d_1\right)\right]}{\sqrt{|k_2|^2+\beta_{\text{TE}}^2}}, \tag{3.4}$$

$$d_2^{\text{TM}} = \frac{\tanh^{-1}\left[\frac{|\varepsilon_2|\sqrt{|k_1|^2+\beta_{\text{TM}}^2}}{|\varepsilon_1|\sqrt{|k_2|^2+\beta_{\text{TM}}^2}}\tanh\left(\sqrt{|k_1|^2+\beta_{\text{TM}}^2}\,d_1\right)\right]}{\sqrt{|k_2|^2+\beta_{\text{TM}}^2}}, \tag{3.5}$$

which imply, due to the single-valued nature of the hyperbolic tangent functions, that one may have a large electrical aperture with still a single mode of operation, traveling at the interface of the ENG–MNG waveguide. This is in some senses related to our earlier

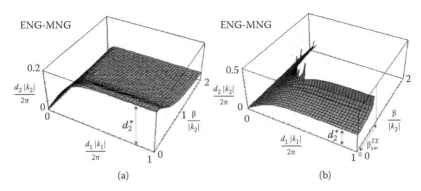

FIGURE 3.2 Dispersion diagram for TE mode in an ENG–MNG waveguide, illustrating the relationship among normalized d_1, d_2, and normalized real-valued β_{TE}, as described in (3.4), for two sets of material parameters for a pair of ENG–MNG slabs at a given frequency: (a) when $\varepsilon_1 = -2\varepsilon_0, \mu_1 = \mu_0, \varepsilon_2 = 3\varepsilon_0, \mu_2 = -2\mu_0$, and (b) when $\varepsilon_1 = -5\varepsilon_0, \mu_1 = 2\mu_0, \varepsilon_2 = 2\varepsilon_0, \mu_2 = -\mu_0$. The set of material parameters chosen in (a) does not allow a TE surface wave at the ENG–MNG interface, while the set chosen in (b) does. (From A. Alù and N. Engheta, in *guided modes in a wavelength jilled with a Pair of Single Negative (SNG), Double Negative (DNG), and/or Double Positive (DPS) Layers IEEE Transactions on Microwave Theory and Techniques*, vol. MTT-52, no. 1, pp. 199–210, January 2004.) (With permission.)

work on the interface resonance between such complementary bilayers [41] and the surface modes guided by such interface.

Figure 3.2, as an example, reports the TE dispersion diagrams (relative to Equation 3.4) with d_1, d_2, and β_{TE} as variable parameters for the single TE mode in two different ENG–MNG waveguides, with geometries described in the caption. It is evident how these waveguides may support a monomodal operation despite their electrically large size, property unique for such metamaterial-loaded waveguides, and of interest for different microwave applications.

As another interesting property that Equations 3.2 and 3.3 imply, guided modes may be squeezed in extremely tiny waveguides, with lateral dimensions below the diffraction limit of half-wavelength. This is related to the discovery that the combination of DPS and DNG materials, or more in general complementary metamaterials, may compensate the phase in a cavity, as first anticipated in Reference [37]. This may be seen, for instance, in the distribution of the fields in the transverse cross section of a DPS–DNG subwavelength waveguide, as the one reported in Figure 3.3. In this case, the waveguide total cross section is taken to be only one hundredth of the wavelength in the first slab, and it is evident that the drastic difference from the conventional standing-wave distribution in a regular (much larger) waveguide loaded with standard materials. This is a first evidence of the possibility of beating the diffraction limit in a cavity or a waveguide with the proper combination of complementary materials.

We have discussed in great details in Reference [6] the anomalous power propagation properties in such waveguides loaded by complementary metamaterials, showing that the power flow is obliged to follow a very atypical behavior, flowing in the opposite directions in the two layers with oppositely signed constitutive parameters. Even though at first sight counterintuitive, this behavior indeed satisfies causality, and

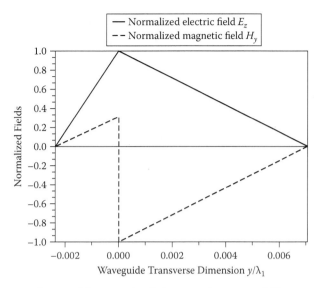

FIGURE 3.3 Distribution of the normalized electric and magnetic field in a DPS–DNG thin waveguide with $\varepsilon_1 = 2\varepsilon_0$, $\mu_1 = \mu_0$, $\varepsilon_2 = -3\varepsilon_0$, $\mu_2 = -3\mu_0$, $\beta = 0.47k_1$. (From G. V. Eleftheriades, and K. G. Balmain [editors], *Negative Refraction Metamaterials: Fundamental Properties and Applications*, IEEE Press, John Wiley & Sons, Hoboken, New Jersey, 2005.)

it corresponds in many senses to the sum of reflected and incident wave in any standard waveguide. The main striking feature in this type of waveguide is that reflected and incident powers are carried by the same unique mode, and they travel in distinct regions of the waveguide.

Somehow related to these anomalous power properties in closed waveguides, we have also studied in details [6] the exotic properties of a coupler composed by complementary open waveguides, as depicted in Figure 3.4. As depicted in the figure, also in this case the power flow is required to be opposite in the two regions of the coupler, implying anomalous "contradirectional" coupling in such geometry.

This may produce very interesting applications, in some senses similar to those of a regular Bragg reflector or grating, but with a reflected wave that is completely isolated in space from the incident one, with interesting applications and anomalies.

3.4 Antennas

As another fascinating application of metamaterials at microwave frequencies, antennas loaded with exotic materials may be considerably squeezed in size. Once again, the key mechanism is the resonance between complementary materials interfacing each other, which may substitute the geometrical resonance typical of antennas with size comparable to the wavelength. Small antennas loaded with metamaterials have been studied by several groups worldwide. The seminal work by Ziolkowski and his group [16] has envisioned the possibility to squeeze the dimensions of a general class of antennas by

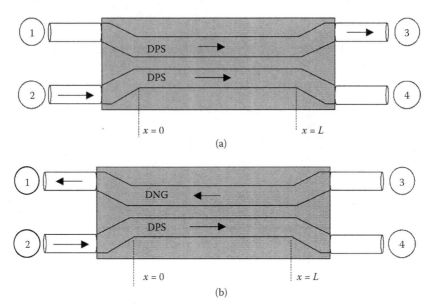

FIGURE 3.4 (a) Standard directional coupler with DPS–DPS open waveguides; (b) Contradirectional (backward) coupler formed by a DPS and a DNG slab, as proposed in Reference 6. (From G. V. Eleftheriades, and K. G. Balmain [editors], *Negative Refraction Metamaterials: Fundamental Properties and Applications*, IEEE Press, John Wiley & Sons, Hoboken, New Jersey, 2005.)

employing DNG, ENG, or MNG metamaterials, and subsequent works by his and other groups have successfully realized these concepts.

The general interest in reducing the size of radiators has been a topic of interest in the engineering community since decades ago and several solutions are currently available, usually at the expenses of other radiation features, such as bandwidth, gain, efficiency, and radiation pattern purity [78–85]. Without entering into the details of these standard techniques' advantages and disadvantages, which are well known in the technical literature, we have been interested in exploring how metamaterials may provide interesting novel phenomena for overcoming some of these limitations. In different antenna setups, materials with exotic electromagnetic properties and negative effective constitutive parameters have indeed shown to be able to work near the limits for antenna reduction, maintaining reasonable values of efficiency and other radiation parameters [75–77]. In particular, one of the problems we have recently tackled refers to the possibility of applying the phase compensation mechanisms to squeeze the size of patch antennas by partially loading them with an MNG material. As we have shown in some recent technical contributions [51–53], the proper choice of filling ratio for the material underneath the patch, partially loaded with a negative-permeability material and partially with a positive-permeability medium that may be even free space, may cause a patch resonance to be in principle independent on the total size of the antenna itself. Even though the

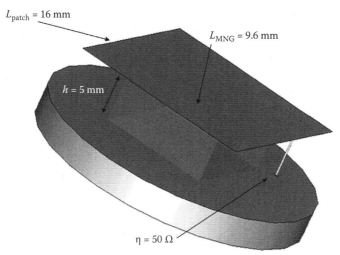

L_patch = 16 mm

L_MNG = 9.6 mm

h = 5 mm

η = 50 Ω

FIGURE 3.5 Geometry for our simulation: a rectangular patch antenna loaded with a nega-tive-permeability metamaterial block that partially fills the volume underneath the patch. The dimensions of the structure are indicated in the figure, together with the feeding probe. The ground plane is assumed infinite extent in these simulations.

bandwidth of such device is still limited by Chu's limit when passive materials are used, a proper choice of the patch geometry may allow obtaining a noticeable gain and effi-ciency from such structures, despite their subwavelength size. We have shown, using analytical techniques and numerical simulations, how the circular geometry may in particular allow good performance once the proper resonant mode is selected.

Here we report some numerical results related to these ideas, in which we propose some innovative concepts that may lead to the possible realistic design and realization of such compact patch antennas. We have simulated with finite integration technique software [86], a related structure as depicted in Figure 3.5, consisting of a rectangular patch of subwavelength dimensions partially loaded with a negative-permeability meta-material with square shape. In contrast with our previous results, here the geometry has been selected to be square, instead of circular, which may arguably be easier to realize with current technology. As we have anticipated earlier, the realization of MNG meta-materials at microwave frequencies has reached a level that may hold good promise for the practical realization of these ideas.

The rectangular block of MNG metamaterial that we have considered here has been assumed to have its plasmonic features around the frequency $f = 1.6$ *GHz*. The Lorentz model for such a material, employed in our simulations, is taken to have the following form:

$$\mu(\omega) = 1 + \frac{56.964 \cdot 10^9}{83.056 \cdot 10^9 - \omega^2 + j\omega 0.036 \cdot 10^9}. \qquad (3.6)$$

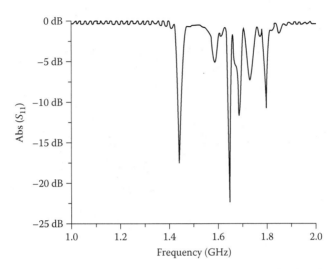

FIGURE 3.6 Reflection coefficient at the input port (coaxial cable) for the antenna of Figure 3.1.

We have also assumed that the substrate over which the patch antenna is printed has a realistic constant permittivity of $\varepsilon(\omega) = 12.2 - j0.08$. This dielectric substrate is assumed to be the substrate over which the SRR inclusions may be realistically printed, analogous to what suggested in Reference [53].

Figure 3.6 reports the simulated s_{11} coefficient at the input port of the probe (with characteristic impedance 50 Ω) feeding the square patch along its diagonal to excite the proper radiating mode of operation. It can be seen that around the expected operation frequency (which was derived based on the filling ratio between the negative-permeability material and the total size of the patch), that is, around $f = 1.6$ GHz, we get a good matching with the coaxial cable. This is confirmed also by full-wave numerical simulations in the frequency domain we have recently performed (not reported here) on the same antenna. These results confirm our theoretical analysis and the design we have carried out, and suggest realistic possibilities for matching such a subwavelength radiator with a realistic feed. As we have verified in our numerical simulations, the first resonant dip on the left side (at lower frequencies) is not of interest, since it is related to a dip mainly due to absorption, related to the high index of refraction and high loss factor of the Lorentzian filling material. Around the frequency of interest, however, the reflection coefficient shows interesting resonances due to the radiation properties of such an antenna.

Figure 3.7 reports the radiation pattern of this antenna at $f = 1.57$ GHz, which is the point where maximum total efficiency from the antenna is expected. The diagram confirms the excitation of the correct radiation mode, similar to a larger patch. These simulations suggest that it is in principle possible to decrease the size of the antenna without altering considerably its radiation pattern.

Figure 3.8 reports the samples of the broadside (red, lighter) and maximum (black, darker) gain (without considering reflections at the input port, as defined by IEEE) for the

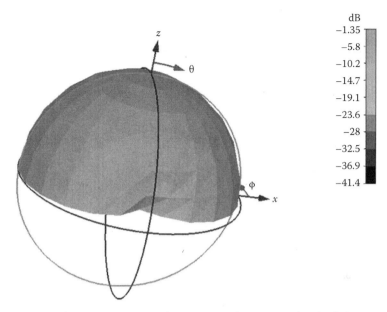

FIGURE 3.7 Radiation pattern at a frequency in the resonant band of the antenna of Figure 3.1.

antenna under analysis, confirming the results of the previous discussion. It is seen that for the first resonant dip in Figure 3.6, the gain is very poor, but in the higher frequency range, reasonable gain may be achieved, despite the electrically small size of the antenna. Due to the relatively high losses in the substrate considered here, the gain is still quite limited, but better results, closer to the theoretical limits of gain for electrically small antennas,

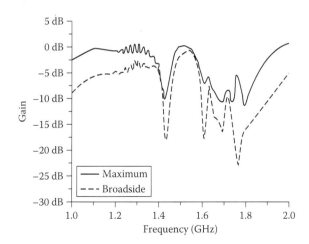

FIGURE 3.8 Broadside and maximum gain (without considering the mismatch) for the antenna of Figure 3.1.

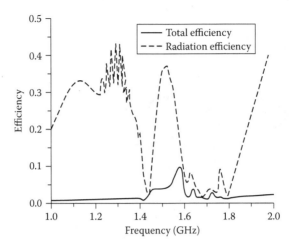

FIGURE 3.9   Radiation and total efficiency for the antenna of Figure 3.1.

may be achieved with lower loss substrates. Notice that when the maximum gain tentatively coincides with the broadside gain, it means that the correct mode of operation has been properly excited, consistent with the radiation pattern in Figure 3.7. This implies that the MNG substrate is effectively and properly modifying the current flow across such a subwavelength patch, so that the radiating mode resembles that of a larger patch. This is consistent with what is described in our earlier papers on this topic [51–53].

Figure 3.9 reports the radiation efficiency η_{rad} and the total efficiency η_{tot} (considering the mismatch at the port) in the frequency domain of interest. Frequency domain solutions of the same problem confirm these results. Qualitatively, these results also confirm the resonant behavior of the antenna around the design frequency.

The field distributions at the resonant frequency of the patch have also been evaluated. Figures 3.10–3.13 report the electric field, magnetic field, and surface current distributions on the patch, showing that, despite its subwavelength size, it is possible to have a resonant mode with proper field distributions, resulting in effective radiation in the surrounding free-space environment. These results are consistent with our earlier theoretical predictions for the circular patch, and extend those results to a rectangular configuration, which may be more practical for the antenna realization.

From Figures 3.7–3.13, one notices that the levels of efficiency and gain are lower than those achieved in our previous publications on metamaterial circular patch antennas. The reason is that here we have assumed realistic losses in the SRR metamaterial and in the substrate in order to consider practically realizable metamaterials within fabrication possibilities. One can speculate that with better technology and lower material loss it may be possible to achieve levels of gain and efficiency comparable with those of a rectangular patch of standard size, but with much reduced dimensions.

As a new set of simulations, to move even closer to the practical consideration of the metamaterial patch antenna, we scaled down the dimensions of the patch antenna to fit a smaller size, that is, we designed a patch antenna with side $L_{patch} = 10$ mm. Our theory confirms the possibility of resonance to be fairly independent on the total size

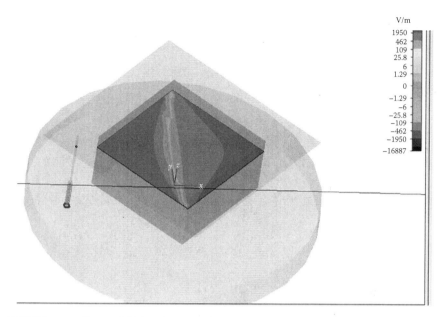

FIGURE 3.10 Electric field distribution underneath the patch at the resonant frequency of the antenna of Figure 3.5.

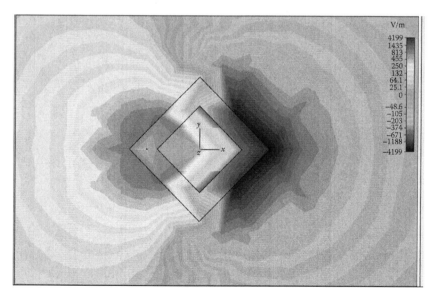

FIGURE 3.11 Electric field distribution over the whole plane underneath the patch at the resonant frequency of the antenna of Figure 3.5 (frequency domain simulation).

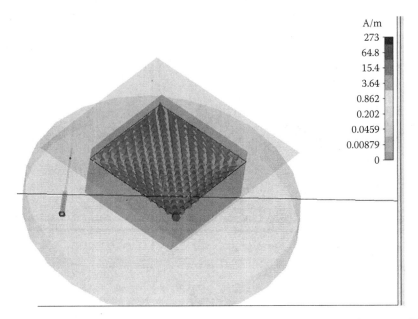

FIGURE 3.12   Magnetic field distribution underneath the patch at the resonant frequency of the antenna of Figure 3.5.

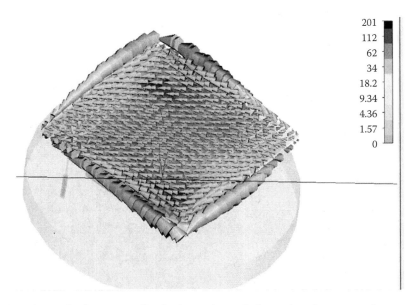

FIGURE 3.13   Surface current distribution underneath the patch at the resonant frequency of the antenna of Figure 3.5.

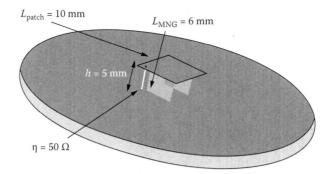

$L_{patch} = 10$ mm
$L_{MNG} = 6$ mm
$h = 5$ mm
$\eta = 50\ \Omega$

FIGURE 3.14 Alternative geometry for the square metamaterial patch antenna: the metamaterial considered here, and the other parameters have been kept the same, but the geometry has been scaled down by a factor of 1.6 with respect to the case of Figure 3.5, and the dielectric substrate has been considered to be finite.

of the patch, but related only to the relative dimensions of the underneath metamaterial block and the patch size, which have been kept with the same ratio. The new antenna is depicted in Figure 3.14 and, in this case, the simulations also consider the presence of a circular ground plane of finite size.

In the following figures, we report the results of our simulations for the radiation parameters in this new configuration. Figure 3.15 reports the reflection coefficient at the port, confirming that the resonant frequency remains stable with this scaling of dimensions. This is quite striking, confirming our earlier theoretical work on metamaterial-loaded antennas and resonators.

Also, the other antenna parameters confirm our theoretical predictions and expectations, namely, slightly reduced efficiency and gain, due to the smaller aperture of the

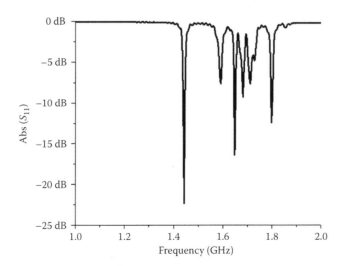

FIGURE 3.15 Reflection coefficient at the input port (coaxial cable) for the antenna of Figure 3.14.

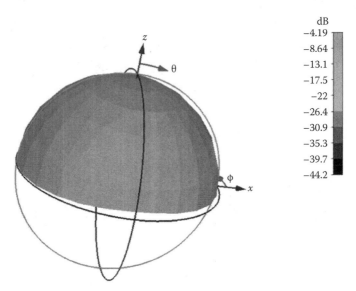

FIGURE 3.16 Radiation pattern at a frequency in the resonant band of the antenna of Figure 3.14.

antenna, but similar properties in terms of excitation of the proper radiating mode on the patch. We envision that the use of lower loss materials may help to further improve the radiation performance of such antennas, and in particular, their overall total efficiency.

We have also extensively worked on leaky-wave antennas with reduced size, utilizing complementary metamaterial loading. We have shown in our recent works how planar

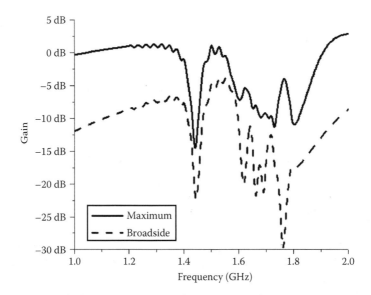

FIGURE 3.17 Broadside and maximum gain for the antenna of Figure 3.14.

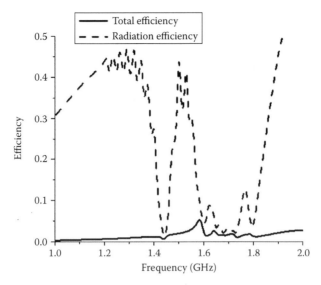

FIGURE 3.18 Radiation and total efficiency for the antenna of Figure 3.14.

[55] and cylindrical [56] configurations with leaky-wave radiation may strongly benefit from the use of ENG and MNG materials, achieving higher directivity and compactness compared to conventional leaky-wave radiators.

3.5 Plasmonic Cloaking

In the previous sections, we presented several applications for which a judicious combination of "complementary" metamaterials may provide anomalous resonances and exotic properties. As another interesting example of how ENZ and ENG metamaterials may be applied at microwave frequencies, in some senses as the "dual" of the compact resonances described earlier, we have shown that the total scattering cross section of a composite object containing materials with oppositely signed polarizability may be drastically reduced, making the object essentially "transparent" or invisible [42–45]. In the quasi-static case (electrically small objects), the conditions on the permeabilities and permittivities of the two materials composing a spherical core-shell system and on the ratio a_1/a between inner and outer radii of the core-shell object for achieving a zero scattering condition are the following:

$$\text{TE: } \gamma \equiv \frac{a_1}{a} \simeq \sqrt[3]{\frac{(\mu_2 - \mu_0)[2\mu_2 + \mu_1]}{(\mu_2 - \mu_1)[2\mu_2 + \mu_0]}}$$

$$\text{TM: } \gamma \equiv \frac{a_1}{a} \simeq \sqrt[3]{\frac{(\varepsilon_2 - \varepsilon_0)[2\varepsilon_2 + \varepsilon_1]}{(\varepsilon_2 - \varepsilon_1)[2\varepsilon_2 + \varepsilon_0]}}$$

(3.7)

for the TE and TM polarizations, respectively, where the subscript 2 refers to the outer shell, subscript 1 refers to core and subscript 0 refers to the background material. This implies

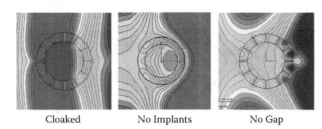

Cloaked   No Implants   No Gap

FIGURE 3.19 (*See color insert.*) Consistent with the results in Reference 44, the total electric field distribution at a snap shop in time for a sample geometry of a metamaterial cloak designed at microwave frequencies in the three cases: (a) with a metamaterial cover that is designed by filling a high permittivity shell with metallic fins; (b) with the same dielectric shell cover, but removing the metallic plates; (c) again with the same dielectric shell cover, but with metallic plates extended to touch the boundaries of the shell, thus modifying the wave interaction and weakening the cloaking effect. (From M. G. Silveirinha, et al., *Phys. Rev. E,* vol. 75, 036603 [16 pages], March 7, 2007.)

that for a given dielectric or metallic sphere it is possible to choose a suitable metamaterial "cover" layer with proper material parameters in order to make the overall system effectively "transparent" [45]. As an example, we show in Figure 3.19 some full-wave numerical simulations for a metamaterial cloak designed to operate at microwave frequencies. The geometry is composed of a dielectric cylinder, the object to be cloaked, with permittivity epsilon = 3*epsilon0 and radius R = 0.8/k0, with k_0 being the wave number in free space, surrounded by a parallel-plate metamaterial made of a shell of high permittivity ($\varepsilon_{diel} = 35.4\ \varepsilon_0$) and twelve metallic fins radially oriented. As detailed in [44], the fins should not touch the boundaries of the metamaterial shell, since they require a suitably designed gap in order to create the effective metamaterial properties of interest.

These numerical simulations of Figure 3.19, consistent with those reported in [44], show how the proper design of a microwave metamaterial shell may produce drastic scattering reduction from an obstacle. In particular, Figure 3.19 shows the amplitude of the near-zone electric field for the three cases of: (a) the dielectric cylinder surrounded by the cover as designed in [44]; (b) the same geometry, but with the metallic fins removed; (c) the same geometry, but with the metallic plates extended also inside the virtual gap required for the metamaterial to have the effective properties of interest. In the three cases, the structure is excited by a plane wave impinging from the right of the figure with electric field of 1*V/m* and polarized normal to the plane of the figure (i.e., the electric field vector is parallel with the axis of the structure). It is impressive to see how the proper design of the metamaterial reduces the scattering, despite the addition of a high-permittivity shell and metallic fins. It is also evident from this figure that the proper design of the gaps for the virtual interface is necessary to obtain the desired scattering reduction. When the gaps are not present, the perturbation of the field distribution is indeed much larger in all directions, due to the additional scattering from these elements.

From the preceding example, one can see how the effective negative polarizability of properly designed microwave metamaterials may modify the scattering properties of a given conducting or dielectric object of moderate size, and produce an effective and elegant way for cloaking applications.

3.6 Supercoupling

Not only do the anomalous characteristics and properties of microwave ENZ metamaterials provide an interesting way for cloaking objects, as outlined in the previous section, but they may also lead to many other interesting potential applications of interest [50]. Applications for directive radiation, enhancement of transmission, and tailoring of the phase pattern are among the most popular examples. As a further anomalous phenomenon associated with the ENZ properties, we have explored the possibility that narrow waveguide channels and bends filled with ENZ metamaterials may provide an anomalous tunneling that results in a drastic increase of transmission. The long wavelength in zero-permittivity metamaterials allows field distribution with little phase variation within such materials filling a tight channel, essentially independent of its length, shape, and geometry [46–48, 87,88]. Such "supercoupling" phenomenon, as we have named it, is particularly interesting in the microwave domain, and it has been recently verified experimentally by different groups. In particular, we have investigated the possibility to demonstrating experimentally this phenomenon in a hollow narrow waveguide channel, exploiting the intrinsic dispersion of the modes supported by rectangular channels.

In a rectangular waveguide, the propagation of the dominant TE_{10} mode may indeed be regarded, for some of its features, as equivalent to that of a transverse electromagnetic *TEM* wave traveling in a "metamaterial" with effective constitutive parameters:

$$\varepsilon_{eff} = \varepsilon_0 [\varepsilon_r - c^2/(4 f^2 w^2)]$$
$$\mu_{eff} = \mu_0, \tag{3.8}$$

where ε_0, μ_0 are the free-space constitutive parameters, ε_r is the relative permittivity of the dielectric filling the waveguide, $c = (\varepsilon_0\mu_0)^{-1/2}$ is the velocity of light in free space, w is the waveguide width, and an $e^{-i2\pi ft}$ time convention is assumed. Zero permittivity may be effectively achieved by working near the cutoff of this mode.

We have applied these concepts in [48] to experimentally demonstrate this supercoupling effect in a setup consistent with the one reported in Figure 3.20, formed by two rectangular waveguides connected by an ultranarrow rectangular channel of much smaller height. The lateral width of the channel has been chosen such that the narrow channel is to operate at cutoff for the design frequency, so that the channel would effectively behave as if filled by a zero permittivity metamaterial, despite the absence of inclusions in the channel. The geometrical parameter ε_{ch} is the effective permittivity of the channel, as defined in (Eq. 3.8). It is evident that when $h_{ch} \ll h$, we should then have $\varepsilon_{ch} \ll \varepsilon_{wg}$ in order to have the waveguides and the narrow channel matched, confirming that in the ideal limit of $\varepsilon_{ch} \rightarrow 0$, that is, at the frequency for which the channel operates at cutoff, we could achieve complete matching and total transmission through the channel even when the height mismatch is extremely high. Even more impressive is the property of this transmission to be essentially independent of the length, bending, and abruptions present inside the supercoupling channel, due to the effect of the "quasistatic" long-wavelength propagation inside the ENZ channel.

In Figure 3.21, consistent with [46], we have shown the full-wave simulation we did in [86] for a rectangular waveguide si $h \gg h_{ch}$ similar to that in Figure 3.20, but here the ultranarrow channel has been bent and twisted both in the E and H plane.

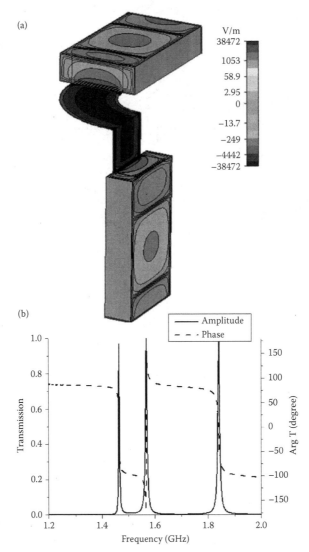

FIGURE 3.20 (a) Geometry and electric field distribution (snapshot in time, normal component) and (b) amplitude and phase of the transmission coefficient through a supercoupler with a 90° bend in the H plane following a 90° bend in the E plane. (From A. Alù, et al., *Phys. Rev. E*, vol. 78, 016604 [10 pages], July 23, 2008.)

Despite the abruptions, cross-sectional mismatch, bending, and twisting, total transmission and zero phase delay are achieved at the design frequency for which the channel is at cutoff. We underline that the whole geometry is filled with air, and the "effective" ENZ operation is achieved by the proper choice of the empty channel width to ensure cutoff, following Equation 3.8. The figure shows the ENZ-based "static-like" propagation inside the channel with uniform phase.

These results show that the metamaterial concepts may shed new light and provide novel ideas in microwave components such as waveguides, and these findings, which do not directly involve the use of complicated technology or intricate designs for meta-material inclusions, may be of importance for several applications, spanning waveguide connections and coupling, filtering, sensing, power conversion, and absorption.

3.7 Conclusions

In this chapter, we have provided an overview of some of the features of microwave metamaterials and a sample of their potential applications. We have focused on various exciting characteristics of microwave metamaterials with exotic electromagnetic properties, and in particular, we reviewed their roles in compact resonators, waveguides and antennas, cloaking, and supercoupling. We believe that metamaterial technology has achieved much development in recent years, particularly for microwave frequencies, and it is therefore possible to predict the integration of this technology in several applied fields of science and engineering in the near future.

Acknowledgment

This work is supported in part by the National Science Foundation (NSF) CAREER award No. ECCS-0953311 to Andrea Alu, and the US Air Force Office of Scientific Research (AFOSR) Young Investigator Research Program (YIP) award No. FA9550-11-1-0009 to Andrea Alu, and the US Air Force Office of Scientific Research (AFOSR) grant number FA9550-08-1-0220 to Nader Engheta.

References

1. J. B. Pendry, A. J. Holden, D. J. Robbins, and W. J. Stewart, Magnetism from conductors and enhanced nonlinear phenomena, *IEEE Trans. Microw. Theory Tech*, vol. 47, no. 11, pp. 2075–2081, November 1999.
2. J. B. Pendry, A. J. Holden, D. J. Robbins, and W. J. Stewart, Low-frequency plasmons in thin wire structures, *J. Phys.: Condensed Matter*, vol. 10, pp. 4785–4809, 1998.
3. R. A. Shelby, D. R. Smith, and S. Schultz, Experimental verification of a negative index of refraction, *Science*, vol. 292, no. 5514, pp. 77–79, 2001.
4. V. G. Veselago, The electrodynamics of substances with simultaneously negative values of ε and μ, *Soviet Physics Uspekhi*, vol. 10, no. 4, pp. 509–514, 1968 [in *Russian Usp. Fiz. Nauk*, vol. 92, pp. 517–526, 1967].
5. C. Caloz, and T. Itoh, *Electromagnetic Metamaterials: Transmission Line Theory and Microwave Applications*, Wiley-IEEE Press, Hoboken, NJ, 2005.
6. G. V. Eleftheriades, and K. G. Balmain (editors), *Negative Refraction Metamaterials: Fundamental Properties and Applications*, IEEE Press, John Wiley & Sons, Hoboken, New Jersey, 2005.
7. N. Engheta, and R. W. Ziolkowski (editors), *Electromagnetic Metamaterials: Physics and Engineering Explorations*, John Wiley & Sons, New York, 2006.
8. A. K. Sarychev, and V. M. Shalaev, *Electrodynamics of Metamaterials*, World Scientific Publishing Company, Singapore, 2007.

9. R. Marqués, F. Martín, and M. Sorolla, *Metamaterials with Negative Parameters: Theory, Design and Microwave Applications*, Wiley Series in Microwave and Optical Engineering, New York, 2008.

10. R. W. Ziolkowski, and E. Heyman, Wave propagation in media having negative permittivity and permeability, *Phys. Rev. E.*, vol. 64, no. 5, 056625, 2001.

11. D. R. Smith, W. J. Padilla, D. C. Vier, S. C. Nemat-Nasser, and S. Schultz, Composite medium with simultaneously negative permeability and permittivity, *Phys. Rev. Lett.*, vol 84, no. 18, pp. 4184–4187, 2000.

12. R. A. Shelby, D. R. Smith, S. C. Nemat-Nasser, and S. Schultz, Microwave transmission through a two-dimensional, isotropic, left-handed metamaterial, *Appl. Phys. Lett.*, vol. 78, no. 4, pp. 489–491, 2001.

13. J. B. Pendry, Negative refraction makes a perfect lens, *Phys. Rev. Lett.*, vol. 85, no. 18, pp. 3966–3969, 2000.

14. R. W. Ziolkowski, Superluminal transmission of information through an electromagnetic metamaterial, *Phys. Rev. E.*, vol. 63, no. 4, 046604, April 2001.

15. R. W. Ziolkowski, Pulsed and CW Gaussian beam interactions with double negative metamaterial slabs, *Opt. Express*, vol. 11, no. 7, pp. 662–681, April 7, 2003.

16. R. W. Ziolkowski, and A. D. Kipple, Application of double-negative materials to increase power radiated by an electrically small antennas, *IEEE Trans. Antennas Propagat., Special Issue on Metamaterials*, vol. 51, no. 10, pp. 2626–2640, October 2003.

17. A. A. Oliner, A planar negative-refractive-index medium without resonant elements, *Digest IEEE MTT Int. Microw. Symp.* (IMS'03), Philadelphia, PA, pp. 191–194, June 8–13, 2003.

18. G. V. Eleftheriades, A. K. Iyer, and P. C. Kremer, Planar negative refractive index media using periodically L-C loaded transmission lines, *IEEE Trans. Microw. Theory Tech*, vol. 50, no. 12, pp. 2702–2712, December 2002.

19. L. Liu, C. Caloz, C.-C. Chang, and T. Itoh, Forward coupling phenomena between artificial left-handed transmission lines, *J. Appl. Phys.*, vol. 92, no. 9, pp. 5560–5565, November 1, 2002.

20. Z. M. Zhang, and C. J. Fu, Unusual photon tunneling in the presence of a layer with a negative refractive index, *Appl. Phys. Lett.*, vol. 80, no. 6, pp. 1097–1099, February 11, 2002.

21. I. V. Lindell, S. A. Tretyakov, K. I. Nikoskinen, and S. Ilvonen, BW media – media with negative parameters, capable of supporting backward waves, *Microw. Opt. Tech. Lett.*, vol. 31, no. 2, pp. 129–133, 2001.

22. S. A. Tretyakov, Metamaterials with wideband negative permittivity and permeability, *Microw. Opt. Tech. Lett.*, vol. 31, no. 3, pp. 163–165, 2001.

23. K. G. Balmain, A. A. E. Luttgen, and P. C. Kremer, Resonance cone formation, reflection, refraction and focusing in a planar anisotropic metamaterial, *IEEE Antennas Wireless Propagat. Lett.*, vol. 1, pp. 146–149, 2002.

24. M. W. McCall, A. Lakhtakia, and W. S. Weiglhofer, The negative index of refraction demystified, *Eur. J. Phys.*, vol. 23, pp. 353–359, 2002.

25. A. Lakhtakia, Reversed circular dichroism of isotropic chiral mediums with negative permeability and permittivity, *Microw. Opt. Tech. Lett.*, vol. 33, no. 2, pp. 96–97, April 20, 2002.

26. A. N. Lagarkov, and V. N. Kisel, Electrodynamics properties of simple bodies made of materials with negative permeability and negative permittivity, *Doklady Phys.*, vol. 46, no. 3, pp. 163–165, 2001.

27. M. W. Feise, P. J. Bevelacqua, and J. B. Schneider, Effects of surface waves on behavior of perfect lenses, *Phys. Rev. B*, vol. 66, 035113, 2002.

28. R. Marques, F. Medina, and R. Rafii-El-Idrissi, (2002) Role of bianisotropy in negative permeability and left-handed metamaterials, *Phys. Rev. B*, vol. 65, no. 14, 144440, 2002.

29. R. Marques, J. Martel, F. Mesa, and F. Medina, A new 2-D isotropic left-handed metamaterial design: Theory and experiment, *Microw. Opt. Tech. Lett.*, vol. 36, pp. 405–408, December 2002.

30. D. R. Smith, D. Schurig, and J. B. Pendry, Negative refraction of modulated electromagnetic waves, *Appl. Phys. Lett*, vol. 81, no. 15, pp. 2713–2715, October 7, 2002.

31. P. Gay-Balmaz, and O. J. F. Martin, Efficient isotropic magnetic resonators, *Appl. Phys. Lett.*, vol. 81, no. 5, pp. 939–941, July 29, 2002.

32. J. A. Kong, J. A., B.-I. Wu, and Y. Zhang, A unique lateral displacement of a Gaussian beam transmitted through a slab with negative permittivity and permeability, *Microwave Opt. Tech. Lett.*, vol. 33, no. 2, pp. 136–139, 2002.

33. R. A. Silin, and I. P. Chepurnykh, On media with negative dispersion, *J. Commun. Tech. Electron.*, vol. 46, no. 10, pp. 1121–1125, 2001.

34. D. R. Fredkin, and A. Ron, Effective left-handed (negative index) composite material, *Appl. Phys. Lett.*, vol. 81, no. 10, pp. 1753–1755, September 2, 2002.

35. B.-I. Wu, T. M. Grzegorczyk, Y. Zhang, and J. A. Kong, Guided modes with imaginary transverse wave number in a slab waveguide with negative permittivity and permeability, *J. Appl. Phys.*, vol. 93, no. 11, pp. 9386–9388, June 1, 2003.

36. C. Caloz, C.-C. Chang, and T. Itoh, Full-wave verification of the fundamental properties of left-handed materials in waveguide configurations, *J. Appl. Phys.*, vol. 90, no. 11, pp. 5483–5486, December 2001.

37. N. Engheta, An idea for thin subwavelength cavity resonators using metamaterials with negative permittivity and permeability, *IEEE Antennas and Wireless Propagation Lett.*, vol. 1, no. 1, pp. 10–13, 2002.

38. C. L. Holloway, D. C. Love, E. F. Kuester, A. Salandrino, and N. Engheta, Subwavelength resonators: On the use of metafilm to overcome the lambda/2 size limit, *Institute of Engineering and Technology (IET) Microwave, Antennas and Propagation*, vol. 2, no. 2, pp. 120–129, 2008.

39. A. Alù, and N. Engheta, Guided modes in a waveguide filled with a pair of single-negative (SNG), double-negative (DNG), and/or double-positive (DPS) layers, *IEEE Trans. Microw. Theory Tech.*, vol. MTT-52, no. 1, pp. 199–210, January 2004.

40. A. Alù, and N. Engheta, Radiation from a traveling-wave current sheet at the interface between a conventional material and a material with negative permittivity and permeability, *Microw. Opt. Tech. Lett.*, vol. 35, no. 6, pp. 460–463, December 20, 2002.

41. A. Alù, and N. Engheta, Pairing an epsilon-negative slab with a mu-negative slab: Resonance, tunneling and transparency, *IEEE Trans. Antennas Propagat., Special Issue on "Metamaterials,"* vol. 51, no. 10, pp. 2558–2571, October 2003.

42. A. Alù, and N. Engheta, Cloaking and transparency for collections of particles with metamaterial and plasmonic covers, *Opt. Express*, vol. 15, no. 12, pp. 7578–7590, June 5, 2007.

43. A. Alù, and N. Engheta, Plasmonic materials in transparency and cloaking problems: mechanism, robustness, and physical insights, *Opt. Express*, vol. 15, no. 6, pp. 3318–3332, March 19, 2007.

44. M. G. Silveirinha, A. Alù, and N. Engheta, Parallel plate metamaterials for cloaking structures, *Phys. Rev. E*, vol. 75, 036603 (16 pages), March 7, 2007.

45. A. Alù, and N. Engheta, Achieving transparency with plasmonic and metamaterial coatings, *Phys. Rev. E*, vol. 72, 016623 (9 pages), July 26, 2005 (erratum in *Phys. Rev. E*, vol. 73, 019906, January 24, 2006).

46. A. Alù, M. G. Silveirinha, and N. Engheta, Transmission-line analysis of ε-near-zero (ENZ)-filled narrow channels, *Phys. Rev. E*, vol. 78, 016604 (10 pages), July 23, 2008.

47. A. Alù, and N. Engheta, Dielectric sensing in ε-near-zero narrow waveguide channels, *Phys. Rev. B*, vol. 78, 045102 (5 pages), July 3, 2008.

48. B. Edwards, A. Alù, M. E. Young, M. G. Silveirinha, and N. Engheta, Experimental verification of epsilon-near-zero metamaterial coupling and energy squeezing using a microwave waveguide, *Phys. Rev. Lett.*, vol. 100, 033903, January 25, 2008.

49. A. Alù, F. Bilotti, N. Engheta, and L. Vegni, Theory and simulations of a conformal omni-directional sub-wavelength metamaterial leaky-wave antenna, *IEEE Trans. Antennas Propagat.*, vol. 55, no. 6, part 2, pp. 1698–1708, June 2007.

50. A. Alù, M. G. Silveirinha, A. Salandrino, and N. Engheta, Epsilon-near-zero metamaterials and electromagnetic sources: Tailoring the radiation phase pattern, *Phys. Rev. B*, vol. 75, 155410 (13 pages), April 11, 2007.

51. A. Alù, F. Bilotti, N. Engheta, and L. Vegni, Sub-wavelength planar leaky-wave components with metamaterial bilayers, *IEEE Trans. Antennas Propagat.*, vol. 55, no. 3, part 2, pp. 882–891, March 2007.

52. A. Alù, F. Bilotti, N. Engheta, and L. Vegni, Metamaterial covers over a small aperture, *IEEE Trans. Antennas Propagat.*, vol. AP-54, no. 6, pp. 1632–1643, June 2006.

53. F. Bilotti, A. Alù, and L. Vegni, Design of miniaturized metamaterial patch antennas with μ-negative loading, *IEEE Trans. Antennas Propagat.*, vol. 56, no. 6, pp. 1640–1647, June 2008.

54. A. Alù, F. Bilotti, N. Engheta, and L. Vegni, Theory and simulations of a conformal omni-directional sub-wavelength metamaterial leaky-wave antenna, *IEEE Trans. Antennas Propagat.*, vol. 55, no. 6, part 2, pp. 1698–1708, June 2007.

55. A. Alù, F. Bilotti, N. Engheta, and L. Vegni, Sub-wavelength planar leaky-wave components with metamaterial bilayers, *IEEE Trans. Antennas Propagat.*, vol. 55, no. 3, part 2, pp. 882–891, March 2007.

56. A. Alù, F. Bilotti, N. Engheta, and L. Vegni, Sub-wavelength, compact, resonant patch antennas loaded with metamaterials, *IEEE Trans. Antennas Propagat.*, vol. 55, no. 1, pp. 13–25, January 2007.

57. A. Alù, and N. Engheta, Polarizabilities and effective parameters for collections of spherical nano-particles formed by pairs of concentric double-negative (DNG), single-negative (SNG) and/or double-positive (DPS) metamaterial layers, *J. Appl. Phys.*, vol. 97, 094310 (12 pages), May 1, 2005 (erratum in *Journal of Applied Physics*, vol. 99, 069901, March 15, 2006).

58. A. Alù, and N. Engheta, An overview of salient properties of planar guided-wave structures with double-negative (DNG) and single-negative (SNG) layers, in *Negative Refraction Metamaterials: Fundamental Properties and Applications*, G. V. Eleftheriades, and K. G. Balmain, eds., chap. 9, pp. 339–380, IEEE Press, John Wiley & Sons, Hoboken, New Jersey, 2005.

59. A. Alù, and N. Engheta, Guided modes in a waveguide filled with a pair of single-negative (SNG), double-negative (DNG), and/or double-positive (DPS) layers, *IEEE Trans. Microw. Theory Techniques*, vol. 52, no. 1, pp. 199–210, January 2004.

60. A. Alù, F. Bilotti, and L. Vegni, Analysis of L-L transmission line metamaterials with coupled inductances, *Microw. Opt. Tech. Lett.*, vol. 49, no. 1, pp. 94–97, January 2007.

61. A. Alù, F. Bilotti, and L. Vegni, Exploring the possibility of enhancing the bandwidth of μ-negative metamaterials by employing tunable varactors, *Microw. Opt. Tech. Lett.*, vol. 49, no. 1, pp. 55–59, January 2007.

62. A. Alù, N. Engheta, A. Erentok, and R. W. Ziolkowski, Single-negative, double-negative and low-index metamaterials and their electromagnetic applications, *Radio Science Bulletin*, vol. 319, pp. 6–19, December 2006.

63. A. Alù, N. Engheta, and R. W. Ziolkowski, Finite-difference time-domain analysis of the tunneling and growing exponential in a pair of ε-negative and μ-negative Slabs, *Phys. Rev. E*, vol. 74, 016604 (9 pages), July 18, 2006.

64. A. Alù, and N. Engheta, Physical insight into the "growing" evanescent fields of double-negative metamaterial lenses using their circuit equivalence, *IEEE Trans. Antennas.Propagat.*, vol. 54, no. 1, pp. 268–272, January 2006.

65. A. Alù, and N. Engheta, Evanescent growth and tunneling through stacks of frequency-selective surfaces, *IEEE Antennas Wireless Propagat. Lett.*, vol. 4, pp. 417–420, 2005.

66. N. Engheta, A. Alù, R. W. Ziolkowski, A. Erentok, Fundamentals of Waveguide and Antenna Applications involving DNG and SNG Metamaterials, in *Metamaterials: Physics and Engineering Explorations*, N. Engheta, and R. Ziolkowski, eds., chap. 2, pp. 43–86, IEEE Press, John Wiley & Sons, New York, NY, USA 2006.

67. F. Bilotti, A. Toscano, L. Vegni, K. Aydin, K. B. Alici, and E. Ozbay, Equivalent circuit models for the design of metamaterials based on artificial magnetic inclusions, *IEEE Trans. Microw. Theory Tech.*, 55, 2865, 2007.

68. Tretyakov, S. A., Maslovski, S., and Belov, P. A., An analytical model of metamaterials based on loaded wire dipoles, *IEEE Trans. Antennas Propagat.*, vol. 51, Issue 10, pp. 2652–2658, October 2003.

69. Ricardo Marqués, Ferran Martín, and Mario Sorolla, *Metamaterials with Negative Parameters: Theory, Design and Microwave Applications*, Wiley Series in Microwave and Optical Engineering, 2009.

70. Sergei Tretyakov, *Analytical Modeling in Applied Electromagnetics*, Artech House Boston, MA, USA, Electromagnetic Analysis Series, 2003.

71. W. Rotman, Plasma simulation by artificial dielectrics and parallel-plate media *IRE Trans. Antennas Propagat.* 10, 82, 1962.

72. R. Marqués, J. Martel, F. Mesa, and F. Medina, "Left-Handed-Media Simulation and Transmission of EM Waves in Subwavelength Split-Ring-Resonator-Loaded Metallic Waveguides" *Phys. Rev. Lett.*, 89, 183901, 2002.

73. J. D. Baena, L. Jelinek, R. Marqués, and F. Medina, "Near-perfect tunneling and amplification of evanescent electromagnetic waves in a waveguide filled by a meta-material: Theory and experiments," *Phys. Rev. B,* 72, 075116, 2005.

74. S. Hrabar, J. Bartolic, and Z. Sipus, "Waveguide miniaturization using uniaxial negative permeability matamaterial," *IEEE Trans. Antennas Propagat.,* 53, 110, 2005.

75. S. F. Mahmoud, A new miniaturized annular ring patch resonator partially loaded by a metamaterial ring with negative permeability and permittivity, *IEEE Antennas Wireless Propagat. Lett.,* vol. 3, pp. 19–22, November 1, 2004.

76. M. E. Ermutlu, and S. Tretyakov, Patch antennas partially loaded with a dispersive backward-wave material, *Proc. 2005 IEEE AP-S USNC/URSI Int. Symp.,* Washington, DC, July 3–8, 2005.

77. J. S. Petko, and D.H. Werner, Theoretical formulation for an electrically small microstrip patch antenna loaded with negative index materials, *Proc. 2005 IEEE AP-S and USNC/URSI Int. Symp.,* Washington, DC, July 3–8, 2005.

78. D. M. Pozar, and D. H. Schaubert, *Microstrip Antennas: The Analysis and Design of Microstrip Antennas and Arrays,* IEEE Press, New York, 1995.

79. J. R. James, and P. S. Hall, *Handbook of Microstrip Antennas,* Peter Peregrinus, London, 1989.

80. S. Maci, G. Biffi Gentili, P. Piazzesi, and C. Salvador, Dual-band slot-loaded patch antenna, *IEEE Proc.-Microw. Antennas Propagat.,* vol. 142, no. 3, pp. 225–232, June 1995.

81. R. Porath, Theory of miniaturized shorting-post microstrip antennas, *IEEE Trans. Antennas Propagat.,* vol. 48, pp. 41–47, no. 1, January 2000.

82. Shun-Shi Zhong, and Jun-Hai Cui, Compact circularly polarized microstrip antenna with magnetic substrate, *Proc. of the 2002 IEEE AP-S Int. Symp.,* vol. 1, pp. 793–796, San Antonio, TX, June 2002.

83. D. Psychoudakis, Y. H. Koh, J. L. Volakis, and J. H. Halloran, Design method for aperture-coupled microstrip patch antennas on textured dielectric substrates, *IEEE Trans. Antennas Propagat.,* vol. 52, pp. 2763–2766, no. 10, October 2004.

84. C. Reilly, W. J. Chappell, J. Halloran, K. Sarabandi, J. Volakis, N. Kikuchi, and L. P. B. Katchi, New fabrication technology for ceramic metamaterials, *Proc. 2002 IEEE AP-S Int. Symp.,* vol. 2, pp. 376–379, June 16–21, 2002.

85. J. L. Volakis, Chi-Chih-Chen, Ming Lee, B. Kramer, and D. Psychoudakis, Miniaturization methods for narrowband and ultrawideband antennas, *Proc. 2005 IEEE Int. Workshop on Antenna Technology: Small Antennas and Novel Metamaterials,* pp. 119–121, March 7–9, 2005.

86. CST Design Studio, www.cst.com.

87. M. G. Silveirinha, and N. Engheta, "Tunneling of Electromagnetic Energy through Subwavelength Channels and Bends using ε-Near-Zero Materials"; *Phys. Rev. Lett.* 97, 157403, 2006.

88. M. G. Silveirinha, and N. Engheta, "Theory of supercoupling, squeezing wave energy, and field confinement in narrow channels and tight bends using ε near-zero metamaterials," *Phys. Rev. B,* 76, 245109, 2007.

4

Dielectric Metamaterials

Dawn Tan,
Kazuhiro Ikeda, and
Yeshaiahu Fainman
Dept. of Electrical and
Computer Engineering
University of California
San Diego

4.1 Introduction

In this chapter, we focus on metamaterials composed primarily of dielectric materials that are engineered on the nanometer scale so as to have emergent optical properties not otherwise present. The increased localization of the optical field as a result of these engineered materials is responsible for phenomena such as form-birefringence, structural dispersion, and enhanced optical nonlinear interactions. Equivalently, characteristics such as the local polarizability of the metamaterial and the dispersion may be controlled by geometry, properties of constituent materials, and their composition. The introduction of periodicity in metamaterials results in a modification in the dispersion relation that can be used to create an artificial bandgap [1–3]. The manipulation and modification of this periodicity allows the bandgap to shift and parts of the bandgap to be accessed by propagating modes. Photonic crystal (PhC) waveguides rely on this concept: a line of defects is introduced into the otherwise periodic structure so as to guide light. The confinement of light within the PhC lattice is also used to realize devices such as super collimators [16,73], super prisms [8], super lenses [9], omnidirectional filters, [10] and lasers [11] through proper design and optimization of Bloch modes.

Similar to the PhC in its periodic nature is a class of metamaterials that exploits the advantages of both continuous free space and discrete guided wave modes. The simplest example involves the propagation of light in a waveguiding slab, where confinement occurs only in the vertical direction; the free space propagation occurs in the plane of the slab. This configuration, aptly termed "free space optics on a chip" (FSOC), enables interaction with discrete optical components that are located along the propagation direction. Functionalizing devices for such integrated systems would require free space implementations of focusing, beam steering, and wavelength selectivity [12]. Realization of these functionalities can exploit periodic, quasi-periodic, or even random nanostructured composites. By altering at nanometer scales, the surface morphology of a dielectric using nanolithography and advanced etching techniques, we can realize these complex structures and control the material's local polarizability. As we shall see in this chapter, these structures fall into the deeply subwavelength regime with spatial features $\ll \lambda/2n$ and require metamaterials engineering with very high spatial resolution.

The engineering of composite dielectrics can continue to larger scales, creating metamaterials that involve feature sizes on a subwavelength scale, for example, just $< \lambda/2n$. The common themes of periodicity/quasi-periodicity and enhanced effects due to light confinement of guided modes will still remain. Continuing the simplification of PhC with periodicity in two dimensions that has been used for planar confinement, an alternative system whereby light confinement in 2D is achieved by total internal reflection and periodicity in 1D is introduced to create a bandgap in the third dimension. One method of implementation involves using a channel waveguide instead of a slab waveguide in the FSOC case for guiding light, and periodically modulating the effective index of the channel waveguide along the propagation direction. This results in a periodic quantum wire akin to a 1D PhC. Thereafter, interesting properties may be engineered by once again introducing defects in the periodic structure to access forbidden modes or slightly changing the periodicity to alter the effective bandgap and dispersive properties [13–17]. The strong confinement of fields in these engineered quantum wire-like structures will also enable exploring interaction with other overlapping fields or discrete components despite the highly guided nature of these modes.

Silicon on insulator materials (SOI) and III-V compound semiconductor materials will be used for most of the discussion in this chapter because SOI is compatible with the well-established CMOS fabrication process, and III-V semiconductors have been frequently used in heterogeneous integrated circuits and systems. In addition, a large index difference between silicon and its oxide exists, which leads to highly confined modes and enables the miniaturization of on-chip silicon-based photonic circuits. Furthermore, silicon is optically transparent and has a very low material absorption coefficient at the wavelengths region around 1.55 μm, which is also used for telecommunications. Waveguiding loss in SOI platforms has a state-of-the-art value of less than 1 dB/cm. In terms of the impact for future systems applications, it is evident that next-generation computing would benefit greatly from all-optical data transfer and processing on a chip. Electrical interconnections inherent in today's computing cannot measure up in terms of both speed and bandwidth [18]. Researchers in the field are aware of the need to bypass any sort of electro-optic process in order to take computing speeds to the next level. Much work is being done in creating both passive and active devices in SOI. Discrete device components such as modulators [19,20], Filters [21], and resonators [16,17] have been demonstrated. Active devices utilizing Raman gain [22] and hybrid silicon lasers [13] that achieve gain from a bonded III-V material have been demonstrated. The momentum of research in this area is the best evidence that silicon photonics is set to revolutionize the field of computing and communications.

At the core of progress made in this field is the ability to realize these structures and test them in experimental settings. Lithography, either via electrons or light at various wavelengths, is instrumental to creating the masks for pattern transfer to SOI wafers. State-of-the-art electron-beam writers boast a resolution of tens of nanometers. Present-day research is pushing the limits of what the e-beam writers can realistically deliver. Experts predict that nanolithography will reach a resolution of 16 nm by the year 2020 and that both top-down and bottom-up integration methods will become feasible [24]. The next generation of devices promises to deliver interesting behavior at the atomic scale if this bottleneck in processing can be overcome. But until the improvement takes place, only theoretical work can be done on this class of devices.

In this chapter, we will divide the analysis of the dielectric metamaterials into two categories, namely those in the deeply subwavelength scale and the subwavelength scale. As we shall see, interesting material emergent properties arise when the materials are engineered to sizes smaller than or comparable to the wavelength of light in the said medium.

4.2 Deeply Subwavelength Scale

In this section, we investigate a class of dielectrics characterized by feature sizes $\delta \ll \lambda/2n$, where λ is the free-space wavelength of device operation and n is the refractive index of the dielectric material. The photonic structures having periodic or quasi-periodic refractive index variation with characteristic distances much smaller than the wavelength of light can be called "metamaterials" (from the Greek word "μετά" = "after," "beyond")—materials that gain their properties from the structure rather than only from constituents. This approach, as we will discover in detail in this section, can be illustrated by the simplest example: form-birefringent materials, one-dimensional periodic structures which have polarization-dependent index of refraction [24–26] and unusual nonlinear properties [27]. Extending this concept to 2D geometry or implementing aperiodicity enables other useful functionalities such as converting a linear polarization state to radial or azimuthal polarization [28] and creating a graded-index medium [29]. It was also shown that the metamaterial approach can help to overcome fabrication difficulties and create a Fresnel lens analogue using binary lithographic fabrication with a deeply subwavelength feature size of less than 60 nm [30].

4.2.1 Form Birefringence in 1D Periodic Structures

Form birefringence occurs in structures that have deeply subwavelength periodicity. The altered surface morphology of the dielectric used to construct such structures results in a large difference between the effective indices of the TE- and TM-polarized optical fields, since they need to satisfy different boundary conditions. Form-birefringent nano-structures (FBNs) are advantageous compared to naturally birefringent materials in that (1) the strength of birefringence, $\Delta n/n$ (where Δn and n are the difference and average refractive indices for the two orthogonal polarizations), is larger in the former; (2) the extent of form birefringence, Δn, may be adjusted by varying the duty ratio as well as the shape of the microstructures; and (3) FBNs may be used to modify the reflection proper-ties of the dielectric boundaries. These features are useful for constructing polarization-selective beam splitters and general-purpose polarization-selective diffractive optical elements such as birefringent computer-generated holograms (BCGHs).

In this section, we will present the theory, fabrication, and experimental character-ization of a form-birefringent computer-generated hologram implemented on a GaAs substrate. The device will be designed to transmit the TE-polarization straight ahead and deflect the TM polarization at an angle.

The layout of a binary phase diffractive structure is shown below (Figure 4.1). Two periodic functions exist within the structure: within a single period T, one pixel consists of a high-spatial frequency grating (HSFG) with a deeply subwavelength period Λ, and the next pixel is the substrate material. The HSFG introduces only the 0th diffraction order owing to its subwavelength nature. The diffractive structure, however, introduces multiple diffraction orders. The phase differences Φ_{TE} and Φ_{TM} between an incident field in the HSFG region (ray 1) and the diffractive region (ray 2) for TE and TM polariza-tions, respectively, are given by:

$$(2\pi/\lambda)(n_s - n_{\mathrm{TE}})d = \Phi_{\mathrm{TE}} \tag{4.1a}$$

$$(2\pi/\lambda)(n_s - n_{\mathrm{TM}})d = \Phi_{\mathrm{TM}}, \tag{4.1b}$$

where λ is the wavelength in vacuum, d is the HSFG layer thickness, n_s is the refractive index of the substrate, and n_{TE} and n_{TM} are the refractive indices of the HSFG for TE and TM polarizations, respectively.

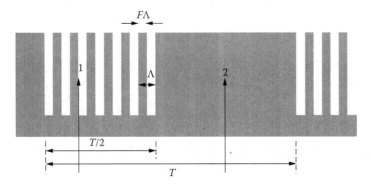

FIGURE 4.1 Schematic diagram of FBCGH.

When the condition $\lambda \gg \Lambda$ is met, the second-order effective medium theory (EMT) [31] may be used to accurately to calculate the effective indices, $n_{2\text{TE}}$ and $n_{2\text{TM}}$, with high accuracy:

$$n_{2\text{TE}} = \left\{ n_{0\text{TE}}^2 + \frac{1}{3}\left[\frac{\Lambda \pi F}{\lambda}(1-F)(n^2 - n_0^2) \right]^2 \right\}^{\frac{1}{2}} \tag{4.2a}$$

$$n_{2\text{TM}} = \left\{ n_{0\text{TM}}^2 + \frac{1}{3}\left[\frac{\Lambda \pi F}{\lambda} n_{0\text{TM}}^3 n_{0\text{TE}} \times (1-F)\left(\frac{1}{n^2} - \frac{1}{n_o^2} \right) \right]^2 \right\}^{\frac{1}{2}}, \tag{4.2b}$$

where

$$n_{0\text{TE}} = \left[Fn^2 + (1-F)n_0^2 \right]^{\frac{1}{2}}$$

and

$$n_{0\text{TM}} = \left[\frac{(n_0 n)^2}{Fn_0^2 + (1-F)n^2} \right]^{\frac{1}{2}}$$

are the effective indices calculated with zero-order EMT, F is the HSFG grating fill factor, and n and n_0 are the refractive indices of the two materials that form the HSFG. GaAs is chosen as the substrate and, therefore, $n = n_s = 3.37$ at our chosen reconstruction wavelength of 1.55 μm. In general, it is evident from the boundary conditions that the indices for two polarizations satisfy $n_{\text{TE}} > n_{\text{TM}}$. This occurs due to the continuity of the electric displacement for the TM-polarized fields, causing most of the electric field to be localized in the low-dielectric-constant material (i.e., air). To design a diffractive polarization beam splitter, we implement a binary phase grating for TE polarization without affecting the TM polarization. Therefore, $\Phi_{\text{TE}} = \pi$ and $\Phi_{\text{TM}} = 2\pi$. The value of Λ chosen should be large enough to be fabricated and small enough to not cause propagating diffraction orders greater than the 0th order. Using rigorous coupled wave analysis (RCWA) simulations [32,33], the criterion required is

$$\Lambda \leq \lambda / n_s. \tag{4.3}$$

Setting $\Phi_{\text{TE}} = \pi$ and $\Phi_{\text{TM}} = 2\pi$ in Equation 4.1, we obtain the following expression:

$$\frac{n_s - n_{\text{TE}}}{n_s - n_{\text{TM}}} = \frac{\Phi_{\text{TE}}}{\Phi_{\text{TM}}} = 1:2. \tag{4.4}$$

Substituting n_{TE} and n_{TM} from Equation 4.2 into Equation 4.4 and choosing $\Lambda = 0.3\ \mu m$ to satisfy Equation 4.3, we obtain for the grating fill factor, $F = 0.3509$. The corresponding effective refractive indices from Equation 4.9 are then calculated to be $n_{2TE} = 2.309$ and $n_{2TM} = 1.2447$. From Equation 4.1, we find the required HSFG etch depth, $d = 0.728\ \mu m$.

For further design confirmation, RCWA simulations are performed to find the phase delay introduced by the HSFG. In the simulation, a single period of the surface relief grating is divided into a large number of planar layers. The optical fields are formulated in terms of spatial harmonics by Fourier series expansions of the dielectric constant of each layer. Boundary conditions are matched and the energy conservation law is employed to solve the resultant coupled diffraction equations. In our simulation, we only try to calculate the phase delay caused by HSFG to confirm the results that we obtained by using the EMT. The actual diffraction efficiency of an FBCGH is estimated later by scalar diffraction theory. The grating parameters are the same as those given earlier. The simulation indicates that the phase delay introduced by a 0.73-mm-thick HSFG is 2.154π for TE polarization and 1.190π for TM. A GaAs layer of the same thickness without HSFG introduces a 3.178π phase delay. Thus, the designed grating will have a 1.024π phase difference between the HSFG pixel and an unetched pixel for TE polarization and 1.987π for TM polarization. This simulation shows the validity of our design. It also indicates that the EMT, if used carefully, can be used in designing FBCGH elements.

For experimental validation, fabrication of the designed device is performed on GaAs. E-beam lithography, followed by chemically assisted ion beam etching, is performed in a total device size of 100 by 100 μm. The fabricated dimensions were $T = 10\ \mu m$, $\Lambda = 0.3\ \mu m$, $F = 0.35$, and $d = 0.75\ \mu m$. Figure 4.2 shows an SEM micrograph of the fabricated device. Measurements were made using the setup shown in Figure 4.3. An He-Ne laser with an

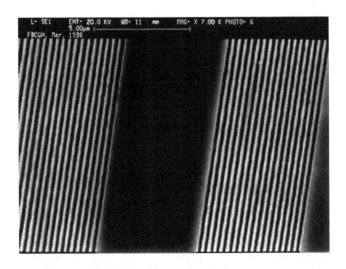

FIGURE 4.2   SEM micrograph of device fabricated on GaAs.

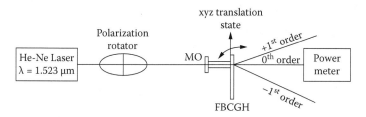

FIGURE 4.3 Experimental setup used to measure FBCGH.

operating wavelength of 1.523 μm was focused onto the fabricated FBCGH using a 6X microscope objective. A Germanium detector was used to measure the far-field diffraction patterns. Using a polarization rotator, we were able to detect separately the TE and TM polarizations of the incident beam.

The aim of the measurement is to confirm the form-birefringent behavior of the FBCGH. This was measured using the difference in diffraction intensity for the detected TE and TM fields. In the binary phase FBCGH reconstruction stage, we anticipate observing ±1 propagating diffraction orders. In our experiments, only two spots on the IR phosphor viewing card under TE polarization and one spot under TM polarization were observed, even though higher orders exist and may be detected with a photodetector. We can optimize the distance between the microscope objective (MO) and the FBCGH by minimizing the measured energy diffracted into the 0th diffraction order at TE-polarized illumination.

The measured diffraction efficiency, excluding reflection, and the polarization contrast ratios, are summarized in Table 4.1. The diffraction efficiency was calculated as the ratio of the intensity measured at a certain diffraction order to that of the total light transmitted through the GaAs substrate without an FBCGH. These measured results show that the FBCGH has good polarization selectivity (large polarization contrast ratios) and diffraction efficiencies close to the theoretical limit. Note that the form-birefringent structure also serves as an antireflection coating, explaining the slightly higher measured diffraction efficiencies compared to that predicted by scalar diffraction theory for a binary phase element (40.5%).

The expected results calculated with scalar diffraction theory are also listed in the table for comparison. The slight asymmetry between the efficiencies of the ±1st diffraction orders is due to imperfect normal incidence. Polarization contrast ratios as large as 275:1 and diffraction efficiencies as high as >40% for the ±1st diffraction orders are experimentally observed.

TABLE 4.1 Measured results for fabricated FBCGH

Performance	0th Order	+1st Order	−1st Order
TE Efficiency	0.86% (0.0%)	41.4% (40.5%)	44.2% (40.5%)
TM Efficiency	75.5% (100%)	0.15% (0%)	0.44% (0%)
Polarization Contrast Ratio	88.2:1	275:1	99.2:1

Note: Results calculated using scalar diffraction theory for a binary phase level diffractive optical element are given in parentheses for comparison.

4.2.2 Inhomogeneous Dielectric Metamaterials with Space-Variant Polarizability

It is also possible to mold the light flow in the planar configuration using the metamaterial approach. Bringing functionality of the table-top optical information processing components to a chip will create compact devices that can benefit from fast data transfer, small form-factor, parallel processing, and low power consumption. Implementing free-space-like propagation for planar optics means that while the light is confined by index difference in the vertical direction (chip plane), the horizontal beam size is regulated by phenomena similar to that seen in 3D free-space optics such as diffraction, reflection, and refraction. This can be seen as a direct and more natural transition of the conventional free space bulk optical components and devices to the chip-scale photonic integrated circuits.

To create a dielectric metamaterial, we use a subwavelength structure that can be fabricated in the high refractive index slab (see Figure 4.4).

The slab has an index of refraction, n_1, and the gaps in the etched subwavelength structure can be filled with a material possessing a low index of refraction, n_2, such as, for example, air with $n_2 = 1$. This slab structure is constructed on the cladding with an overall lower index of refraction $n_c < n_1$, to ensure confinement in the vertical direction. For some material systems such as SOI, the cladding with the guiding slab is located on the top of a thick substrate with refractive index n_s. Consider a grating with a period Λ, with $F\Lambda$ being a fraction of the unit cell filled with high-index material. It can be shown that the second-order effective medium theory approximation [34] is accurate for small grating periods $\Lambda < \lambda/n$ [35] and for grating thicknesses larger than $\lambda/3$ [36]. Other approaches in design and analysis of these subwavelength grating metamaterial structures include numerical methods such as RCWA, finite element method (FEM), and the finite-difference time-domain (FDTD) approach.

This concept can be used, for example, in creating new materials with refractive indices different from these of the constituent materials. For example, for SOI material systems, we usually have silicon with index of refraction of $n_{Si} = 3.48$ and silicon dioxide with $n_{SiO2} = 1.46$ as the only materials available for structure design. In table-top free-space optics, we, on the other hand, have a variety of materials such as different glasses, crystals, polymers, etc. This fact makes it difficult to directly transfer table-top optical

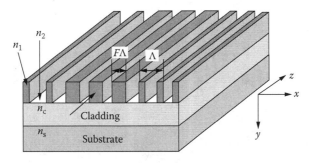

FIGURE 4.4   Schematic diagram of a subwavelength planar metamaterial.

setups for on-chip implementations. Metamaterials can provide a solution to overcome this difficulty. For example, by implementing the scheme of subwavelength grating, the index achievable for an SOI material system varies from 1.5 to 3.4, thus covering almost fully the range between high-index silicon and low-index oxide. This range was estimated for TE-polarized fields in structures with a period of $\Lambda = 400$ nm satisfying the $\Lambda < \lambda/n = 1500$ nm$/3.5 \approx 400$ nm condition, with filling factors varying from 0.1 to 0.9 to comply with the state-of-the-art nanofabrication capabilities (e.g., feature size ~ 40 nm), of the optical field was assumed.

4.2.2.1 Graded Index Structures

Next we examine a subwavelength structure with a linearly varying filling factor, corresponding to introducing a linear spatial chirp. Such a slab will equivalently act as a graded index metamaterial media, where the effective index of refraction in the transverse direction decreases or increases linearly. It is well known that in such a "graded" index material, the incident beam of light will bends toward the higher index of refraction. We performed numerical simulations of light propagation in such a spatially chirped subwavelength grating structure with initial periods of $\Lambda = 150$ nm and $\Lambda = 300$ nm. The numeric simulation results for the SOI material platform are summarized in Figure 4.5.

The propagation of light for the structure modeled with a short initial period (i.e., deeply subwavelength limit at $\Lambda = 150$ nm) shows truly graded index behavior (see Figure 4.5a), while for a larger initial period of $\Lambda = 300$ nm we can observe some reflection (see Figure 4.5b). In the latter case, we observe a "snake"-like propagation trajectory due to a Bragg reflection: as the light beam propagates close to the normal to the gradient of refractive index, its effective wavelength becomes smaller (as the index increases) until the Bragg matching condition is satisfied, that is, $\lambda_{eff} = \lambda/n = 2\Lambda \sin\theta$ (for the first diffraction order). After the Bragg-matched reflection, the light beam propagations in a graded index medium, but this time toward the lower effective refractive index. Consequently, the "total internal reflection" condition will be satisfied, and the

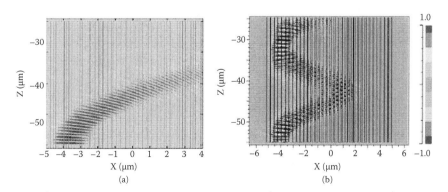

FIGURE 4.5 *See color insert.* Numerical modeling results showing light propagation in subwavelength SOI gratings with spatial chirp introduced by linearly varying the filling factor (increasing from left to right) and initial periods of (a) $\Lambda = 150$ nm and (b) $\Lambda = 300$ nm.

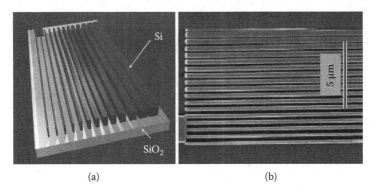

(a)   (b)

FIGURE 4.6 (a) Schematic diagram and (b) SEM micrograph of the fabricated graded index dielectric metamaterial.

light beam will be reflected back propagating toward the increasing effective index of the metamaterial slab. This type of snake-like light propagation occurring due to a combination of nonresonant and resonant transient metamaterial behavior is quite unusual and cannot be observed in natural materials.

To experimentally demonstrate this possibility, we fabricated a silicon subwavelength structure with an initial period $\Lambda = 500$ nm and linearly varying filling factor (see Figure 4.6).

The propagation of light in the graded index medium was studied experimentally using our heterodyne near-field scanning optical microscope (HNSOM) system (see Appendix A). Amplitude distribution over a larger area shows the snake-like propagation trajectory of the light beam as it propagated in the nanostructure (Figure 4.7a), while smaller area scans clearly demonstrate (Figure 4.7b,c) light bending toward the higher refractive index. The phase image (Figure 4.7c) also illustrates the decrease of the effective wavelength as the light propagates to the higher effective refractive index region. The spacing between the phase fronts at the input to the device corresponds to a wavelength of approximately $\lambda_{eff} = 1.4$ μm and at the higher index medium $\lambda_{eff} = 1$ μm.

The presented experimental results prove that the concept of dielectric metamaterial realized with subwavelength structures works well, creating a graded index medium that cannot be found in natural materials. A similar device can be used for wavelength division multiplexing (WDM), in which two signals with different carrier wavelengths can be spatially separated after several periods of snake-like mode propagation utilizing the spectral dependence of the Bragg reflection phenomena.

4.2.2.2 Quadratic Effective Index Profile and Applications

The aperiodic deeply subwavelength nanostructure can be used for focusing the propagating fields in the slab structure. The most common element to implement focusing functionality in free space is a lens that implements a quadratic phase function. For on-chip planar optics realization, we can create a dielectric metamaterial similar to that described earlier, implementing such a quadratic phase modulation in the transverse direction. To design such a material, equations (Equation 4.5) can be used to provide a

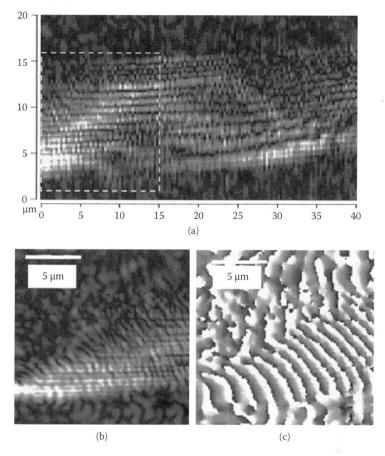

(a)

(b)   (c)

FIGURE 4.7 H-NSOM measurements of the light propagating in the graded index metamaterial: (a) amplitude distribution shows propagation consisting of periodic reflections with bending toward refractive index gradients; (b) and (c) near-field amplitude and phase distributions of the boxed area showing light bending and decreasing the values of the effective wavelength in the two regions of high and low effective refractive index.

quadratic phase profile of the effective refractive index by varying the filling factor of the initial subwavelength grating. Assuming TE-polarized fields, the quadratic index can be described by

$$n(x) = n_o\left(1 - \frac{1}{2}\alpha x^2\right) \tag{4.5}$$

with the desired parameters of $n_0 = 3.2$ and $\alpha = 0.01$ in the SOI material system without SiO_2 overcoating (i.e., using air as a top cladding), which gives $n_1 = 3.5$ and $n_2 = 1$; we also choose wavelength $\lambda = 1550$ nm and grating period $\Lambda = 400$ nm. Using these values and assumptions, we calculate and plot in Figure 4.8 the duty ratio variation of the initial subwavelength grating versus the transverse coordinate.

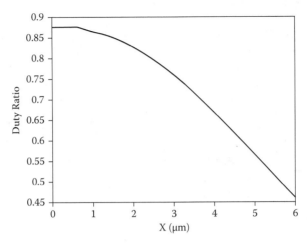

FIGURE 4.8 The calculated dependence of the filling factor versus the transverse spatial coordinate to implement a quadratic phase modulation (or quadratic effective index profile) in the transverse direction of the optical beam propagating in the slab.

Light propagation in such metamaterials structures can be analyzed using various numerical methods. Specifically, we use FDTD analysis employing the Rsoft FullWAVE software package. Our design results assume an initial grating period of $\Lambda = 400$ nm and filling factors ranging from 0.25 to 0.75 that can be implemented using currently available nanofabrication tools that can provide the smallest feature size of 100 nm (Figure 4.9a) [29]. The device length in the simulations was set to 10 μm, and the resulting obtained focusing distance was 8 μm.

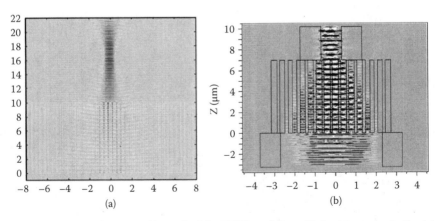

FIGURE 4.9 *See color insert.* (a) Result of the FDTD modeling of light propagating in the slab with the embedded subwavelength grating with variable filling factor providing quadratic dependence of refractive index on the transversal direction; (b) Numerical modeling of mode matching using the planar subwavelength metamaterial effectively realizing a lens functionality.

Such mode-matching functionality can be used for chip-scale integration of various nanophotonic components and devices. In contrast to commonly used waveguide tapers, the discussed metamaterials approach can be used for chip-scale mode matching between various devices and components. Such planar mode-matching elements that can realize arbitrary phase functions in the transverse direction of the propagating fields can provide a more natural and general approach for nanophotonics integration. Additional advantages of this approach are more flexible engineering capabilities and the smaller footprint of such mode-matching components. Numerical modeling in Figure 4.9b shows that a planar lens made of dielectric metamaterial can provide mode matching between 5-μm-wide channel waveguide and 1-μm-wide waveguide with an efficiency of better than 90% and the device length in this case is smaller than 10 μm. Note that adiabatic waveguide tapers need to be significantly longer to provide similar efficiency under these conditions.

4.2.2.3 Experimental Investigation of the Planar Metamaterial for Mode Matching

To demonstrate the foregoing concepts, we design and fabricate a specific device using an SOI material system. A silicon slab with a height of 250 nm integrated on top of a 3 μm SiO_2 layer on a silicon wafer provides light confinement in the vertical direction. The proposed mode-matching metamaterial device or a metamaterial "lens" is used to demonstrate matching the modes between two similar 2-μm-wide silicon channel waveguides located 25 μm apart (see Figure 4.10). The light beam emerging from the waveguide (on the left) will excite "free space modes" in the 25 μm silicon slab spacing, and due to diffraction effects the optical field coupled into the second waveguide will

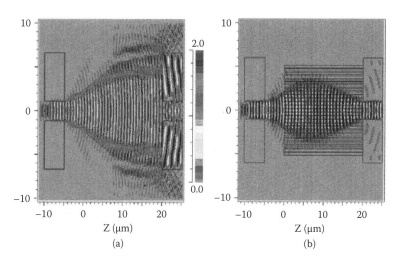

FIGURE 4.10 Results of FDTD modeling showing (a) inefficient coupling between the waveguides due to the diffraction in the silicon slab and (b) efficient coupling due to spatially chirped subwavelength dielectric grating acting as a planar lens compensating for the diffraction in a 5 μm silicon slab, and focusing the beam into the second waveguide.

not be efficient (see Figure 4.10a). The FDTD analysis estimates a coupling efficiency of approximately 20%. Next we investigate our metamaterial mode-matching structure by leaving the 5-μm-long silicon slab unperturbed to allow excitation of "free space modes" and in the remaining 20 μm silicon slab, we implement our metamaterial "lens." Here, we chose the initial period of the grating structure as $\Lambda = 400$ nm and the largest filling factor $F_{max} = 0.75$ to ensure that the smallest feature size will not exceed the practical 100 nm resolution limit. The smallest filling factor cannot exceed $F_{min} = 0.35$ to provide confinement of the mode in the vertical direction (i.e., to assure that effective index of the modulated slab has an effective index larger than that of the silicon dioxide). The numerical FDTD simulation results shown in Figure 4.10b show that with the metamaterial "lens," coupling efficiency can be increased to approximately 95%.

Fabrication of the metamaterial-based graded index "lens" consists of patterning the device using electron beam lithography followed by reactive ion etching with chlorine-based chemistry. An SEM micrograph showing the layout of the entire fabricated device is depicted in Figure 4.11a. A slanted view of a magnified portion of the device is shown in Figure 4.11b.

We estimated the accuracy of the fabricated structure by acquiring a sequence of magnified top-view SEM images from which we estimated the local duty cycle. In general, the obtained local duty cycle was slightly larger than the designed one. This was particularly noticeable around the central portion of the lens. For example, the central airgap was only ~50–60 nm wide, as opposed to the designed value of 100 nm. We believe that this deviation can be attributed mostly to the proximity effect. Proximity correction [37] can be applied to further improve the fabrication process.

To better understand the effect of structural deviation on the propagation of light within the device, we conducted an additional FDTD simulation taking into account the geometry of the fabricated structure. The result, depicted in Figure 4.12, shows that focusing occurred within the structure, about 6 μm before the output. The "overfocusing" effect is expected, as the higher values of duty cycle around the central section create an effective material with larger quadratic parameter α compared to that of the designed material.

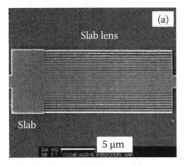

FIGURE 4.11 Scanning electron micrograph showing the fabricated device: (a) a top view of the entire structure, (b) magnified slanted view showing part of the slab lens and the output waveguide.

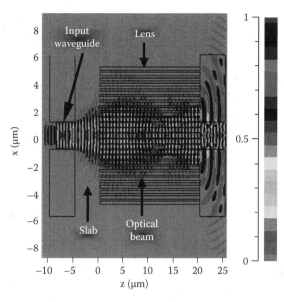

FIGURE 4.12   *See color insert.* Finite difference time domain simulation shows the beam propagation within the device, where the deviation of the fabricated structure from the design is taken into account. The result shows that focusing occurs about 5 μm before the output waveguide.

Typically, characterization of nanophotonic devices is performed by analyzing the light intensity measured at the output of the device. Unfortunately, this approach cannot probe the amplitude and, even more importantly, the phase profile of the optical beam as it propagates within the structure. To overcome this deficiency, the fabricated samples are characterized using our heterodyne near-field scanning optical microscope (H-NSOM) [38], capable of measuring both amplitude and phase of the propagating optical field with a resolution of about 100 nm. The H-NSOM is ideal for characterization of our component, because it allows direct observation of the curvature of the spatial phase front of the field propagating in the slab lens. Figure 4.7 shows the measured amplitude and phase of the optical field propagating through the device at a wavelength of $\lambda = 1550$ nm. Figure 4.13 shows the near-field amplitude distribution in the whole structure of the fabricated device. Figure 4.14a shows the amplitude of the optical field in the region that includes the input waveguide, the nonpatterned slab ("S"), and large portion of the slab lens section ("L"). The dashed vertical white lines mark the boundaries between the various sections of the device. Light propagates from left to right. Figure 4.14b shows the measured phase in the same region. Figure 4.14c shows several cross sections of the phase front calculated from Figure 4.14b at several planes along the z-axis. The obtained results clearly show the expanding of the optical beam in the region of the slab. As expected, the phase front is diverging in this section. As the beam enters the metamaterial, the curvature of the phase front gradually decreases and becomes planar after about 5 μm propagation in the slab. Then, the phase front begins to converge toward the focus. As anticipated from Figure 4.12, the new focus is generated slightly

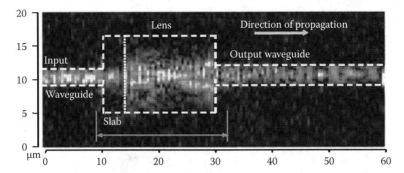

FIGURE 4.13 Near-field amplitude distribution showing light propagation in the device. The geometry of the device is indicated by the dashed line.

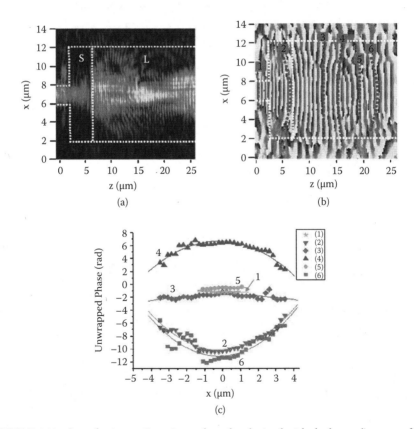

FIGURE 4.14 *See color insert.* Experimental results obtained with the heterodyne near-field scanning optical microscope (H-NSOM). (a) The amplitude and (b) the phase of the optical field in the region that includes the input waveguide, the nonpatterned slab ("S"), and large portion of the slab lens section ("L"). The dashed vertical white lines mark the boundaries between the various sections. Light is propagating from left to right. (c) Cross sections showing the phase profile at several planes along the device. The planes are marked in Figure 4.14b.

before the output waveguide. This can be observed both from the phase image showing a planar phase profile about 6 μm before the end of the slab lens (Figure 4.14b,c), and from the amplitude image showing a strong peak at the same location (Figure 4.14a). As the beam continues to propagate, the phase front starts to diverge again, and the optical beam expands.

The investigated metamaterial-based graded index slab "lens" device is the first step toward the realization of the FSOC concept. Our heterodyne-based near-field measurements clearly demonstrate the focusing effect. This experimental demonstration opens new possibilities in the field of on-chip integrated photonic devices, as the demonstrated component can be integrated with other building blocks (some of these are yet to be developed) to create future devices and systems based on the concept of FSOC. We believe that this new concept may become essential for applications such as optical interconnections, information processing, spectroscopy, and sensing on a chip.

4.3 Quantum Wires: Subwavelength-Scale Metamaterials

We refer to subwavelength structures as structures with features that are smaller than the wavelength of light but comparable to it. An example of such periodic structures is a general family of resonant structured materials such as Photonic Crystal (PhC) lattices and the whole family of devices that can be implemented in PhC lattice. In this section, we explore a novel practical approach that we call Quantum Wires metamaterials, which can be used to implement both subwavelength as well as deeply subwavelength nanostructures. It should be noted that subwavelength-scale devices are characterized by feature sizes $\Lambda < \lambda/2n$, where λ is the vacuum wavelength of operation of the device, and n is the effective index of the specific mode in the device. The effective index of the mode in the material $n_{eff} = \beta/k$, where β is the propagation constant of the waveguide mode and k is the wave number in vacuum. For example, for an SOI materials system with silicon refractive index $n_{Si} = 3.48$ and silicon oxide refractive index $n_{SiO2} = 1.46$, we can construct a typical single-mode silicon waveguide with n_{eff} of around 2.5 for operation at the telecommunications wavelength of 1.55 μm. In the following text, we will mainly focus on the subwavelength regime to demonstrate the unique capabilities of this approach and demonstrate experimentally example devices. We first investigate a periodic 1D PhC quantum wire, and extend the investigation to a quasi-periodic quantum wire. Finally, we study the characteristics of a filter created by coupling two such quantum wires together.

4.3.1 Vertical Gratings in Silicon Wires

The conventional approach to implementing subwavelength periodic structures such as, for example, distributed Bragg reflectors (DBRs) relies on a two-step lithographic process involving first the definition of a waveguide, followed by definition of the horizontal stripes on the waveguide to bring about the periodic index modulation satisfying the Bragg requirement. In contrast, our approach is using single-step lithography to make it practical and compatible with the current silicon photonics fabrication processes. In our method, we use vertical gratings that are created by periodically corrugating

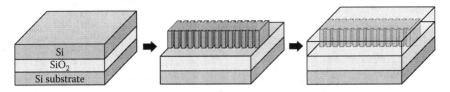

FIGURE 4.15   Single-step lithography to define the vertical gratings.

the sidewalls of a single-mode silicon waveguide wire with fixed amplitude, thus significantly simplifying the fabrication process (see Figure 4.15). In general, the width of the waveguide, W, will determine the effective refractive index of the waveguide mode, whereas the sidewall modulation, ΔW_s, will determine the strength of the coupling in the quantum wire. Typical values of W are on the order of few hundred nanometers, and the typical values of ΔW_s are on the order of tens of nanometers, dramatically altering the behavior of the waveguide. The capability to control the sidewall resolution and amplitude of the modulation enables precision control of the effective index and the modal structures in the quantum wires to create functionalities that cannot be achieved in currently available photonic nanostructures.

Next, we show a specific example of realizing a Bragg grating using our approach. The Bragg requirement is governed by the equation $\lambda_B = 2 \cdot n_{eff} \cdot \Lambda_B$ [39], where λ_B is the Bragg wavelength and Λ_B is the period of the grating. The refractive indices of Si and SiO$_2$ are assumed to be 3.5 and 1.4 in this example. The proposed structure implementing a DBR device is shown in Figure 4.16, it consists of a 500 by 250 nm single-mode silicon waveguide that has an effective refractive index of $n_{eff} = 2.56$. The corresponding Bragg period is thus $\Lambda_B = 305$ nm for operation of the device at the wavelength of 1.55 µm.

For the transverse electric (TE) fundamental mode, where the electric field is parallel to the wafer surface, the coupling coefficient κ for the first-order sidewall modulated quantum wire is given by

$$\kappa = \Gamma \left(\frac{\pi}{\lambda} \cdot \frac{\Delta W_s}{W_{eff}} \cdot \frac{n_s^2 - n_{eff}^2}{n_{eff} q} \right) \left(\frac{n_{eff}^2}{n_s^2} - \frac{n_{eff}^2}{n_{SiO_2}^2} + 1 \right), \qquad (4.6)$$

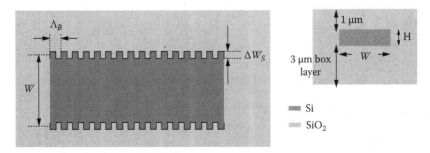

FIGURE 4.16   Schematic diagram of a sidewall-modulated quantum wire realized in SOI with $W = 500$ nm, $H = 250$ nm, ΔW_s, and $\Lambda_B = 305$ nm.

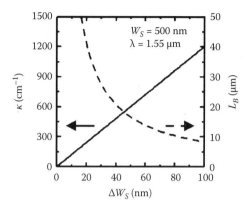

FIGURE 4.17 Calculated coupling coefficient (κ) and the corresponding Bragg length ($L_B = 1/\kappa$) versus ΔW_S.

where n_s is the effective index of the symmetric slab waveguide; Γ is the fraction of the TE mode confined in the silicon core (~85% for this structure) [40]; $q \equiv (n/n_s)^2 + (n/n_{SiO_2})^2 - 1$; and the waveguide effective width $W_{eff} \equiv Ws + 2/(q.\gamma)$, where $\gamma \equiv \sqrt{\beta^2 - k^2 n_{SiO_2}^2}$. The coupling coefficient, κ, increases linearly with ΔW_s (see Equation 4.6) as described by the plot in Figure 4.17. The Bragg length, $L_B = 1/\kappa$, is also plotted in Figure 4.17 as a function of ΔW_s.

From Figure 4.17, we obtain ΔW_s values of 50, 75, and 100 nm corresponding to values κ of 600, 900, and 1200/cm. According to coupled mode theory (CMT), the bandwidth of a Bragg grating is given by

$$\Delta \lambda = \frac{\lambda^2}{\pi . n_{eff}} |\kappa|. \tag{4.7}$$

Therefore, the sidewall modulation values, ΔW_s of 50, 75, and 100 nm result in a stopband of 18, 27, and 36 nm, respectively. It should be noted that CMT is merely an approximation in this case since inherent to CMT is the assumption that the grating under investigation is in the weak coupling regime. Since the sidewall-modulated quantum wires in these example values correspond to operation in a strong coupling regime, the slowly varying envelope approximation used to eliminate the second-order derivative in the propagation direction in the coupled-mode equations no longer holds. Nevertheless, the CMT gives us a useful albeit approximate result, but for more accurate results FDTD numeric simulations should be used. Examples of such simulations will be shown later in the chapter.

4.3.2 Resonant Transmission Filter in Sidewall-Modulated Quantum Wires

The first example device implemented in a sidewall-modulated quantum wire is a Fabry-Perot (FP)-type filter consisting of two sidewall-modulated quantum wire gratings with

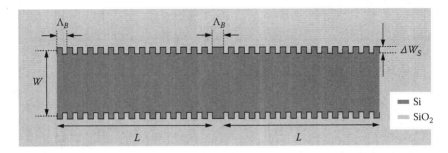

FIGURE 4.18 Top view of resonant transmission filter in sidewall-modulated quantum wire, where W_S is the average width of the waveguide, ΔW_S is the full depth of a single sidewall of the waveguide, and L and Λ_B are the total length and the period of a single Bragg reflector, respectively. For SOI realization, we use $d = \Lambda_B/2$, $\Lambda_B = 305$ nm, $W = 500$ nm, and $\Delta W_s = 50$ nm.

reflection and transmission amplitude coefficients, r and t, separated by a spacer of length, $d = \Lambda_B/2$, shown in Figure 4.18. The quarter wave spacer leads to a phase shift, $\varphi = \pi/2$, implementing a single-mode resonator. The transmission amplitude of the resonant filter, t_{RF}, is given by

$$t_{RF} = \frac{t^2.\exp(i\varphi)}{1-r^2.\exp(2i\varphi)}, \tag{4.8}$$

where $t = 1/\{\cosh(\mu L) - (i\delta - \alpha/2)[\sinh(\mu L)]/\mu\}$; $r = i\kappa t[\sinh(\mu L)]/\mu$; $\mu \equiv \kappa^2 + (i\delta - \alpha/2)2$, with a detuning parameter $\delta \equiv \beta - \beta_B \sim (d\beta/d\omega)(\omega - \omega_B) = (\omega - \omega_B)/v_g$; α is the loss of the Bragg reflector; $\omega \equiv 2\pi c/\lambda$ is angular frequency; $\omega_B \equiv 2\pi c/\lambda_B$ is the Bragg angular frequency; and $v_g \equiv c/n_g$ is the group velocity of the waveguide mode with speed of light in vacuum c.

The group index of the waveguide mode, n_g, can be experimentally determined by measuring the period of the Fabry–Perot oscillations, $\Delta\lambda_{FP}$, in a straight waveguide of identical width W. These oscillations occur due to refractive index mismatch at the cleaved facets of the fabricated waveguide structure. The equation governing the relationship between n_g and $\Delta\lambda_{FP}$ is given by

$$n_g = n_{eff} - \lambda_B.\left(\frac{\Delta n}{\Delta\lambda_{FP}}\right). \tag{4.9}$$

The group index of ~3.8 was measured experimentally in a fabricated 500-nm-wide waveguide in a 250-nm-thick Si slab on SOI.

The transmitted power, $|t_{RF}|^2$, is calculated using Equation 4.3 and plotted as a function of the Bragg detuning, $(\lambda-\lambda_B)$ in Figure 4.19. In general, $|t_{RF}|^2$ is periodic in ϕ with period π, supporting transmission resonances occurring in the center of the stopband in odd multiples of $\pi/2$. However, this resonance can be tuned by varying the length of the spacer (d). The sidewall modulation amplitude determined the coupling strength, which

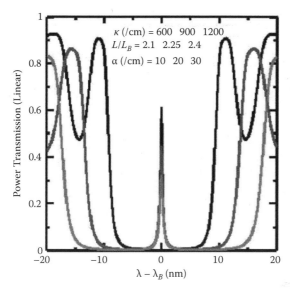

FIGURE 4.19 Calculated power transmission of a Fabry–Perot filter constructed with two side-wall-modulated quantum wires with a single-mode spacer.

in turn determined the width of the stop band in the transmission spectrum. The curves with bandwidths of black, blue, and red lines correspond to filters constructed using vertical gratings with $\kappa = 600/cm$, $\alpha = 10/cm$, $L = 70\ \mu m$ $(2.1L_B)$ (Filter 1); $\kappa = 900/cm$, $\alpha = 20/cm$, $L = 50\ \mu m$ $(2.25L_B)$ (Filter 2); and $\kappa = 1200/cm$; $\alpha = 30/cm$, and $L = 40\ \mu m$ $(2.4L_B)$ (Filter 3), respectively. We used a grating period, $\Lambda_B = 305$ nm, and a spacer, $d = 152.5$ nm.

For experimental validation of the quantum wire filters, we fabricated three devices with sidewall modulation of $\Delta W_s = 50$, 75, and 100 nm. An SOI wafer consisting of a 250 nm silicon top layer with a 3 μm buried SiO_2 (BOX) layer on a silicon wafer was used. An e-beam resist, polymethylmethacrylate (PMMA), was spin-coated on the sample, followed by E-beam lithography. A scanning electron microscope (SEM) with a pattern generation software was used to pattern the whole structure on an area of ~150 μm × 300 μm. Following a standard lift-off process to transfer the pattern to a nickel mask, the silicon layer was selectively etched in a reactive ion etcher (RIE) using a mixture of BCl_3/Ar with gas flow rates BCl_3/Ar = 10/10 sccm, pressure ~ 30 mTorr, power density ~ 0.1 W/cm², and temperature ~ 20°C. It results in an etching rate of ~75 nm/min for a sample size of ~0.5 cm². Figures 4.20a,b show top and tilted SEM images of the etched structures, respectively. After removing nickel, a buffer layer of silicon dioxide ~1 μm thick is deposited to obtain symmetry in the vertical direction. In addition, a single-step antireflection coating (Si_3N_4) was sputtered on the cleaved facets of the device to reduce Fabry–Perot resonance at the air–silicon interfaces.

Experimental characterization of the fabricated devices was performed using a TE-polarized broadband source. The light was coupled into a polarization-maintaining tapered fiber with an output spot diameter of ~2.5 μm, producing ~20 dB of polarization

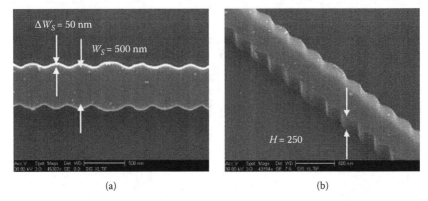

(a)   (b)

FIGURE 4.20 Scanning electron micrographs of a dry etched quantum wire filter device. (a) Top view, (b) tilted view.

extinction. Another polarization-maintaining tapered fiber is used to collect light from the fabricated devices, and its relative power transmission over the wavelength is analyzed. Figure 4.21 shows the measured spectrum the fabricated devices.

The experiments show that good agreement is achieved between the calculated and measured transmission spectra. For filters with $\Delta W_s = 50$ nm, the experimentally measured stopband of 19 nm is found to be in good agreement with the calculated stopband of 18 nm. The transmission line introduced by the defect (i.e., cavity) has a bandwidth $\Delta \lambda$ of about 0.5 nm, corresponding to the quality factor of the resonant mode of the filter, $Q = \lambda_B/\Delta\lambda = 3000$.

By thermally tuning the refractive indices of the silicon quantum wire, the resonant peak can be designed to shift at a frequency limited only by the thermal response of the material. Modulators may be designed based on this principle and have been demonstrated to have a kilohertz modulation speed [41]. Similar to PhC lattice concepts,

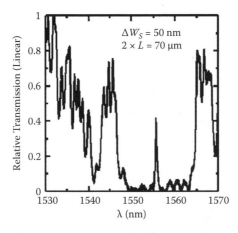

FIGURE 4.21 Measured transmission spectrum for filter with $\Delta W_s = 50$ nm.

the sidewall-modulated quantum wires have the potential to serve as practical building blocks of more complex devices that can be integrated into circuits and systems.

4.3.3 Chirped Sidewall-Modulated Grating Quantum Wires for Dispersion Engineering

In this section, we further extend the sidewall-modulated quantum wire gratings to engineer dispersion for chip-scale integration. The device is a ubiquitous chirped fiber Bragg grating (CFBG), used to compensate for group velocity dispersion (GVD) in fiber-optic applications. In this application, the sidewall modulating gratings are quasi-periodic, in contrast to the previous section, where they had a fixed period [42]. The device is an example of how the local phase and, equivalently, dispersive properties can be modified by engineering the material at the nanoscale. Our GVD device extends the simple sidewall modulation technique used in the past to construct lasers [43,44], filters [14], modulators [41], and resonators such as the one shown in the previous section. We show numerically and experimentally that by controlling device parameters such as type of apodization, and magnitude and sign of the chirp, the GVD properties of our device such as bandwidth, wavelength of operation, sign, and magnitude of dispersion can be tailored to meet the needs of various chip-scale applications.

Our GVD device uses the same SOI technology that was demonstrated in the previous section. A schematic diagram of the device is shown in Figure 4.22. The refractive indices of Si and SiO_2 are assumed to be 3.48 and 1.46, respectively, for the purposes of this study. The mean width, $W = 500$ nm, and height, $H = 250$ nm, of the waveguide resulted in an effective mode index, $n_{eff} = 2.63$, and are chosen for single-mode operation at the chosen Bragg wavelength, $\lambda_o = 1.55$ μm. The corresponding Bragg grating period period $\Lambda_o = \lambda_o/(2n_{eff}) = 295$ nm (which will be called later the mean period). Next, a linear chirp is applied to the Bragg grating to introduce a quadratic phase across the optical spectrum centered at the resonant Bragg frequency. Consequently, a linear group delay will appear across the reflection band—this leads to the term, chirped Bragg grating (CBG). The CBG period at any point z is given by $\Lambda(z) = \Lambda_o + \frac{F}{2\pi} \cdot \frac{\Lambda_o^2}{L} \cdot \frac{2z}{L}$, where F is the chirp parameter and L is the device length [45]. The total chirp in period for a fixed F is $\Delta\Lambda = \frac{F}{\pi} \cdot \frac{\Lambda_o^2}{L}$.

The large sidewall modulation of 50 nm on our CBG devices implies that in contrast to weakly coupled CFBGs in silica, they are operating in a strong coupling regime.

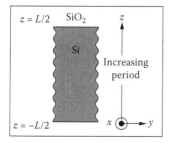

FIGURE 4.22 (a) Top view of chirped Bragg grating (CBG) device geometry realized with a sidewall-modulated quantum wire and (b) transverse profile of the quantum wire structure.

Moreover, the relatively short length of a strongly chirped device may not meet the slowly varying envelope approximation inherent in CMT. Therefore, we perform finite difference time domain (FDTD) simulations to study their characteristics. The effective index of a slab of height H infinite in the y-direction was used in place of the material refractive index of silicon to reduce the problem to two dimensions (2D). Although the reduction of the 3D problem to a 2D one is approximate, this method retains the salient spectral characteristics [46] and has shown good experimental agreement for similar structures [47].

A well-known phenomenon that occurs in Bragg filters is the ripple which occurs from the abrupt start and end of the periodic structures in the filter. The windowing effect, which leads to degradation in the filter response, may be alleviated by gradually increasing the amplitude of index modulation, and equivalently κ, from 0 at the edge to its maximum value at the center of the grating. Therefore, the effectiveness of different apodization filters [48] in achieving a flat-top response and in-band ripple suppression is first investigated. The reflection and group delay spectra for $L = 100\,\mu m$ and $\Delta\Lambda = 7.5$ nm are shown in Figure 4.23. While all apodization filters are effective in suppressing group delay ripple, asymmetric Blackman apodization [48] is the most effective in maximizing the CBG bandwidth while maintaining a flat response. The apodization applied is asymmetric in that the coupling strength of the input half of the CBG increases from 0 at the CBG edge to its maximum value at the center of the CBG. The rear half of the CBG has a constant κ. Therefore, we adopt asymmetric Blackman apodization for our CBGs for further studies of the effect of $\Delta\Lambda$ on the CBG devices.

Figure 4.24 shows that increasing $\Delta\Lambda$ from 4 nm, 7.5 nm, to 12 nm for a CBG with $L = 100\,\mu m$ results in a wider reflection bandwidth, that is, 38, 60, and 88 nm, respectively, and correspondingly, a smaller calculated GVD, that is, 7.0×10^5 ps/nm/km, 3.3×10^5 ps/nm/km, and 2.1×10^5 ps/nm/km, respectively. This trend is found to be in agreement with observations in weakly coupled CFBGs [49].

The designed CBG device was fabricated on an SOI wafer with a 250 nm layer of silicon on top of a 3 μm buried oxide layer on a silicon wafer. We use e-beam lithography followed by reactive ion etching. SiO_2 cladding was deposited over the fabricated Si structure

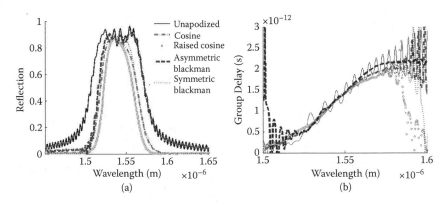

FIGURE 4.23 2D FDTD simulation results in $\Delta\Lambda = 7.5$ nm and $L = 100$ μm for different apodization filters: (a) Reflection response and (b) group delay response.

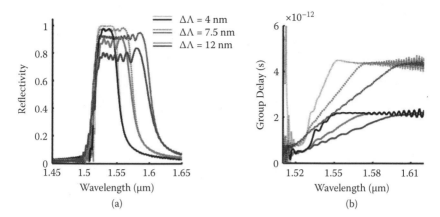

FIGURE 4.24 Effect of chirp on CBG: (a) Reflection response and (b) group delay response. Dotted lines denote $L = 200\ \mu m$ whereas solid lines denote $L = 100\ \mu m$.

using plasma-enhanced chemical vapor deposition (PECVD). SEM micrographs of fabricated structures before PECVD deposition of SiO_2 are shown in Figure 4.25.

The setup used for characterization of fabricated devices is shown in Figure 4.26. Characterization of the fabricated CBG devices was performed using TE-polarized light from a broadband source (1.52 μm to 1.62 μm) connected to a circulator. Light reflected from the CBG device is rerouted through the circulator and enters the input port of an optical spectrum analyzer, where the reflection spectrum is obtained.

We fabricated test samples with an access waveguide around 200 μm in length in front of the CBG structure. For the dispersion characterization, we use Fabry–Perot (FP) resonance oscillations in the measured reflection spectrum of the device. The resonator is defined by two reflectors, the cleaved input facet of the Si waveguide and the CBG. A method similar to that reported in Reference [50] is used: the length traversed by each

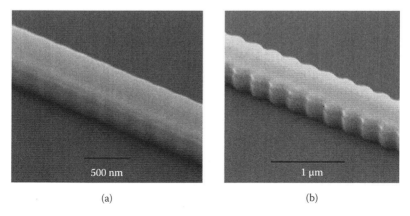

FIGURE 4.25 SEM micrographs of (a) apodized and (b) unapodized parts of the CBG before deposition of SiO_2 overcladding.

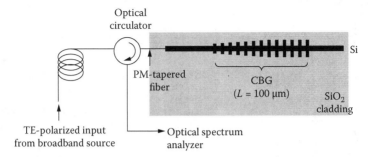

FIGURE 4.26 Setup used for experimental measurement of the CBG device.

wavelength component from the cleaved input facet to the point of reflection in the CBG may be experimentally determined using the relation

$$l_c = \frac{\lambda^2}{2n_g(\Delta\lambda_{FP})}, \qquad (4.10)$$

where $\Delta\lambda_{FP}$ is the free spectral range of the FP resonator and the value of n_g is fixed at 4.2. The total group delay is subsequently found using $\frac{2l_c \times n_g}{c} = \frac{\lambda^2}{c(\Delta\lambda)}$, which is independent of the value of n_g. Finally, the GVD is obtained using the derivative of the group delay with respect to wavelength. Since the GVD of the access waveguide is two orders of magnitude lower than that of our device [51], the measured dispersion properties are dominated by those of our CBG devices. Typical experimental data of the reflection spectrum is shown in Figure 4.27 for a fabricated CBG with $\Delta\Lambda = -7.5$ nm and $L = 100$ μm. The reflectivity data is compared with the numeric modeling result obtained for the same CBG geometry but with the average waveguide width adjusted to $W = 520$ nm to account for fabrication inaccuracies causing a 20 nm shift of the center wavelength of operation

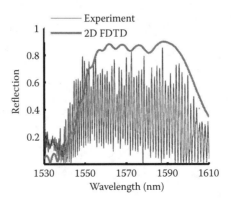

FIGURE 4.27 Comparison between simulated and measured reflection spectra envelopes. The high-frequency FP-oscillations in the measured spectrum corresponding to $L = 100$ μm and $\Delta\Lambda = -7.5$ nm are created to measure the GVD of the device.

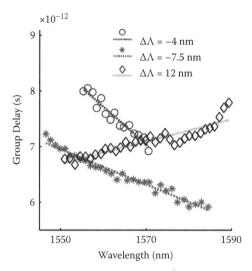

FIGURE 4.28 Measured group delay for CBGs with $L = 100$ µm for different values of $\Delta\Lambda$. Solid lines denote linear fit to measured data.

determined experimentally. The modeled result shows good agreement with the envelope and bandwidth of the measured spectrum.

Within the device bandwidth in Figure 4.27, the period of the FP oscillations increases as wavelength increases, implying negative (i.e., normal) dispersion. Three fabricated CBG devices with $L = 100$ µm and $\Delta\Lambda = -4$ nm, -7.5 nm, and 12 nm were characterized in terms of their group delay (Figure 4.28). The GVD is extracted from the slope of a linear fit applied to the experimental data. The expected GVD values for $L = 100$ µm, $\Delta\Lambda = -4$ nm, -7.5 nm and 12 nm from the simulations are -0.070 ps/nm, -0.033 ps/nm, and 0.021 ps/nm respectively; the measured GVD values are -0.067, -0.032, and 0.020 ps/nm, respectively, showing good agreement between the numeric and experimental results. Note that, as expected, changing the sign of the chirp changes the sign of the dispersion (Figure 4.28).

In this work, we investigate CBGs limited in length to 100 µm, because the high-resolution writing window of our e-beam system is 100 µm. However, it should be noted that longer CBGs are highly desirable for several reasons. First, with longer CBGs, we can achieve close to 100% reflectivity even for large $\Delta\Lambda$. Second, the ripple in both the reflection and group delay spectra can be significantly reduced for longer asymmetrically apodized CBGs [48]. The strong coupling coefficient inherent in the device ensures that reflectivity is close to 100% even for short device lengths. Referring to Figure 4.24, it appears that the ripple within the reflection band worsens toward higher wavelengths. Since the CBGs are asymmetrically apodized, there exists a discontinuity at the end of the CBG. As chirp increases, the "effective value" of $\kappa.L$ for each wavelength component decreases. Therefore, some light reaches the end of the CBG where a broadband reflection occurs. Interference of the distributed reflection of spectral components from the main part of the CBG with the broadband reflected light gives rise to ripple in the spectra. The group delay ripple worsens for the smaller wavelengths since these components

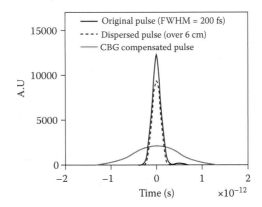

FIGURE 4.29 Simulated CBG dispersion compensation of a 200 fs pulse propagating in a 6-cm-long silicon waveguide (300 by 500 nm cross-sectional area) with a GVD of 1100 ps/nm/km.

propagate further into the CBG, owing to the smaller coupling at the apodized input end, and a larger proportion of it reaches the end of the CBG. Therefore, we expect that for larger κ or L, the ripple arising from this phenomenon can be reduced. This is confirmed by the increased extent of ripple for $L = 100$ μm compared to $L = 200$ μm for fixed chirp (Figure 4.24).

By introducing the sidewall modulation to waveguides of arbitrary dimensions, a CBG may be integrated with the waveguide for dispersion compensation. As an example, a 300 by 500 nm single-mode waveguide exhibits a GVD of 1100 ps/nm/km [51]. Figure 4.29 shows by simulation that the broadening of a pulse with temporal full width at half maximum (FWHM) of 200 fs, which has propagated over a length of 6 mm in the waveguide, is well compensated for by a CBG with $L = 200$ μm and $\Delta\Lambda = -7.5$ nm. Third-order dispersion (TOD) is neglected here since the TOD length [52] is longer than the dispersion length of the waveguide. The bandwidth of the pulse is well within that of the CBG. It may be seen from Figure 4.29 that the effective suppression of ripple in the reflection and group delay spectra results in minimal distortion. Note that depending on the geometry, silicon waveguides can exhibit either normal or anomalous dispersion. The introduced CBG device enables compensation of positive and negative sign of dispersion by simply changing the sign of the chirp. Moreover, since the operating wavelengths of the device are determined by the mean Bragg period, this approach can be used in an extremely wide range of optical frequencies.

In summary, the CBG device proposed here provides a platform to engineer normal or anomalous dispersion for photonic systems applications, including GVD compensation in silicon photonic structures on a chip. The experimental tests of the designed and fabricated devices validate the 2D FDTD method used for modeling of the CBG structures. Asymmetric Blackman apodization was found to be most effective in suppressing group delay ripple while maximizing the usable bandwidth. The experiments demonstrated a dispersion value of 7.0×10^5 ps/nm/km over a wide spectral range of about 40 nm in the near-infrared spectral range of 1.55 μm. Both normal and anomalous dispersion have been demonstrated. The operating bandwidth may be adjusted and arbitrary amounts of dispersion achieved by adjusting the sign and magnitude of the chirp. The tunability of the CBG bandwidth makes it highly suitable for accommodating ultrashort pulses with high spectral content.

4.2.4 Wavelength-Selective Coupler Based on Coupled Vertical Gratings

The sidewall modulation quantum wires have an additional advantage due to the fact that the sidewall modulation allows creation of spatially coupled structures in the transverse direction of the slab. This approach is expected to further advance our ability to create optical functionalities that cannot be easily achieved with conventional optical materials and devices. Our next example is based on using a pair of sidewall-modulated quantum wires that are brought together close enough to allow coupling between the modes in this structures. In general, rigorous analysis of such coupled structures should be performed by studying supermodes, but in this section we will assume relatively weak coupling, allowing the use of CMT. Our example device is a pair of coupled sidewall-modulated quantum wires that are used to realize a wavelength-selective coupler (WSC) [53]. The WSC functionality is similar to that of an add/drop spectral filter widely used in wavelength division multiplexed (WDM) optical communication systems. It should be noted that this approach can be extended to a multigrating layout for realization of large counts of WDM I/O ports.

The current state of the art on-chip WDM schemes involve the use of high-order coupled ring resonators [54] to implement an add/drop filter. These high-order ring resonators have the advantages of both good extinction ratio and cross-talk suppression. Their main drawback is in the limited free spectral range (FSR) of ~20 nm for ring radii of ~5 μm; this limits the total number of channels that is possible to achieve using the ring-resonator implementation of the WDM add/drop filter device. As a means of circumventing the drawback of limited FSR, we present a pair of coupled sidewall-modulated quantum wires, WG1 and WG2 (see Figure 4.30), implementing a basic unit cell WSC for WDM add/drop filter device.

In our design, the two coupled waveguides WG1 and WG2 with widths W_1 and W_2, respectively, and sidewall-modulating grating depths ΔW_1 and ΔW_2 are separated by a coupling gap distance, G. The fixed period in both gratings, Λ, will be chosen for operation at a specific drop wavelength, λ_c (in our example $\lambda_c = 1.55$ μm). By using CMT [39], we obtain for our WSC device structure three Bragg matching conditions:

$$2\beta_1 = \frac{2\pi}{\Lambda} \tag{4.11a}$$

$$2\beta_2 = \frac{2\pi}{\Lambda} \tag{4.11b}$$

$$\beta_1 + \beta_2 = \frac{2\pi}{\Lambda}, \tag{4.11c}$$

where β_1 and β_2 are the propagation constants for the waveguides with widths W_1 and W_2, respectively. Equations 4.11a–4.11c represent the backward coupling in waveguide W_1,

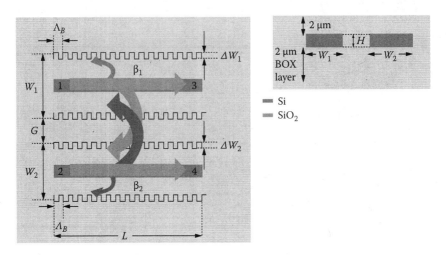

FIGURE 4.30   Schematic diagram of WSC (top view and cross section). The 2×2 device consists of the I/O ports 1 through 4 in the two waveguiding structures WG1 and WG2.

waveguide W_2, and the cross-coupling between waveguides WG1 and WG2. Therefore, we expect to have three stop-bands existing for this device.

The Bragg condition of interest to us is represented by Equation 4.11c. The cross-coupling between waveguides WG1 and WG2 will enable implementing the wavelength-selective add/drop functionality. Therefore, our first step in designing the WSC device is to allocate the three Bragg conditions to the desired wavelengths by a suitable choice of geometric parameters W_1, W_2, and Λ. Specifically, we need to ensure that Equation 4.11c is satisfied at our chosen drop wavelength of $\lambda_c = 1.55\,\mu m$. The device will utilize an SOI material platform with a 250-nm-thick layer of silicon on a 2-μm-thick BOX layer. An SiO_2 overcladding will be also deposited on the final device. As an example, we use a value of $W_1 = 550$ nm and $W_2 = 430$ nm. The corresponding value of β for TE-polarization is calculated using 2D analysis with reduction using the effective index method. The grating's effect on β is neglected in these calculations. Plots of the calculated values of β are shown in Figure 4.31. The left-hand side of each Bragg condition is plotted in solid lines, whereas the right-hand side is plotted in dotted lines. The points of intersection between the solid and dotted lines provide the three Bragg conditions. In this way, the wavelengths corresponding to Equations 4.11a–c for our specific values of W_1 and W_2 may be inferred from the graph.

Next, we calculate the specific value, Λ_B, to set a specific wavelength at cross-coupling $\lambda_c = 1.55\,\mu m$ for this example using Equation 4.11c as $\Lambda_B = 2\pi/\{\beta_1(\lambda=1.55\,\mu m)+\beta_2(\lambda=1.55\,\mu m)\} = 301$ nm. With this value of Λ_B, we find the corresponding wavelengths, λ_1 and λ_2, for the other two Bragg conditions given by Equations 4.11a and 4.11b and check if these are at the desired wavelengths. In our design example, we tried to allocate these backward couplings outside the telecommunications C band, and indeed these wavelengths are located outside the C band with this design ($\lambda_1 \sim 1.52\,\mu m$ and $\lambda_2 \sim 1.59\,\mu m$). Note that these stopbands may also be used for bandstop filters together

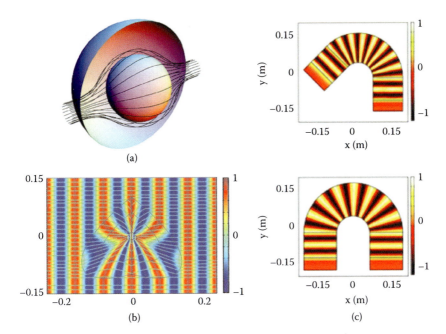

(a)

(b)

(c)

COLOR FIGURE 1.11 Example of the structures designed using the transformation optics approach. (a) A cloak. (From J. B. Pendry et al., *Science* 312, 1780, 2006. With permission.) (b) A wave concentrator. (From W. X. Jiang et al., *Appl. Phys. Lett.* 92, 264101, 2008.) (c) A reflectionless waveguide filled with anisotropic and inhomogeneous material. (From W. X. Jiang et al., *Phys. Rev. E* 78, 066607, 2008; M. Rahm et al., *Phys. Rev. Lett.* 100, 063903, 2008; M. Rahm et al., *Opt. Express* 16, 11555, 2008. With permission.)

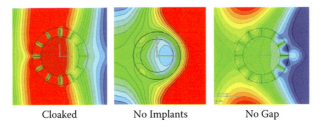

Cloaked   No Implants   No Gap

COLOR FIGURE 3.19 Consistent with the results in Reference 44, the total electric field distribution at a snap shop in time for a sample geometry of a metamaterial cloak designed at microwave frequencies in the three cases: (a) with a metamaterial cover that is designed by filling a high permittivity shell with metallic fins; (b) with the same dielectric shell cover, but removing the metallic plates; (c) again with the same dielectric shell cover, but with metallic plates extended to touch the boundaries of the shell, thus modifying the wave interaction and weakening the cloaking effect. (From M. G. Silveirinha, et al., *Phys. Rev. E*, vol. 75, 036603 [16 pages], March 7, 2007.)

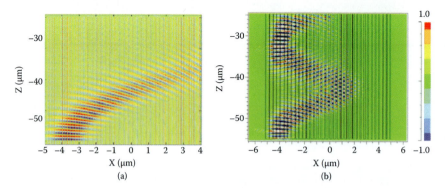

COLOR FIGURE 4.5 Numerical modeling results showing light propagation in subwavelength SOI gratings with spatial chirp introduced by linearly varying the filling factor (increasing from left to right) and initial periods of (a) $\Lambda = 150$ nm and (b) $\Lambda = 300$ nm.

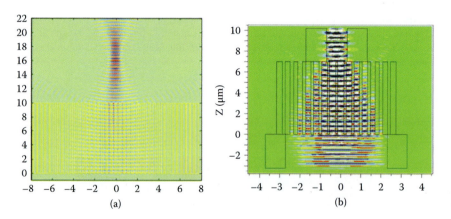

COLOR FIGURE 4.9 (a) Result of the FDTD modeling of light propagating in the slab with the embedded subwavelength grating with variable filling factor providing quadratic dependence of refractive index on the transversal direction; (b) Numerical modeling of mode matching using the planar subwavelength metamaterial effectively realizing a lens functionality.

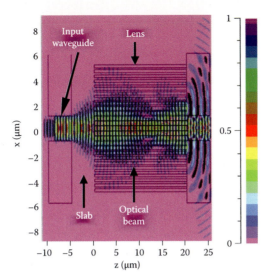

COLOR FIGURE 4.12 Finite difference time domain simulation shows the beam propagation within the device, where the deviation of the fabricated structure from the design is taken into account. The result shows that focusing occurs about 5 μm before the output waveguide.

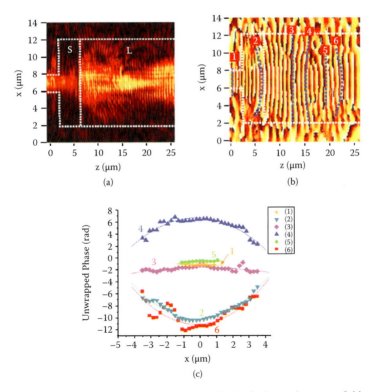

COLOR FIGURE 4.14 Experimental results obtained with the heterodyne near-field scanning optical microscope (H-NSOM). (a) The amplitude and (b) the phase of the optical field in the region that includes the input waveguide, the nonpatterned slab ("S"), and large portion of the slab lens section ("L"). The dashed vertical white lines mark the boundaries between the various sections. Light is propagating from left to right. (c) Cross sections showing the phase profile at several planes along the device. The planes are marked in Figure 4.14b.

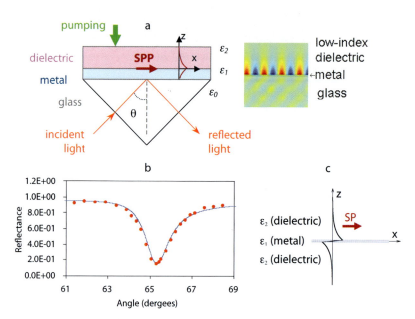

COLOR FIGURE 5.2 (a) Kretchman geometry for excitation of surface plasmon polaritons and distribution of electric field; θ is the incidence angle; pumping was applied to the dye-doped polymeric film in the experiments described in Sections 5.3.2.1 and 5.3.2.2. (b) Angular dependence of the reflectance detected in silver film in the setup of (a). Dots—experiment, solid line—calculations according to Equation 5.1. (c) Schematic of the field distribution in long-range surface plasmon propagating in a thin metallic waveguide.

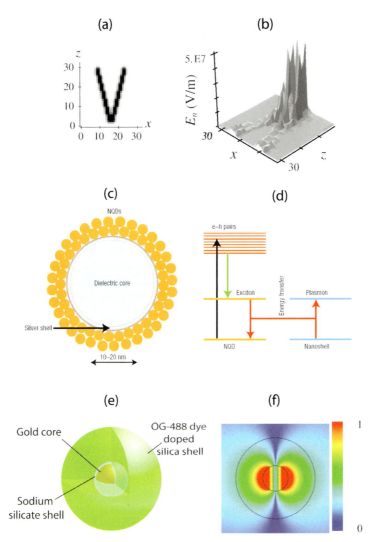

COLOR FIGURE 5.5 (a) V-shaped metallic nanostructure and (b) corresponding distribution of the field amplidude in a SPASER mode. (Adapted from Bergman, D. and M. Stockman. 2003. *Phys. Rev. Lett.* 90:027402.) (c) Hybrid spherical nanoparticle architecture of a spaser, employing quantum dots as a gain medium, and (d) schematic diagram of the energy transfer in a spaser. (Adapted from Stockman, M. I., 2008. *Nat. Photonics* 2:327–29.) (e) Hybrid spaser nanoparticle consisting of a gold core and dye-doped silica shell and (f) the corresponding mode. (Adapted from Noginov, M. A. et al. 2009. *Nature* 460:1110–12.)

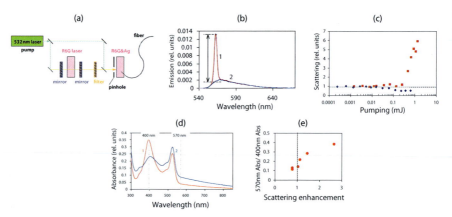

COLOR FIGURE 5.12 (a) Pump-probe experimental setup for Rayleigh scattering measurements. (b) Spectra of scattered light and spontaneous emission (1) and spontaneous emission only (2). (c) Intensity of Rayleigh scattering as a function of pumping energy in two different dye–Ag aggregate mixtures. (d) Absorption spectra of dye–Ag aggregate mixtures; R6G—2.1 × 10⁻⁵ M, Ag aggregate—8.7 × 10¹³ cm⁻³. Squares in (c) and trace 1 in (d) correspond to one particular mixture, and diamonds in (c) and trace 2 in (d) correspond to another mixture. (e) The ratio of absorption coefficients of dye–Ag aggregate mixtures at 570 nm and 400 nm plotted versus enhancement of the Rayleigh scattering measured at 0.46 mJ. (Adapted from Noginov, M. A. 2008. *J. Nanophotonics* 2: DOI:10.1117/1.3073670.)

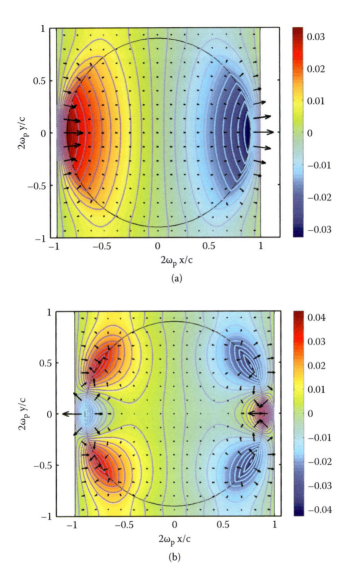

(a)

(b)

COLOR FIGURE 9.2 The eigenvalue potentials of the two strongest dipole-active resonances of the plasmonic structure with the parameters from Figure 9.2 (color and contours) and the corresponding electric field (arrows). (a): The strongest dipole-active resonance at $s_1 = 0.14$ (or $w_1 = 0.38w_p$), and (b) the second-strongest dipole-active resonance at $s_1 = 0.40$ (or $w_1 = 0.63w_p$). (Reprinted with full permission from Shvets, G. and Urzhumov, Y., *Phys. Rev. Lett.* 93, 243902, 2004. Copyright [2004] by the American Physical Society.)

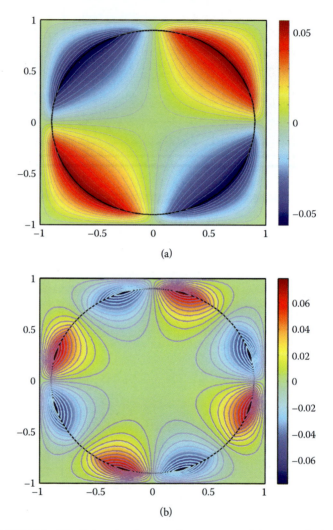

(a)

(b)

COLOR FIGURE 9.3 Dipole-inactive electrostatic resonances of the square lattice plasmonic metamaterials with the parameters as in Figure 9.2. Left: quadrupole resonance at $s_Q = 0.29$ ($w_Q = 0.54w_p$); right: octupole resonance at $s_O = 0.36$ ($w_O = 0.6w_p$).

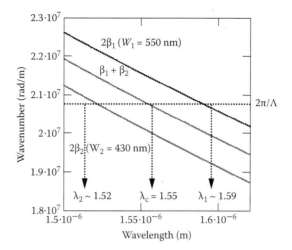

FIGURE 4.31 Propagation constants β calculated for TE-polarized optical fields using 2D analysis with the effective index method, assuming a silicon slab thickness of 250 nm and that both sides of the silicon slab are surrounded by silica claddings. The solid lines are the plots of the left-hand sides of the Bragg conditions in Equations 4.11a–c versus the wavelength of light in vacuum. The dotted horizontal line indicates the wavenumber of the Bragg grating to realize the drop filter for a wavelength $\lambda_c = 1.55$ μm, corresponding to the right-hand sides of Equations 4.11a–c. The cross sections between the solid lines and the dotted line correspond to the three Bragg conditions.

with the cross-coupling. If the designed Bragg wavelengths are not appropriate, then we repeat the foregoing procedure until the desired Bragg conditions are satisfied.

In our next step, we calculate the coupling coefficients to achieve the desired bandwidths for the Bragg conditions using the modulations ΔW_1, ΔW_2, and coupling gap G. The coupling coefficients for each of the three Bragg conditions are given by the CMT [39]:

$$\kappa_1 = \frac{\varepsilon_0 \omega}{4} \left\langle E_1 \left| \Delta n_g^2 \right| E_1 \right\rangle \tag{4.12a}$$

$$\kappa_2 = \frac{\varepsilon_0 \omega}{4} \left\langle E_2 \left| \Delta n_g^2 \right| E_2 \right\rangle \tag{4.12b}$$

$$\kappa_c = \frac{\varepsilon_0 \omega}{4} \left\langle E_1 \left| \Delta n_g^2 \right| E_2 \right\rangle \tag{4.12c}$$

with

$$\Delta n_g^2 = \begin{cases} -(n_{core}^2 - n_{clad}^2)\dfrac{f(z)+1}{2} & \text{in grating region} \\ 0 & \text{otherwise,} \end{cases} \tag{4.12d}$$

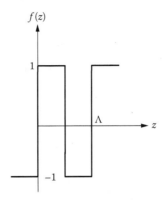

FIGURE 4.32　Periodic function $f(z)$ used for calculating the coupling coefficients.

where κ_1, κ_2, and κ_c are the coupling coefficients for the backward coupling in wave-guides 1 and 2, and the cross-coupling between waveguides 1 and 2, respectively; ε_o is the permittivity of vacuum, ω is the angular frequency of light, and E_1 and E_2 are the field distributions in waveguides WG1 and WG2 for TE polarization, again using the 2D effective index method; Δn_g is the index perturbation, which has values only at the grating region, whereas the silicon core effective index, $n_{core} = 2.96$, and the silica cladding refractive index, $n_{clad} = 1.4$; $f(z)$ is the periodic function shown in Figure 4.32, where z is the coordinate along the grating waveguide.

Intuitively, the larger the modulation (i.e., larger ΔW_1 and ΔW_2) and the field overlap (i.e., smaller W_1, W_2, and G), the larger the coupling coefficients. For known coupling coefficients, we can estimate the coupling bandwidths, $\Delta\lambda_i = \lambda^2|\kappa_i|/\pi n_{eff}$, and the corresponding Bragg lengths, $L_{Bi} = 1/|\kappa_i|$, where $i = 1, 2, c$ and n_{eff} is the effective index for WG1 or WG2 for the backward couplings, but the average effective index is used for the cross-coupling. Figure 4.33 shows plots of $\Delta\lambda_i$ and L_{Bi} as a function of G for several grating

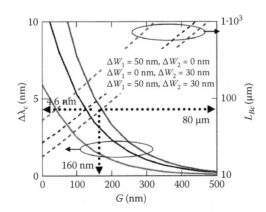

FIGURE 4.33　Plots of the bandwidth $\Delta\lambda_c$ and Bragg length L_{Bc} for cross-coupling with relation to the gap G, for several grating depths: (i) $\Delta W_1 = 50$ nm and $\Delta W_2 = 0$ nm, (ii) $\Delta W_1 = 0$ nm and $\Delta W_2 = 30$ nm, and (iii) $\Delta W_1 = 50$ nm and $\Delta W_2 = 30$ nm.

modulation depths: (i) $\Delta W_1 = 50$ nm and $\Delta W_2 = 0$ nm, (ii) $\Delta W_1 = 0$ nm and $\Delta W_2 = 30$ nm, and (iii) $\Delta W_1 = 50$ nm and $\Delta W_2 = 30$ nm. We observe that a larger grating depth and smaller G give a wider bandwidth and a shorter Bragg length. We select $\Delta W_1 = 50$ nm and $\Delta W_2 = 30$ nm from this point forward. Since the maximum writing field for high resolution of our e-beam system is 100 µm by 100 µm, we fix the WSC length to be $L_{BC} = 80$ µm for this example. Inferring from Figure 4.33, we have a value of $G = 160$ nm and $\Delta\lambda_c = 4.6$ nm for $L_{BC} = 80$ µm. The bandwidth may be reduced by increasing G or decreasing ΔW_1 and/or ΔW_2. In both cases, a longer interaction length is required for the same extinction. Note that a higher extinction and suppressed sidelobes will require a longer grating length than L_{BC} and apodization of the grating [48,42] similar to that shown in the previous section.

Numerical simulations using 2D FDTD are performed to confirm the designed WSC device. The 2D FDTD device layout is shown in Figures 4.34a and 4.35a. Due to the limited accuracy of the analytic effective index method, we have adjusted the grating period to 305 nm to ensure operation at 1.55 µm. Figures 4.34b and 4.35b show the calculated spectra in the output ports for light incident in Ports 1 and 2, respectively, with the designed bands indicated by the green area. We can see that the backward couplings in waveguides WG1 and WG2 and cross-coupling occur almost at the designed wavelengths with a small error due to inaccuracy of the analytic effective index method. We also note that the fabricated grating profile will take on a sinusoidal corrugation rather than a square-wave corrugation. Therefore, a sinusoidal-shaped grating is assumed in the FDTD simulation. The resultant coupling coefficients are slightly higher than these of the square-shaped grating in the analysis, as is evidenced by the wider bandwidths and higher extinction of cross-coupling than that for the Bragg length in the modeling results.

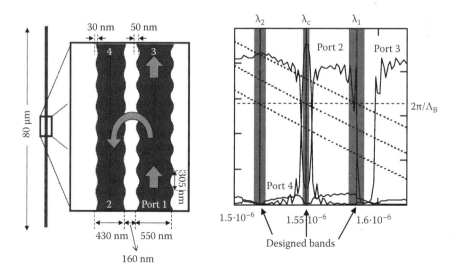

FIGURE 4.34 (a) Layout used for 2D FDTD simulation and (b) spectral response for light incident at Port 1.

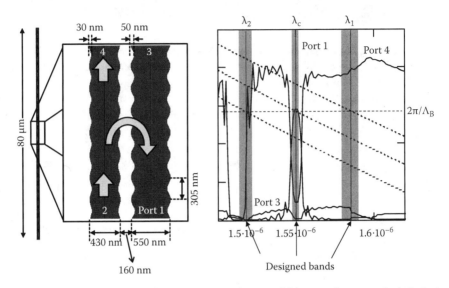

FIGURE 4.35 (a) Layout used for 2D FDTD simulation and (b) spectral response for light incident at Port 2.

Fabrication of the designed WSC was performed in an SOI material system using e-beam lithography followed by reactive ion etching. Figure 4.36 shows optical image and SEM micrographs of the fabricated device. The transmission spectrum of the light incident on Port 1 was measured using the same setup as that in Section 4.2.2. The result is shown in Figure 4.37. The backward coupling in waveguide 1 and the cross-coupling

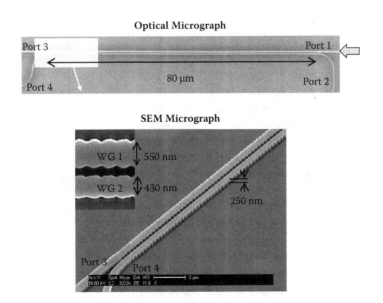

FIGURE 4.36 Optical and SEM micrographs of fabricated WSC device.

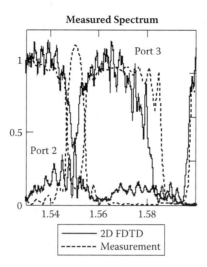

FIGURE 4.37 Measured transmission spectrum with light incident at Port 1 compared with the modeling results.

occur at the designed wavelengths, whereas the extinction ratio is less than that predicted by the simulation. We attribute the smaller extinction to the reduced coupling coefficient due to the incomplete uniformity (e.g., period fluctuation and sidewall roughness) and the incomplete dimensions of the gratings and the waveguides.

The FP oscillation in the measured spectrum corresponds to the e-beam stitching error introduced during the fabrication process. Better extinction can be achieved by improving the fabrication process and/or increasing the WSC length. Note that our device has a very large FSR supporting the total bandwidth of about 70 nm, which makes it very attractive for WDM systems applications.

The introduced WSC device concept can be further extended to configure a multiport add–drop device cascading several coupled sidewall-modulated quantum wires (see Figure 4.38). When another sidewall-modulated quantum wire with a different waveguide width, $W3$, is added to the central waveguide, another cross-coupling effect will occur at a wavelength satisfying the Bragg matching condition. In this case, $W3 = 400$ nm.

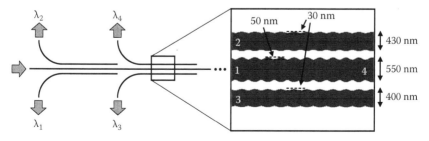

FIGURE 4.38 Schematic diagram of multiport add–drop device by adding another waveguide and cascading the WSC.

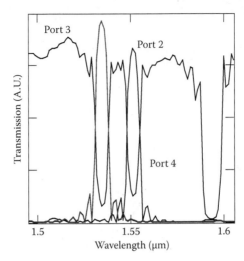

FIGURE 4.39   2D FDTD simulation results for the transmission spectrum to the output ports.

The 2D FDTD simulation results for the device shown in Figure 4.38 are summarized in Figure 4.39. We observe an additional cross-coupling at ~1.53 µm. The WSC structure may also be used in a ring resonator to select a single resonant mode from multiple resonant modes.

The WSC implemented using coupled vertical gratings on an SOI platform offers a good platform for applications in WDM. The operation and analytic design procedure presented allows multiple channels to be designed for. The transmission spectra expected by the procedure shows good agreement with the FDTD simulation results. Measurements of the fabricated WSC demonstrated the expected operation. The available WDM bandwidth of this device is not limiting within the C band, unlike the high-order ring-resonator implementation for WDM applications.

4.4 Discussions and Future Perspectives

This chapter has given an introduction to the emerging field of dielectric metamaterials. The main points include our ability to use standard and emerging nanofabrication tools to create complex geometries of composite materials, varying the spatial distributions periodically, quasi-periodically, or even at random. The feature sizes of these nanocomposites can be deeply subwavelength, which enables creation of materials and devices with unique functionalities that are not possible to achieve with existing technologies. These structures can also be created on subwavelength scale, where the individual features are coupled and can be operated in spatial as well as spectral resonance. We have described the localized effects and alterations in optical behavior of such dielectric metamaterials with space variant polarizability (form birefringence, snake-like propagation, and metamaterials lens), and various functionalities achieved by single and spatially coupled resonant quantum wires.

The current trend of using an SOI material platform will continue because it can utilize well-established CMOS-compatible fabrication processes in the future as microelectronics is projected to reach resolutions on the order of 10–16 nm. In addition, future computing systems will increasingly rely on optical interconnects. These optical interconnects will be used on shorter and shorter length scales, with silicon photonics playing a central role by enabling efficient manipulation, transformation, filtering and detection of light for the purpose of meeting the needs of future information systems. It should be noted that efficient generation of light on a silicon chip is still in its infancy and may not be able to overcome the fundamental issues prohibiting efficient generation of light in indirect bandgap semiconductors; however, alternative solutions similar to delivery of electrical power from off-the-chip sources will bring optical fields into Si chips. Advancement of heterogeneous metamaterials is certainly another valid alternative, but it may require major technological and manufacturing advances. It is evident that the fundamental limits of scale and composition in present-day technology will continue to improve, reaching out to smaller and smaller features in multitudes of materials compositions. It is possible that current trends in engineering electronic bandgaps are not merely a pipe dream, and may one day merge with engineering nanophotonic metamaterials, leading to new device concepts that will serve as the backbone of future information systems technologies.

Engineering the polarizability of dielectric nanostructures is still in its nascent stages, and numerous novel functionalities are still to come.

Appendix A: Dielectric Material Characterization Using Near-Field Scanning Optical Microscopy

The conventional microscopy investigation of nanophotonic phenomena is very limited due to the diffraction limit. It is very hard to resolve features separated by less than half a wavelength in classical microscopy. The idea of improving the resolution of optical measurements by bringing a subwavelength aperture close to the object of interest was first introduced by Synge in 1928 and experimentally realized only in 1983 by two independently working research groups: Dieter W. Pohl and his colleagues working in IBM laboratories [55] and Aaron Lewis with colleagues in Cornell University [56]. Since then, Near-field Scanning Optical Microscopy (NSOM)—the technique realizing this idea— became a very popular characterization and investigation tool. It is used in a number of applications where measurements of local properties with nanoscale resolution are crucial. Several review articles [57–60] and books [61,62] were published on this topic describing the concept of near-field microscopy in detail, as well as improvements introduced to this technique over the years.

Aperture NSOM probes typically have a conical or pyramidal shape with an opening of 50–200 nm at the apex. The resolution of the image is approximately equal to the diameter of the aperture, so we are no longer limited to the diffraction limit and can resolve features as small as $\lambda/30$. There is also a possibility of working with apertureless probes. In this case, a sharp edge of a tip or a nanoparticle works as a local scatterer or field concentrator depending on the configuration, and both the illumination and the collection of light are performed in the far-field, but the interaction of the probe and sample occurs in the near-field region. The resolution of such apertureless probes is determined by the sharpness of the tip, and it is usually better than that of aperture probes, but the recovery of signal from the background is very complicated, and the signal-to-noise ratio is quite poor.

The efficiency of light transmission through a small aperture ($d \ll \lambda$) rapidly drops as the aperture size decreases: $T \sim (d/\lambda)^4$ [63]; thus, for a realistic aperture of 100 nm and visible wavelengths, the transmission efficiency only reaches 10^{-6}–10^{-8} [64]. Such small transmission coefficients demand using powerful optical sources (often lasers), and efficient detection schemes and detectors. Heterodyne detection is an example of an interferometric technique that not only improves the detection efficiency but also allows measurement of the optical phase.

The concept of heterodyne detection is to mix the signal of interest with a coherent reference beam possessing a slightly shifted optical frequency. This can be done by implementing the Mach–Zehnder interferometer scheme with one arm (signal arm) including NSOM and the other arm (reference arm) providing a frequency-shifted reference. The signal collected by the near-field probe depends on its position and can be described as:

$$E_S(x,y) = A_S(x,y)e^{j[\omega_0 t + \phi_S(x,y) + \beta_S]}. \tag{4.13}$$

A pair of acousto-optic modulators (AOMs) spectrally shifts the field in the reference arm by the heterodyne frequency $\Delta\omega$. The modulators are driven by phase-locked radio frequency (RF) oscillators and arranged to operate in the positive and negative diffraction orders such as $\Delta\omega = 40.07\ \text{MHz} - 40\ \text{MHz} = 70\ \text{kHz}$, yielding the optical field in the reference arm:

$$E_R = A_R e^{j[\omega_0 t + \Delta\omega t + \beta_R]}. \tag{4.14}$$

A frequency difference of 70 kHz is sufficient to avoid the significant $1/f$ noise, but simultaneously lies within the bandwidth of the photodetector. The two fields given by Equation 4.13 and Equation 4.14 are added coherently, yielding intensity:

$$I(x,y) \sim A_R^2 + 2A_S(x,y)A_R \sin[\phi_S(x,y) + \Delta\omega t + \beta_0] + A_S^2, \tag{4.15}$$

where the first term represents a constant bias; the second term contains the desired interference signal of interest oscillating at the heterodyne frequency, $\Delta\omega$; and the term proportional to $A_S^2(x,y)$ can be neglected since $A_R \gg A_S$. The coherent gain of the heterodyne detection can be explained comparing the averaged interference cross-term to the intensity term characteristic of direct detection, $A_S A_R \gg A_S^2$. A lock-in amplifier referenced from the RF mixer combining the two AOM-driving signals places a narrow bandpass filter around the heterodyne frequency of 70 kHz—this rejects a large portion of noise and a constant term of A_R^2 that carries no information about the light collected by the near-field probe. The electronic signal proportional to the cross-term $I_E(x,y) \sim 2A_S(x,y)A_R \sin[\phi_S(x,y) + \Delta\omega t + \beta_0]$ is then processed in the lock-in to extract the optical amplitude $A_S(x,y)$ and phase $\phi_S(x,y)$ contributions.

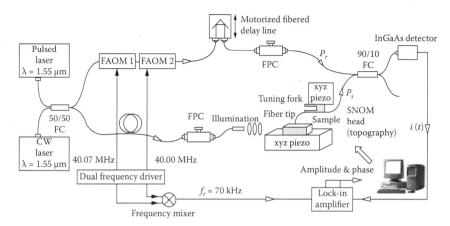

FIGURE 4.40 Scheme of the H-NSOM setup with the collection system for effect of tip analysis.

An example of the heterodyne NSOM (HNSOM) [38] arrangement is shown in Figure 4.40. The Mach–Zehnder interferometer here is composed of the readily available telecom components: 10/90 directional couplers, fiber-pigtailed acousto-optic modulators (AOMs), and polarization controllers. The couplers are arranged in such way that most of the optical power is coupled to and from the signal arm.

This in-fiber realization provides better interferometric stability; polarization-maintaining fibers can also be used to maximize the interference term. To reduce the environmental effects, the microscope head and the coupling unit are covered with a custom-made plastic box (outlined with a dashed line).

The scanning process provides simultaneously three images: sample topography, deduced from the AFM feedback system which keeps the probe at a constant height above the sample surface; and amplitude and phase distributions of the evanescent and propagating optical fields. An example of HNSOM characterization of a microring resonator with a 10 μm diameter is given in Figure 4.41.

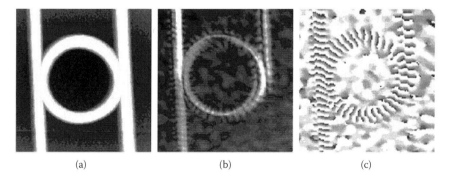

(a)	(b)	(c)

FIGURE 4.41 (a) Microring resonator topography obtained in NSOM measurements; (b) and (c) near-field amplitude and phase distributions (respectively), obtained for resonance at a free-space wavelength of $\lambda = 1535$ nm.

References

1. J. D. Joannopoulos, S. G. Johnson, J. N. Winn, and R. D. Meade, *Photonic Crystals: Molding the Flow of Light*, 2nd ed., Princeton University Press, Princeton, NJ, 2008.
2. M. Abashin, P. Tortora, I. Marki, U. Levy, W. Nakagawa, L. Vaccaro, H. Herzig, and Y. Fainman, Near-field characterization of propagating optical modes in photonic crystal waveguides, *Opt. Express* 14, 1643–1657, 2006.
3. S. John, Strong localization of photons in certain disordered dielectric superlattices, *Phys. Rev. Lett.* 58, 2486–2489, 1987.
4. R. D. Meade, A. Devenyi, J. D. Joannopoulos, O. L. Alerhand, D. A. Smith, and K. Kash, Novel applications of photonic band gap materials: Low-loss bends and high Q cavities, *J. Appl. Phys.* 75, 4753, 1994.
5. S. Lin, E. Chow, V. Hietala, P. R. Villeneuve, and J. D. Joannopoulos, Experimental demonstration of guiding and bending of electromagnetic waves in a photonic crystal, *Science* 282, 274–276, 1998.
6. H. Kosaka, T. Kawashima, A. Tomita, M. Notomi, T. Tamamura, T. Sato, and S. Kawakami, self-collimatting phenomena in *Appl. Phys. Lett.* 74, 1212, 1999.
7. E. Schonbrun, Q. Wu, W. Park, T. Yamashita, C. J. Summers, M. Abashin, and Y. Fainman, Wave front evolution of negatively refracted waves in a photonic crystal, *Appl. Phys. Lett.* 90, 041113, 2007.
8. H. Kosaka, T. Kawashima, A. Tomita, M. Notomi, T. Tamamura, T. Sato, and S. Kawakami, Superprism phenomena in photonic crystals, *Phys. Rev. B* 58, R10096, 1998.
9. C. Luo, S. G. Johnson, and J. D. Joannapoulos, Subwavelength imaging in photonic crystals, *Phys, Rev. B* 68, 045115, 2003.
10. Y. Fink, J. N. Winn, S. Fan, C. Chen, J. Michel, J. D. Joannopoulos, and E. L. Thomas, A dielectric omnidirectional reflector, *Science* 282, 1679–1682, 1998.
11. O. Painter, R. K. Lee, A. Scherer, A. Yariv, J. D. O'Brien, P. D. Dapkus, and I. Kim, Two-dimensional photonic band-gap defect mode laser, *Science* 284, 1819–1821, 1999.
12. U. Levy, M. Abashin, K. Ikeda, A. Krishnamoorthy, J. Cunningham, and Y. Fainman, Inhomogenous dielectric metamaterials with space-variant polarizability, *Phys. Rev. Lett.* 98, 243901, 2007.
13. J. S. Foresi, P. R. Villeneuve, J. Ferrera, E. R. Thoen, G. Steinmeyer, S. Fan, J. D. Joannopoulos, L. C. Kimerling, H. I. Smith, and E. P. Ippen, Photonic-bandgap microcavities in optical waveguides, *Nature* 390, 143–145, 1997.
14. J. T. Hastings, M. H. Lim, J. G. Goodberlet, and H. I. Smith, Optical waveguides with apodized sidewall gratings via spatial-phase-locked electron-beam lithography, *J. Vac. Sci. Technol. B* 20, 2753–2757, 2002.
15. T. E. Murphy, J. T. Hastings, and H. I. Smith, Fabrication and characterization of narrow-band bragg-reflection filters in silicon-on-insulator ridge waveguides, *J. Lightwave Technol.* 19, 1938–1942, 2001.
16. H. C. Kim, K. Ikeda, and Y. Fainman, Resonant waveguide device with vertical gratings, *Opt. Lett.* 32, 539–541, 2007.
17. H. C. Kim, K. Ikeda, and Y. Fainman, Tunable transmission resonant filter and modulator with vertical gratings, *J. Lightwave Technol.* 25, 1147–1151, 2007

18. Miller, D. A. B., Optical interconnects to silicon, *IEEE J. Sel. Top. Quantum Electron.* 6, 1312–1317, 2000.

19. V. R. Almeida, C. A. Barrios, R. R. Panepucci, and M. Lipson, All-optical control of light on a silicon chip, *Nature* 431, 1081–1084, 2004.

20. A. Liu, R. Jones, L. Liao, D. Samara-Rubio, D. Rubin, O. Cohen, R. Nicolaescu, and M. Paniccia, A high-speed silicon optical modulator based on a metal-oxide-semiconductor capacitor, *Nature* 427, 615–618, 2003.

21. B. E. Little, S. T. Chu, H. A. Haus, J. Foresi, and J.-P. Laine, Microring resonator channel dropping filters, *J. Lightwave. Technol.* 15, 998–1005, 1997.

22. O. Boyraz and B. Jalali, Demonstration of a silicon Raman laser, *Opt. Express* 12, 5269–5273, 2004.

23. A. W. Fang, H. Park, O. Cohen, R. Jones, M. J. Paniccia, and J. E. Bowers, Electrically pumped hybrid AlGaInAs-silicon evanescent laser, *Opt. Express*, 14, 9203–9210, 2006.

24. F. Xu, R. Tyan, P. C. Sun, C. Cheng, A. Scherer, and Y. Fainman, Fabrication, modeling, and characterization of form-birefringent nanostructures, *Opt. Lett.* 20, 2457–2459, 1995.

25. R. Tyan, A. Salvekar, Cheng, A. Scherer, F. Xu, P. C. Sun, and Y. Fainman, Design, fabrication, and characterization of form-birefringent multilayer polarizing beam splitter, *J. Opt. Soc. Am. A* 14, 1627–1636, 1997.

26. F. Xu, R.-C. Tyan, P.-C. Sun, Y. Fainman, C.-C. Cheng, and A. Scherer, Form-birefringent computer-generated holograms, *Opt. Lett.* 21(18):1513, 1996.

27. W. Nakagawa, R. Tyan, and Y. Fainman, Analysis of enhanced second-harmonic generation in periodic nanostructures using modified rigorous coupled-wave analysis in the undepleted-pump approximation, *J. Opt. Soc. Am. A* 19, 1919–1928, 2002.

28. U. Levy, C. H. Tsai, L. Pang, and Y. Fainman, Engineering space-variant inhomogeneous media for polarization control, *Opt. Lett.* 29, 1718–1720, 2004.

29. U. Levy, M. Nezhad, H.-C. Kim, C.-H. Tsai, L. Pang, and Y. Fainman, Implementation of a graded-index medium by use of subwavelength structures with graded fill factor, *J. Opt. Soc. Am. A* 22, 724–733, 2005.

30. J. N. Mait, A. Scherer, O. Dial, D. W. Prather, and X. Gao, Diffractive lens fabricated with binary features less than 60 nm, *Opt. Lett.* 25, 381–383, 2000.

31. S. M. Rytov, Electromagnetic properties of a finely stratified medium, *Sov. Phys. JETP 2* 466–475, 1956.

32. I. Richter, P-C. Sun, F. Xu, and Y. Fainman, Design considerations of form birefringent microstructures, *Appl. Opt.* 34, 2421–2429, 1995.

33. E. N. Glytsis and T. K. Gaylord, High-spatial-frequency binary and multilevel stairstep gratings: Polarization-selective mirrors and broadband antireflection surfaces, *Appl. Opt.* 31, 4459–4470, 1992.

34. S. M. Rytov, Electromagnetic properties of a finely stratified medium, *Sov. Phys. JETP 2* 466–475, 1956.

35. I. Richter, P.-C. Sun, F. Xu, and Y. Fainman, Design considerations of form birefringent microstructures, *Appl. Opt.* 34, 2421–2429, 1995.

36. P. Lalanne and D. L. Lalanne, Depth dependence of the effective properties of subwavelength gratings, *J. Opt. Soc. Am. A* 14, 450–458, 1997.

37. M. Parikh, Corrections to proximity effects in electron-beam lithography.1. Theory, *J. Appl. Phys.* 50, 4371–4377, 1979.
38. A. Nesci and Y. Fainman, Complex amplitude of an ultrashort pulse with femtosecond resolution in a waveguide using a coherent NSOM at 1550 nm, in *Wave Optics and Photonic Devices for Optical Information Processing II*, P. Ambs and F. R. Beyette, Jr., eds., *Proc. SPIE* 5181, 62–69, 2003.
39. A.Yariv and P. Yeh, *Optical Waves in Crystals*, Wiley, Hoboken, NJ, 2003.
40. G. P. Agrawal and N. K. Dutta, Analysis of ridge-waveguide distributed feedback lasers, *IEEE J. Quantum Electron.* QE-21, 534–538, 1985.
41. H.-C. Kim, K. Ikeda, and Y. Fainman. Tunable transmission resonant filter and modulator with vertical gratings. *J. Lightwave Technol.* 25 1147–51, 2007.
42. D. T. H. Tan, K. Ikeda, R. E. Saperstein, B. Slutsky, and Y. Fainman, Chip-scale dispersion engineering using chirped vertical gratings, *Opt. Lett.* 33, 3013–3015, 2008.
43. J. Miyazu, T. Segawa, S. Matsuo, T. Ishii, H. Okamoto, Y. Kondo, H. Suzuki, and Y. Yoshikuni, Wavelength tunable mode-locked semiconductor lasers, *Lasers and Electro-Optics, 2005.* CLEO/Pacific Rim 2005. Pacific Rim Conference on , 624–625, 30-02 Aug. 2005
44. K. Sato, A. Hirano, and H. Ishii, Chirp-compensated 40-GHz mode-locked lasers integrated with electroabsorption modulators and chirped gratings, *IEEE J. Quantum Electron.* 5, 590–595, 1999.
45. H. Kogelnik, Filter response of nonuniform almost-periodic structures, *Bell Sys. Tech. J.* 55, 109–126, 1975.
46. M. Gnan, G. Bellanca, H. M. H Chong, P. Bassi, and R. M. De la Rue, Modeling of photonic wire Bragg gratings, *Opt. Quantum Electron.* 38, 133, 2006.
47. V. R. Almeida, R. R. Panepucci, and M. Lipson, Nanotaper for compact mode conversion, *Opt. Lett.* 28, 1302–1304, 2003.
48. R. Kashyap, *Fiber Bragg Gratings*, Academic Press, 1999.
49. K. O. Hill, F. Bilodeau, B. Malo, T. Kitagawa, S. Thériault, D. C. Johnson, J. Albert, and K. Takiguchi, Chirped in-fiber Bragg gratings for compensation of optical-fiber dispersion, *Opt. Lett.* 19, 1314–1316, 1994.
50. M. Notomi, K. Yamada, A. Shinya, J. Takahashi, C. Takahashi, and I. Yokohama, Extremely large group velocity dispersion of line-defect waveguides in photonic crystal slabs, *Phys. Rev. Lett.* 87, 253902, 2001.
51. A. C. Turner, C. Manolatou, B. S. Schmidt and M. Lipson, Tailored anomalous group-velocity dispersion in silicon channel waveguides, *Opt. Express* 14, 4357–4362, 2006.
52. G. P. Agrawal, *Nonlinear Fiber Optics*, 2nd ed., Academic, New York, 1995.
53. K. Ikeda, M. Nezhad, and Y. Fainman, Wavelength selective coupler with vertical gratings on silicon chip, *Appl. Phys. Lett.* 92, 201111, 2008.
54. F. Xia, M. Rooks, L. Sekaric, and Y. Vlasov, Ultra-compact high order ring resonator filters using submicron silicon photonic wires for on-chip optical interconnects, *Opt. Express* 15, 11934–11941, 2007.
55. D. W. Pohl, W. Denk, and M. Lanz, Optical stethoscopy: image recording with resolution $\lambda/20$, *Appl. Phys. Lett.* 44, 651, 1984.

56. A. Lewis, M. Isaacson, A. Harootunian et al., Development of a 500Å spatial resolution light microscope I. Light is efficiently transmitted through $\lambda/16$ diameter apertures, *Ultramicroscopy* 13, 227, 1984.

57. H. Heinzelmann and D. W. Pohl, Scanning near-field optical microscopy, *Appl. Phys. A* 59, 89-101, 1994.

58. Bert Hecht et al., Scanning near-field optical microscopy with aperture probes: Fundamentals and applications, *J. Chem. Phys.* 112, 2000.

59. Julia W. P. Hsu, Near-field scanning optical microscopy studies of electronic and photonic materials and devices, *Mater. Sci. Eng.* 33, 1–50, 2001.

60. L. Novotny and S. J. Stranick, Near-field optical microscopy and spectroscopy with pointed probes, *Annu. Rev. Phys. Chem.* 57, 303–331, 2006.

61. P. N. Prasad, *Nanophotonics*, Wiley Interscience, Hobeken, N. J., 2004.

62. L. Novotny and B. Hecht, *Principles of Nano-Optics*, Cambridge University Press, Cambridge, 2006.

63. H. A. Bethe, Theory of diffraction by small holes, *Phys. Rev.* 66, 163–182, 1944.

64. G. A. Valaskovic, M. Holton, and G. H. Morrison, Parameter control, characterization, and optimization in the fabrication of optical fiber near-field probes, *Appl. Opt.* 34, 1215–1228, 1995.

5

Metamaterials with Optical Gain

M. A. Noginov
*Center for Materials
Research,
Norfolk State University,
Norfolk, Virginia*

5.1 Introduction

Metamaterials—engineered materials with rationally designed geometry, composition, and arrangement of nanostructured building blocks—are predicted to play an increasingly important role in high technologies of the twenty-first century. Unfortunately, most metamaterials, their applications, and metamaterials-based devices suffer from a common problem: absorption loss in metal. Examples include but are not limited to a superlens, which resolution is highly sensitive to optical loss; negative index materials, in which light waves decay at the distances comparable to one wavelength; and optical cloak, whose performance can fail due to losses. All these materials and devices are discussed in different chapters of this book.

On the other hand, numerous applications require *active* metamaterials capable of amplifying light and plasmonic waves, generating stimulated emission of photons or surface plasmons, logic switching (I/0) between two stable states, and many more.

Both problems have a common solution: introduction to a metamaterial of an optical gain or nonlinearity. To date, most of the progress in this area has been made in metamaterials with gain, which is the focus of this chapter.

It is not the purpose of this chapter to provide a comprehensive review of the state of the art in the field. Instead, our goal is to discuss major relevant concepts and experimental results in the form of a tutorial, giving preference to simple hand-waving arguments over mathematical derivations.

The chapter is organized as follows. The first section is Introduction. The concepts of surface plasmons, metamaterials, and optical gain are discussed in Section 5.2. Theoretical proposals of using gain in metamaterials and plasmonic systems are reviewed in Section 5.3. The corresponding experimental realizations are discussed in Section 5.4. Systems and phenomena presented in Sections 5.3 and 5.4 include but are not limited to localized surface plasmons in metallic nanoparticles, surface plasmon polaritons, a spaser, and a negative-index material with optical gain. Finally, the major concepts and results discussed in the chapter are summarized in Section 5.5.

5.2 Surface Plasmons, Metamaterials, and Optical Loss

The unique properties of many metamaterials and their unparalleled responses to electromagnetic waves critically depend on *surface plasmons*.

Localized surface plasmon (SP) is an oscillation of free electrons in a metallic particle, whose resonance frequency is the plasma frequency adjusted by the size and, mainly, the shape of the particle [1–3]. Most commonly, localized SPs are manifested by a band in the extinction spectrum, which is centered at the resonant frequency (Figure 5.1).

Localized surface plasmons have been found on rough surfaces [1–3], in clusters and aggregates of nanoparticles [4–6], as well as in engineered nanostructures [7–10]. SPs

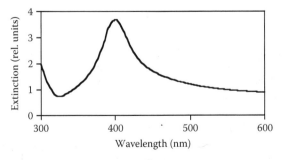

FIGURE 5.1 Extinction spectrum of the Tetrahydrofuran (THF) suspension of Ag nanoparticles featuring the surface plasmon band at 394 nm.

can enhance electric fields in nanoscopic volumes—"hot spots," in which both linear and nonlinear optical responses of molecules and atoms are tremendously enhanced. This leads to a number of important applications. The most matured of them is Surface Enhanced Raman Scattering (SERS) [3], which enables rapid molecular assays for detection of biological and chemical substances [11] and observation of Raman scattering from a single molecule [12,13].

A relevant phenomenon—surface plasmon polariton (SPP, also known as propagating SP)—is a surface electromagnetic wave accompanied by an oscillation of free electrons traveling along the interface between two media, which possess electric permittivities with opposite signs, such as metal and dielectric [14]. The wavevector of SPP propagating at the interface between metal with permittivity ε_1 and dielectric with permittivity ε_2

$$k_x^0 = \frac{\omega}{c} \sqrt{\frac{\varepsilon_1 \varepsilon_2}{\varepsilon_1 + \varepsilon_2}}$$

is larger than the wavevector of a photon propagating in the same dielectric, $\frac{\omega}{c}\sqrt{\varepsilon_2}$ (where ω is the light frequency and c is the speed of light). Because of this mismatch, SPP cannot be excited by a light wave incident from the dielectric with permittivity ε_2. However, SPP propagating in a thin, metallic film at the interface between media ε_1 and ε_2 can be excited from the side of a high-index dielectric prism characterized by $\varepsilon_0 > \varepsilon_2$ (Figure 5.2a), when the projection of the photon wavevector onto the metal–dielectric interface, $k_x(\theta) = \frac{\omega}{c}\sqrt{\varepsilon_0} \sin\theta_0$, matches k_x^0. (Above, all electric permittivities are treated as real.) At the corresponding critical angle, the energy of a light wave is transferred to the SPP, causing a dip in the angular reflectance profile (Figure 5.2b). The distribution of the z component of electric field in SPP is characterized by exponential decay to both metal and dielectric, as shown in Figure 5.2a.

An account of other types of propagating surface plasmons as well as ways to excite them can be found in References [14–16].

Propagating surface plasmons are broadly used in photonic and optoelectronic devices, including waveguides, couplers, splitters, add/drop filters [17–21], quantum cascade lasers [22–24], and sensing [14–16].

Other applications of surface plasmons and related phenomena, dubbed by the common name *nanoplasmonics*, include near-field scanning microscopy, spectroscopy and photomodification with nanoscale resolution [25–27], extraordinary transmission of light through subwavelength holes [28], enhanced response of *p-n* junctions [29], and negative index metamaterials [30,31].

Electric fields of localized and propagating SPs penetrate both metal and dielectric and have their maxima in the vicinity of a metal–dielectric interface (Figure 5.2a). Metals have strong absorption loss (also known as Joule loss or internal loss), which reduces strengths of local electric fields, causes damping of SP oscillations, and shortens the propagation length of SPPs. The absorption loss hinders most existing and potential future applications of nanoplasmonics and metamaterials based on metallic inclusions. Applications relying on resonant enhancement of electric fields, for example,

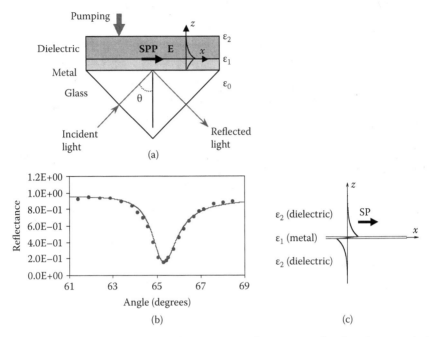

(a)

(b)   (c)

FIGURE 5.2 *See color insert.* (a) Kretchman geometry for excitation of surface plasmon polaritons and distribution of electric field; θ is the incidence angle; pumping was applied to the dye-doped polymeric film in the experiments described in Sections 5.3.2.1 and 5.3.2.2. (b) Angular dependence of the reflectance detected in silver film in the setup of (a). Dots—experiment, solid line—calculations according to Equation 5.1. (c) Schematic of the field distribution in long-range surface plasmon propagating in a thin metallic waveguide.

superlens [30,32], suffer from losses more than nonresonant metamaterials and devices, for example, hyperlens [33–36] or optical cloak [37–39].

5.3 Gain—One Solution to the Loss Problem and Active Metamaterials: Theoretical Proposals and Concepts

Over years, several proposals have been made on how to conquer the plasmon loss. The surface plasmon electric field resides partly in metal and partly in dielectric (Figure 5.2a). As the field in metal experiences strong absorption loss, the field in an adjacent dielectric can have amplification if the dielectric is embedded with atoms, molecules, or quantum dots providing for population inversion and optical gain. The implementation of this idea in systems featuring propagating surface plasmon polaritons (SPPs) and localized surface plasmons (SPs) is discussed in Sections 5.3.1 and 5.3.2.

Stimulated emission of localized and propagating surface plasmons is predicted when SP gain exceeds overall SP loss. This opens a way to active nanoplasmonics and metamaterials as described in Section 5.3.3.

5.3.1 Surface Plasmon Polaritons with Gain

In 1989, Sudarkin and Demkovich proposed to increase the propagation length of SPP by creating population inversion and gain in a dielectric medium adjacent to a metallic film [40]. The proposed experimental test was based on the observation of increased reflectivity of a metallic film in the frustrated (attenuated) total internal reflection setup of Figure 5.2a. The authors of Reference [40] have also briefly discussed the possibility of creating an SP-based laser.

In more recent years, gain-assisted propagation of SPPs in planar metal waveguides of different geometries has been studied by Nezhad et al. in Ref. [41]. SPPs at the interface between metal and dielectric with strong optical gain have been analyzed theoretically by Avrutsky [42]. In particular, it has been shown [42] that the proper choice of optical indices of metal and dielectric can result in an infinitely large effective refractive index of surface waves. Such resonant plasmons have extremely low group velocity and are localized in very close vicinity to the interface. Gain-assisted propagation of a long-range surface plasmon polariton—an SP mode propagating in a thin-film metallic waveguide surrounded by dielectric layers with similar electric permittivities (Figure 5.2c)—has been studied theoritically in Reference [43]. It has been shown that net amplification is possible at visible wavelengths using reasonable pumping power and concentration of laser dye molecules in a gain medium.

In order to better understand how the compensation of SPP loss by gain can be observed experimentally, let us analyze the angular profile of reflectance $R(\theta)$ measured in the setup of Figure 5.2a. According to Reference [4], it is described by the formula

$$R(\theta) = \left| \frac{r_{01} + r_{12} \exp(2ik_{z1}d_1)}{1 + r_{01}r_{12} \exp(2ik_{z1}d_1)} \right|^2, \tag{5.1}$$

where $r_{ik} = (k_{zi}\varepsilon_k - k_{zk}\varepsilon_i)/(k_{zi}\varepsilon_k + k_{zk}\varepsilon_i)$ and

$$k_{zi} = \pm\sqrt{\varepsilon_i\left(\frac{\omega}{c}\right)^2 - k_x(\theta)^2}, \quad i = 0,1,2. \tag{5.2}$$

The parameter $k_z c/\omega$ defines the electric field distribution along the z direction. Its real part can be associated with a tilt of phase-fronts of the waves propagating along the interface [41], while its imaginary part defines the wave attenuation or growth. It has been argued [44] that the correct selection of the sign in Equation 5.2 can be achieved by the cut of the k_z^2 complex plane along the negative imaginary axis (meaning that the phase of complex numbers is defined between $-1/2\pi$ and $3/2\pi$). In particular, in the case of total internal reflection and optical gain in the dielectric medium above the prism, k_z^2 belongs to the third quadrant and k_z belongs to the second quadrant of the corresponding complex planes.

In the limit of small plasmonic loss or gain (when the decay length of SPP, L, is much greater than $2\pi/k_{SPP}$), and in the vicinity of θ_0, Equation 5.1 can be simplified, revealing the physics behind the gain-assisted plasmonic loss compensation [44]:

$$R(\theta) \approx \left|r_{01}^0\right|^2 \left[1 - \frac{4\gamma_i\gamma_r + \delta(\theta)}{(k_x - k_x^0 - \Delta k_x^0)^2 + (\gamma_i + \gamma_r)^2}\right],\tag{5.3}$$

where

$$r_{01}^0 = r_{01}(\theta_0), \delta(\theta) = 4\left(k_x - k_x^0 - \Delta k_x^0\right)\mathrm{Im}(r_0)\mathrm{Im}(e^{i2k_z^0 d_1})/\xi,$$

and

$$\xi = \frac{c(\varepsilon_2' - \varepsilon_1')}{2\omega}\left(\frac{\varepsilon_2' + \varepsilon_1'}{\varepsilon_2' \varepsilon_1'}\right)^{3/2}.$$

The shape of the angular reflectance profile $R(\theta)$ is dominated by the Lorentzian term in Equation 5.3, whose width is determined by the propagation length of SPP:

$$L = [2(\gamma_i + \gamma_r)]^{-1},\tag{5.4}$$

which, in turn, is defined by the sum of the *internal* (absorption) loss,

$$\gamma_i = k_x^{0''} = \frac{\omega}{2c}\left(\frac{\varepsilon_1' \varepsilon_2'}{\varepsilon_1' + \varepsilon_2'}\right)^{3/2}\left(\frac{\varepsilon_1''}{\varepsilon_1'^2} + \frac{\varepsilon_2''}{\varepsilon_2'^2}\right).\tag{5.5}$$

and the *radiation* loss caused by SPP leakage into the prism,

$$\gamma_r = \mathrm{Im}(r_{01}e^{i2k_z^0 d_1})/\xi.\tag{5.6}$$

The relatively small term δ in Equation 5.3 results in a slight asymmetry of $R(\theta)$. The exact equation (Equation 5.1) and approximate equation (Equation 5.3) are in excellent agreement with each other for silver films, whose thickness exceeds ≈ 50 nm [44].

The gain in the dielectric medium reduces the internal loss γ_i of SPP (Equation 5.5). In *thick* metallic films (where $\gamma_i > \gamma_r$ in the absence of gain), the "dip" in the reflectance profile becomes deeper when gain is first added to the system, reaching its minimum value

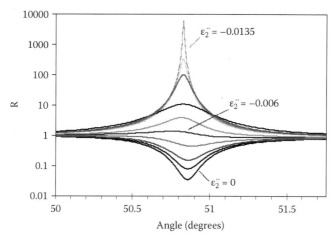

FIGURE 5.3 Reflectance profiles $R(\theta)$ calculated for 55 nm silver film at increasing values of optical gain (negative values ε_2''). Calculations were done for $\varepsilon_0' = 3.18$ $\varepsilon_1' = -16.06$, $\varepsilon_1'' = 0.44$, and $\varepsilon_2' = 1.77$. (Adapted from Noginov, M. A. 2008. *J. Nanophotonics* 2: DOI:10.1117/1.3073670.)

$R_{\min} = 0$ at $\gamma_i \neq \gamma_r$ (Equation 5.3; Figure 5.3). With further increase of the gain, γ_i becomes smaller than γ_r, leading to an increase of $R_{\min}$. This is the starting point for relatively *thin* metallic films such as the 55 nm silver film in Figure 5.3, in which $\gamma_i < \gamma_r$ at $\varepsilon_2'' = 0$ and the resonant value of R monotonically grows with increase of gain.

The resonant value of R is equal to unity when the *internal* loss is completely compensated by gain ($\gamma_i = 0$) at

$$\varepsilon_2'' = -\frac{\varepsilon_1'' \varepsilon_2'^2}{\varepsilon_1'^2}. \qquad (5.7)$$

When the gain is increased to even higher values, γ_i becomes negative, and the dip in the reflectance profile changes to a peak, consistent with the predictions of References [40,46]. The peak has a singularity when the gain compensates *total* SPP loss ($\gamma_i + \gamma_r = 0$) (Figure 5.3). Past the singularity point, the system becomes unstable and cannot be described by stationary equations (Equations 5.1 and 5.3) [47].

The minimum or maximum values of $R(\theta)$ are plotted versus ε_2'' for different thicknesses of the metallic film in Figure 5.4a, and the corresponding values of the full width at half maximum (FWHM) of a dip or a peak are depicted in Figure 5.4b. One can see that in accordance with Equation 5.3, the minimum and the maximum values of R are equidistant from the point corresponding to $\gamma_i = 0$ (Figure 5.4a). At the same time, the width of a dip or a peak W (which according to Equations 5.3 and 5.4 is inversely proportional to the SPP propagation length ($W \propto L^{-1}$)) decreases linearly with an increase of gain (negative ε_2'') from its starting value at $\varepsilon_2'' = 0$ to $W = 0$ in the singularity point, when the total gain compensates the total loss, $\gamma_i + \gamma_r = 0$. The latter condition corresponds to the infinite propagation length of SPP.

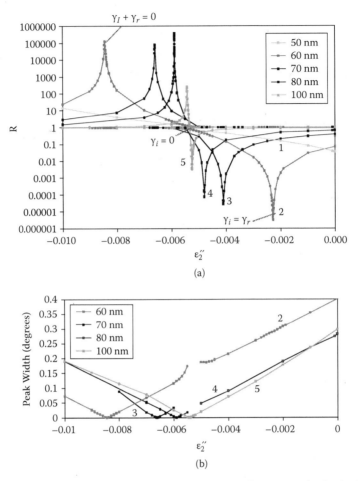

FIGURE 5.4 (a) Dependence of the minimum (maximum) reflectance in the dip (peak) of the reflectance profile $R(\theta)$ versus ε_2'' calculated for the following thicknesses of the silver film: 1–50 nm, 2–60 nm, 3–70 nm, 4–80 nm, 5–100 nm. All other system parameters are as in Figure 5.3. (b) Corresponding FWHM values. (Adapted from Noginov, M. A. 2008. *J. Nanophotonics* 2: DOI:10.1117/1.3073670.)

The gain required to overcome an *internal* loss, independent of the thickness of the metallic film, is given by $\gamma_i = 0$. For the system parameters considered above, this corresponds to the imaginary part of electric permittivity $\varepsilon_2'' = -0.0053$ and gain $g = 421$ cm^{-1}. The critical gain evaluated for actual parameters of the experimental sample studied in Reference [44] is approximately three times larger, $\sim1.3 \times 10^3$ cm^{-1}.

These gain values are approximately equal to those required to compensate for *total* SPP loss in thick metallic films (≥100 nm). In thinner films exhibiting large radiative losses, the critical gain can be several times larger.

5.3.2 Conquering Optical Loss in Localized Surface Plasmons

In 2004, Lawandy has predicted the localized SP resonance in metallic nanospheres to exhibit a singularity when the surrounding dielectric medium has a critical value of optical gain [48]. This singularity, resulting from canceling both real and imaginary terms in the denominator of the polarization of metallic nanospheres, $\propto (\varepsilon_2 - \varepsilon_1)/(2\varepsilon_2 + \varepsilon_2)$, can be evidenced by an increase of the Rayleigh scattering within the plasmon band and lead to a low-threshold random laser action, light localization effects, and enhancement of surface-enhanced Raman scattering [48]. (Here, ε_2 and ε_1 are complex electric permittivities of dielectric and metal, respectively).

This approach has been further developed in Reference [49], in which a three-component system consisting of (1) metallic nanoparticle, (2) shell consisting of molecules with optical gain, and (3) surrounding dielectric (solvent) has been studied. In particular, it has been shown that depending on the thickness of the gain medium, the overall absorption of the complex can increase or decrease with the increase of gain in the dye shell [49].

The critical gain causing a singularity in the nanoparticle's polarization and scattering can be calculated as follows. The polarizability (per unit volume) for isolated spheroidal metallic nanoparticle is given by

$$\beta = (4\pi)^{-1}[\varepsilon_1 - \varepsilon_2]/[\varepsilon_2 + p(\varepsilon_1 - \varepsilon_2)], \qquad (5.8)$$

where p is the depolarization factor [50]. (Note that fractal aggregates of spherical nanoparticles, similar to those used in Reference [50], can be treated in the first approximation as collections of spheroids formed by nanoparticle chains of different lengths.) The resonance wavelength λ_0 is defined by the requirement that the real part of the denominator in Equation 5.8 is equal to zero, $\varepsilon'_2 + p(\varepsilon'_1 - \varepsilon'_2) = 0$. If the dielectric has an optical gain, such that

$$\varepsilon''_2 = -p\varepsilon''_1/(1 - p) \qquad (5.9)$$

at the resonant wavelength $\lambda = \lambda_0$, then the imaginary part of the denominator in Equation 5.8 becomes zero, leading to extremely large local fields limited, in the cw regime, only by saturation effects [48,51].

The gain coefficient $g = (4\pi/\lambda)n''$ can be written as

$$g = (2\pi/\lambda_0)\varepsilon''_2/\sqrt{\varepsilon'_2}, \qquad (5.10)$$

where $n = n' + in'' = \sqrt{\varepsilon' + i\varepsilon''}$ and $n' \approx \sqrt{\varepsilon'_2}$. Thus, the critical gain, which is needed to compensate the loss of *localized* SP, is given by

$$g = -(2\pi/\lambda_0 n)[p/(1 - p)]\varepsilon''_1 . \qquad (5.11)$$

For a fractal aggregate of silver nanoparticles studied in Reference [50], the resonant wavelength $\lambda_0 = 0.56$ μm corresponds to $p \approx 0.114$ (determined by ε_1' and ε_2') and $\varepsilon_1'' \approx 0.405$, so that the required ε_2'' is equal to -0.052 and the gain coefficient g is equal to 4×10^3 cm^{-1} (we used $n = 1.33$ and known optical constants from Reference [52]). As is shown in Section 5.4.2, this value of gain is achievable in the mixture of rhodamine 6G laser dye and aggregated silver nanoparticles.

5.3.3 Spaser

In 2003, Bergman and Stockman have theoretically proposed a novel device capable of generating localized surface plasmons [53]. Its coined name—SPASER—echoes the famous acronym LASER, with the letters SP standing for *surface plasmon* replacing the letter L standing for *light*. Two necessary components of a laser are the gain medium, amplifying oscillating electric field, and the resonator, providing for the stimulated emission feedback. In a spaser, oscillation of light in a photonic cavity (resonator) is replaced by the oscillation of a local surface plasmon field in close proximity to a metallic nanostructure. This local field induces stimulated transitions in a gain medium, which in turn amplify the SP field, thus generating the stimulated emission of surface plasmons. The analogy between the physical properties of spasers and lasers becomes complete because both photons and surface plasmons are bosons and electrically neutral quasi-particles.

The originally proposed spaser consisted of a metallic V-shaped nanostructure surrounded by quantum dots with gain [53]; Figure 5.5a,b. In more recent years, hybrid spherical shell nanoparticle architectures, which can be synthesized using a wet chemistry routine [54,55], have been proposed and studied theoretically [55–58]; Figure 5.5c,d.

A conceptually similar device—the plasmonic nanolaser—theoretically studied in Reference [59]—benefits from the feedback supported by a magnetic plasmonic resonance. The horseshoe geometry of the proposed metallic nanostructures resembled split-ring resonators, core nanostructure used in many negative index metamaterials, optical cloaks etc.

Spasers can be used as generators of coherent SPs (operating in both pulsed and cw regimes) as well as ultrafast amplifiers and bistable logic elements [60]. Originally, spasers were not designed to generate light. However, out-coupling of surface plasmon oscillations to photonic modes (which can be more or less efficient depending on the nature of the SP mode) constitutes a nanolaser [53,55]. In 2008, Zheludev et al. proposed the concept of a lasing spaser: a device consisting of an array of coupled, slightly asymmetrical double-slit split-ring SP resonators providing for stimulated emission of coherent light in a direction perpendicular to the plane of the array [61].

Plasmonic nanostructures can support both luminous and dark modes, which, respectively, can/cannot be excited and observed from the far-field zone [62]. Depending on the shapes and mutual arrangements of nanostructures, the stimulated emission threshold can be first obtained in a luminous mode (as in the V-shaped structure of Reference [53], Figure 5.5a,b) or in a dark mode (as in a lasing spaser of Reference [61]). In spherical nanoparticles consisting of metallic shells deposited onto dielectric cores, Figure 5.5c,d, the two lowest-order modes are dipole and quadrupole, with the former one exhibiting the lowest threshold [56,58,60].

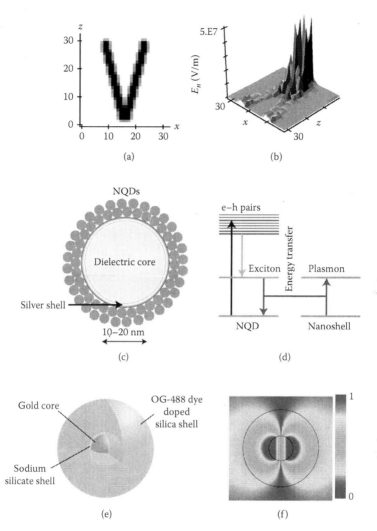

FIGURE 5.5 *See color insert.* Concept of a spaser. (a) V-shaped metallic nanostructure and (b) corresponding distribution of the field amplidude in a SPASER mode. (Adapted from Bergman, D. and M. Stockman. 2003. *Phys. Rev. Lett.* 90:027402.) (c) Hybrid spherical nanoparticle architecture of a spaser, employing quantum dots as a gain medium, and (d) schematic diagram of the energy transfer in a spaser. (Adapted from Stockman, M. I., 2008. *Nat. Photonics* 2:327–29.) (e) Hybrid spaser nanoparticle consisting of a gold core and dye-doped silica shell and (f) the corresponding mode. (Adapted from Noginov, M. A. et al. 2009. *Nature* 460:1110–12.)

To calculate the mode in the first experimentally demonstrated spaser, Figure 5.5e, the three-dimensional wave equation was solved, resulting in a sequence of localized plasmon modes with different values of total angular momenta ℓ and their projections m ($m = -\ell,..., 0, ..., \ell$) [55]. The spectral band of the gain in the experiment [55] corresponded to the lowest-frequency triple-degenerate dipole modes of this sequence ($\ell = 1$, $m = -1$, 0, 1), Figure 5.5f. This degeneracy (similar to that in the p state of a hydrogen atom) can be lifted by small deviations from spherical symmetry in the particle geometry.

5.3.4 Critical Gain: Concepts and Estimates

The critical gain required to compensate optical loss in metal as well as the threshold gain in a spaser are determined by a variety of factors, including shape, sizes, and architectures of a nanostructures; types of metal; and wavelengths. Typical values of the threshold gain range between 10^3 and 10^5 cm^{-1}. Thus, the critical gains estimated in simple plasmonic nanostructures in Sections 5.3.1 and 5.3.2 are on the order of 10^3 cm^{-1}. In the architecture similar to that depicted in Figure 5.5c (with the gain medium *inside* the nanoshell), the threshold gain has been predicted to be equal to 30,800 cm^{-1} for Ag at $\lambda = 510$ nm and 52,800 cm^{-1} for Au at $\lambda = 810$ nm [56]. An even larger value of critical gain, 1.04×10^5 cm^{-1}, has been calculated in a spaser nanoparticle [55]; Figure 5.5e.

The question arises as to whether these extremely high values of gain can be achieved and maintained in real materials and devices. Let us estimate the maximum gain that can be realized in laser dyes. The largest dimension of a rhodamine 6G molecule (an efficient laser dye, which is often used in loss compensation experiments) is ~0.6 nm [63]. Such molecules can be packed with the density exceeding one molecule per cubic nanometer, resulting in a dye concentration $>10^{21}$ cm^{-3}. Multiplying this value by the emission cross-section of the dye, 4×10^{-16} cm^2, one obtains $g > 4 \times 10^5$ cm^{-1}. If molecules are packed with even higher density, then a gain of $\geq 10^6$ cm^{-1} does not seem to be impossible.

Dimerization of dye molecules, causing broadening of the emission band (threefold to fivefold) and nonradiative quenching of emission, present a serious problem for the high gain. Nevertheless, we infer that the gain of ~10^5 cm^{-1} can be achieved in short-pulse experiments (e.g., when gain is excited and probed with laser pulses, which are shorter than the luminescence lifetime of the dye), or if dye molecules are separated by spacer or host molecules (e.g., polymer molecules) preventing the dye from dimerization [64,55].

Quenching of dye molecules due to dimerization is not the only problem hindering high gain. Another big challenge is presented by the stimulated emission, and the as amplified spontaneous emission. Even if the size of the pumped spot is only $l \approx 0.1$ mm, the gain values discussed above would correspond to an enormously large single-path amplification of ~exp (gl) and very high efficiency of stimulated emission, which would wipe out the population inversion and reduce the gain. That is why high gain can be observed in photonic modes in dielectric media only at ultrafast pump and probe, $gct \ll 1$ (where t is the characteristic time of the process and c is the speed of light), or if the size of the pumped volume l is very small, $gl \ll 1$.

The situation becomes radically different in the vicinity of metallic nanostructures or metal–dielectric interfaces, where molecules emit to lossy plasmonic modes. As long as the overall loss in the mode exceeds the gain, a high value of population inversion, which without metal would cause a very strong amplified spontaneous emission in a purely dielectric medium, will be preserved in and not be depleted by the stimulated emission.

Laser dyes are not the only media that can support optical gain. Thus, a gain of 2000 cm^{-1} can be obtained in quantum wells [65], a gain of 80 cm^{-1} in quantum dots [66], a gain of 50 cm^{-1} in semiconducting polymers [67], and a gain of ~1 cm^{-1} in glasses doped with rare earth ions [68]. Even if in each of the categories above the gain can be increased by an order of magnitude, laser dyes remain champions in terms of the maximum gain. The disadvantage of organic dyes is that they photo-bleach. Inorganic semiconductor media are more practical for metamaterials and device applications, since they are robust and potentially suitable for direct electrical pumping.

In a number of publications [53,61,68], gain is described in terms of the negative (and constant in time) imaginary part of electric permittivity ε''. Such an approach gives a fairly accurate description of the system behavior below the stimulation emission threshold (Gain < Loss) and fails completely above the threshold (Gain > Loss). This can be explained by the fact that physically one cannot maintain an overall cw gain in infinitely large systems [44]. Similar to lasers, the population inversion and gain in plasmonic systems will lock at their threshold values, the latter being determined by the condition Gain = Loss. An accurate description of the system's dynamics should be based on the solution of rate equations for excited and ground states of the gain medium and taking into account the finite size of the sample [60,69,70].

5.4 Experimental Demonstrations of Metamaterials and Plasmonic Structures with Gain

5.4.1 Compensation of Loss and Stimulated Emission of Surface Plasmon Polaritons

5.4.1.1 Compensation of Optical Loss by Gain

Experimentally, SPPs were studied in the attenuated total internal reflection setup of Figure 5.2a [44]. The 90° prism was made of glass having an index of refraction $n_0 = 1.784$. Metallic films were produced by evaporating 99.99% pure silver. Rhodamine 6G dye (R6G) and polymethyl methacrylate (PMMA) were dissolved in dichloromethane. The solutions were deposited onto the surface of silver and dried to a film. The concentration of dye in dry PMMA was equal to 10 g/L (2.1×10^{-2} M) and the thickness of the polymer film was on the order of 10 μm.

The prism was mounted on a rotating stage. The reflectance R was probed with p-polarized He-Ne laser beam at $\lambda = 594$ nm. The reflected light was detected by a photomultiplier tube (PMT) connected to an integrating sphere, which was moved during the scan to follow the walk of the beam. The permittivity of the metallic film was determined by fitting the experimental reflectance profile $R(\theta)$ with Equation 5.1.

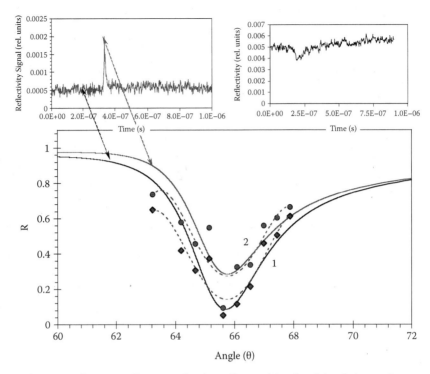

FIGURE 5.6 Reflectance $R(\theta)$ measured without (diamonds) and with (circles) optical pumping in the glass-silver-R6G/PMMA structure. Dashed lines – guides for the eye. Solid lines – fitting with Equation 5.1 at $\varepsilon_0' = n_0^2 = 1.784^2 = 3.183$, $\varepsilon_0'' = 0$, $\varepsilon_1' = -15$, $\varepsilon_1'' = 0.85$, $d_1 = 39$ nm, $\varepsilon_2' = n_2^2 = 1.5^2 = 2.25$, $\varepsilon_2'' \approx 0$ (trace 1), and $\varepsilon_2'' \approx -0.006$ (trace 2). Reflectance kinetics recorded under pumping in 39-nm-thick silver film (left inset) and 90 nm silver film (right inset) under resonant excitation conditions. (Adapted from Noginov, M. A. et al. 2008. *Opt. Express* 16:1385–92.)

In the measurements with optical gain, the R6G/PMMA film was pumped from the back side of the prism (Figure 5.1a) with Q-switched pulses of a frequency-doubled Nd:YAG laser (t_{pulse} ~10 ns). The pumped spot, with diameter equal to ~3 mm, completely overlapped the smaller spot of the probe He-Ne laser beam. The reflected He-Ne laser light was steered with the set of mirrors to the entrance slit of a monochromator, set at $\lambda = 594$ nm, with a PMT attached to the monochromator's exit slit. Experimentally, the reflectance kinetics $R(\theta,t)$ were recorded under short-pulsed pumping at different incidence angles (left inset of Figure 5.6).

The results of the reflectance measurements in the 39 nm silver film are summarized in Figure 5.6. Two sets of data points correspond to the reflectance without pumping (measured in flat parts of the kinetics before the laser pulse) and with pumping (measured in peaks of the kinetics), as shown with arrows in Figure 5.6. By dividing the values of R measured in the presence of gain by those without gain, the relative enhancement of the reflectance signal was calculated to be as high as 280%, a significant improvement in comparison to Reference [71], where the change of the reflectance in the presence of gain did not exceed 0.001%. Fitting both reflectance curves with Equation 5.1 and knowing $\varepsilon_1 = -15$, $\varepsilon_1 = 0.85$, and $\varepsilon_2' = n_2^2 = 2.25$, yields $\varepsilon_2'' \approx -0.006$; this corresponds to an optical gain of 420 cm^{-1} (at $\lambda = 594$ nm) and ~35% reduction in the internal SPP loss [44].

In thicker silver films, calculations predict an initial reduction of the minimum reflectance $R(\theta)$ at small values of gain followed by its increase (after passing the minimum point $R = 0$) at larger gains (Figure 5.4a). The reduction of R was experimentally observed in the 90-nm-thick film, where, instead of a peak in the reflectance kinetics, a dip was detected (right inset of Figure 5.6).

Note that laser dye is not the only gain medium that can elongate the propagation length of the SPP. Thus, in Reference [68], the (relatively low) loss of the long-range SPP in a thin, metallic waveguide was partly compensated by Er^{3+}-doped phosphate glass.

5.4.1.2 Stimulated Emission in Open Paths

In the series of experiments described below [72], the experimental setup was almost identical to that in the previous subsection. Silver films with a thickness of 39 nm–81 nm were deposited on the glass prism with an index of refraction $n_0 = \sqrt{\varepsilon_0} = 1.7835$ and coated with 1–3 μm PMMA films doped with R6G dye in concentration 2.2×10^{-2} M. The smaller (than in Section 5.4.1.1) thickness of the PMMA/R6G film allowed for a stronger pumping of dye molecules in the layer adjacent to the silver surface.

Three different sets of experiments have been performed:

1. In the first set of measurements, used primarily for calibration purposes, SPPs were excited from the bottom of the prism (Figure 5.2a), and the reflectance of the sample R was measured as a function of the incidence angle θ. Fitting the reflectance profile $R(\theta)$ with Equation 5.1 allowed one to determine the dielectric constants of the particular silver films studied.

   In the remaining experiments, the samples were pumped with a pulsed frequency-doubled Nd:YAG laser (t_{pulse} ~10 ns, $\lambda = 532$ nm) from the PMMA side at a nearly normal angle of incidence.

2. In the second set of measurements, SPPs were excited via emission of R6G molecules. The laser light at $\lambda = 532$ nm corresponds to the absorption maximum of R6G. It excites dye molecules in the PMMA/R6G volume, in particular, in the vicinity of the silver film where the pumping-induced SPP is confined. Excited dye molecules, in turn, partly emit to the SPP modes at corresponding frequencies. The SPPs excited by optically pumped molecules (reported earlier in References [73,74]) get decoupled from the prism at angles corresponding to the SPPs' wavenumbers. At low pumping intensity, the SPP spectra recorded at different angles θ qualitatively resembled the R6G spontaneous emission spectrum modulated by the SPP decoupling function (Figure 5.7) [72].

3. The character of SPP emission excited by optically pumped dye molecules changed dramatically at high pumping intensity.
   i. The emission spectra considerably narrowed in comparison to those at low pumping; Figure 5.8a.
   ii. The narrowed emission spectra peaking at ~602 nm became almost independent of the observation angle.

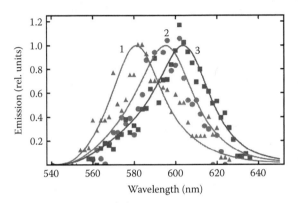

FIGURE 5.7 Spectra of SPP spontaneous emission decoupled at different angles θ; triangles—θ = 67.17°, circles—θ = 66.14°, squares—θ = 65.62°. The thickness of the silver film is 57 nm. (Adapted from Noginov, M. A. et al. 2008. *Phys. Rev. Lett.* 101:226806.)

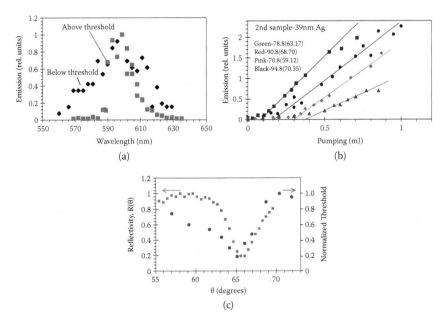

FIGURE 5.8 (a) Spectra of the SPP emission recorded at θ = 68.7° at (diamonds) low pumping density 10.9 mJ/cm² and (squares) high pumping density 81.9 mJ/cm². (b) Input–output curves of SPP emission recorded at different angles θ. The thickness of the silver film is 39 nm. The diameter of the pumped spot is 2.16 mm, λ = 602 nm. Squares—θ = 63.17°, circles—θ = 68.70°, diamonds—θ = 59.12°, triangles—θ = 70.35°. Resonance angle — θ = 65.56°. (c) Dependence of the SPP stimulated emission threshold versus θ (circles) and the reflectance profile (squares) measured at λ ≈ 604 nm. (Adapted from Noginov, M. A. et al. 2008. *Phys. Rev. Lett.* 101:226806.)

iii. The dependence of the emission intensity (recorded in the maximum of the emission spectrum) on the pumping intensity was strongly nonlinear with the distinct threshold; Figure 5.8b.

iv. The value of the threshold I_{th} depended on the observation angle θ. The angular dependence $I_{th}(\theta)$ resembled the reflectance profile $R(\theta)$; Figure 5.8c.

The experimental results above suggest the stimulated character of emission decoupled from SPP modes. Knowing the concentration of R6G molecules in the film (1.3×10^{19} cm^{-3}), absorption and emission cross sections of R6G ($\sigma_{abs}^{\lambda=532nm} = 4.3 \times 10^{-16}$ cm^2, $\sigma_{em}^{\lambda=600nm} = 1.9 \times 10^{-16}$ cm^2), and lifetime of excited R6G molecules at a given concentration of dye, $\tau \approx 1$ ns [75], we were able to evaluate the experimental threshold gain to be of the same order of magnitude as the theoretically predicted ones satisfying the condition $\gamma_i + \gamma_r = 0$—gain compensates total loss.

To prove that the stimulated emission seen in the foregoing experiments came from direct generation of SPPs rather than from random lasing in PMMA/R6G film decoupled from the prism via SPPs, the emission collected from the back side of the prism as well as from glass slides with similar deposited films was studied close to the normal angle of incidence and at a grazing angle. At high pumping energy, the collected emission spectrally narrowed and demonstrated an input-output behavior with a threshold. However, the values of the thresholds (higher at nearly normal direction than at a grazing angle) were significantly larger than the threshold in the case of the SPP decoupled from the prism, and the narrowed emission spectra had their maxima at shorter wavelengths (as short as ~584 nm). Thus, the emission decoupled from the glass prism was generated by SPPs rather than originating in purely photonic modes within the PMMA/R6G film.

The demonstrated phenomenon adds a new stimulated emission source to the toolbox of nanophotonic materials and devices and proves that total compensation by metamaterials' loss by gain is indeed possible.

5.4.1.3 Stimulated Emission in Microring Cavities

The stimulated emission effect described in Section 5.4.1.3 was, most likely, an SPP amplification in open paths. Whether a feedback provided by the SPP scattering (similar to that in random lasers [76,77]) existed in the system, it did not have a distinctive signature in the emission behavior. In the present section, we discuss the stimulated emission of SPPs with feedback provided by a microring cavity.

Experimentally, 10 μm thick gold wire was dipped in a dichloromethane solution of polymethyl methacrylate (PMMA) and Rhodamine 6G dye (R6G) and then dried in air. As a result, the wire was coated with ≈10 μm thick film (inset of Figure 5.9a). The concentration of R6G in dry PMMA was equal to 10 g/L (2.1×10^{-2} M). The samples were excited with ≈10 ns pulses of a frequency-doubled Q-switched Nd:YAG laser ($\lambda = 532$ nm) focused on a 0.25 mm spot. Above a certain pumping energy threshold, a series of narrow spectral lines, characteristic of microring laser cavities [83,84], appeared in the emission spectra (Figure 5.9a) [80].

The energy spacing between emission lines is governed by the product of the real part of the refraction index n' and the diameter of the lasing mode d. In whispering gallery mode microring lasers, the latter is equivalent to the outer diameter of the ring-shaped or

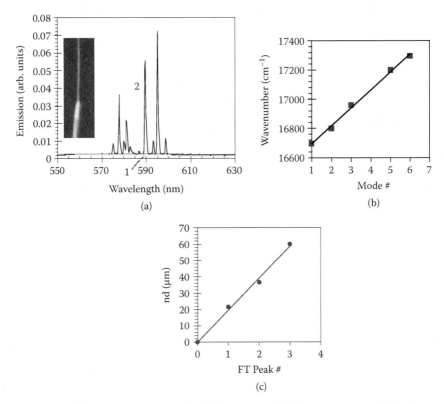

FIGURE 5.9 (a) Emission spectrum of R6G/PMMA-coated gold wire; trace 1—0.026 mJ, trace 2—0.052 mJ. Inset: microphotograph of the wire; top part—10 μm bare gold wire, bottom part—~30 μm wire coated with R6G/PMMA (~30 μm). (b) Mode energy versus mode number (calculated from trace 2 of (a); $n'd$ = 19.9 μm. (c) Position of the FT peak (in units of $n'd$) versus FT peak number (calculated from trace 2 of (a); $n'd$ = 19.6 μm. (From Kitur, J. K. et al. 2009. In *Frontiers in Optics 2009/Laser Science XXV/Adaptive Optics: Methods, Analysis and Applications/Advances in Optical Materials/ Computational Optical Sensing and Imaging/Femtosecond Laser Microfabrication/Signal Recovery and Synthesis on CD-ROM* (Optical Society of America, Washington, DC, 2009), FTuB5.)

cylinder-shaped gain medium. From the spacing between the lasing modes (Figure 5.9b), it has been calculated that $n'd$ = 19.9 μm [80]. Alternatively, the Fourier transform (FT) of the emission spectrum was calculated, revealing a series of equidistant peaks corresponding to multiples of $n'd$. Not surprisingly, this measurement resulted in nearly the same value of $n'd$ equal to 19.6 μm (Figure 5.9c) [80]. The index of refraction of R6G-doped PMMA is equal to n'_{PMMA} = 1.50 [81] and that of SPP propagating at the interface between gold and PMMA is equal to $n'_{SPP} = \text{Re}\left(\sqrt{(\varepsilon_{PMMA}\varepsilon_{Au})/((\varepsilon_{PMMA}+\varepsilon_{Au}))}\right) = 1.74$ at λ = 590 nm [14,52]. The diameter of the SPP mode is approximately equal to 10 μm, and the corresponding product, $n'_{SPP}d$ = 17.4 μm, is very close to that measured experimentally. On the other hand, if it existed, the whispering gallery mode concentrated at the outer diameter of ≈30 μm gold-PMMA cylinder would have $n'_{PMMA}d$ = 45 μm. This indicates that the observed stimulated emission modes of a ring cavity were due to surface plasmon polaritons propagating at the surface of the gold wire. A ballpark estimate shows that the gain in the system was of the same order of magnitude as required to compensate the absorption loss of the SPP. The fact that whispering gallery modes concentrated at the outer surface of the R6G/PMMA cylinder were not observed is explained by a roughness of the PMMA–air interface.

5.4.1.4 Semiconductor Lasers Enabled by Surface Plasmon Polaritons

Semiconductors are arguably the most promising gain media for applications in nano-plasmonics and metamaterials since they are suitable for direct electrical pumping. Over the years, several semiconductor lasers, in which laser mode has been supported by SPPs, have been reported in the literature.

SPP-based configurations have proved to be particularly useful in quantum cascade semiconductor lasers operating in the mid-infrared (mid-IR) range of the spectrum. In typical dielectric waveguides based on refractive index contrast and composed of a high-index core sandwiched between two lower index cladding layers, the transverse dimension of the confined mode is proportional to the effective emission wavelength. At long (≥10 μm) wavelengths, this requires the core and the cladding to be so thick that they are difficult to handle using state-of-the-art fabrication techniques. However, the mode may be more confined, and the gain layer can be significantly thinner if the SPP mode is used instead of a regular photonic mode [22].

The metallic layer deposited on top of the semiconductor structure can both support SPPs and serve as an electrode supplying carriers to the semiconductor gain medium. This idea was used in several works, in which mid-IR quantum cascade lasers with advanced properties operating at 8–11.5 μm [22], ~17 μm [23], and ~19 μm [82] have been demonstrated. Note that the SPP loss is very small at long wavelengths, because of a very small skin depth, making SPPs an ideal solution for long-wavelength laser applications.

In Reference [83], SPPs were employed to design miniature semiconductor lasers (operating at telecom wavelengths of ~1.5 μm) with the width of the core layer smaller than the diffraction limit. The laser mode was supported by *gap plasmons* propagating in a waveguide formed by two parallel metallic walls surrounding a heterostructure of a semiconductor laser; Figure 5.10. The smallest width of the waveguide (Figure 5.10) providing for the stimulated emission was 90 nm. Though the loss in the demonstrated laser was larger than that in the mid-IR quantum cascade laser described earlier (because of a shorter wavelength), it still could be overcome by the electrically pumped gain.

In the visible wavelength range, SPP loss becomes significantly larger; however, it still could be conquered by a gain optically excited with short laser pulses. In Reference [84], the $\lambda = 489$ nm laser, whose mode was nanoscopic in two dimensions, consisted of a CdS nanowire separated from the silver substrate by a 5-nm-thick MgF_2 layer. The stimulated emission in the demonstrated laser was supported by SPPs propagating along the metal–dielectric interface beneath the nanowire and reflecting off its ends. The very high field in the 5 nm gap separating CdS wire and silver surface makes the proposed design promising for future extreme intracavity spectroscopy applications.

A very substantial step toward the demonstration of a lasing spaser, theoretically predicted in Reference [61], has been accomplished in ref. [85]. Experimentally, a layer of

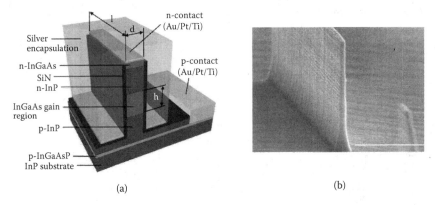

(a)   (b)

FIGURE 5.10   Structure of cavity formed by a rectangular semiconductor pillar encapsulated in silver. (a) Schematic showing the device layer structure. (b) Scanning electron microscope image showing the semiconductor core of one of the devices. The scale bar is 1 μm. (Adapted from Hill, M. T. et al. 2009. *Opt. Express* 17:11107–112.)

PbS quantum dots was placed on the top of the array of double-slit split-ring resonators. Although the optical gain, pumped with a frequency-doubled CW Nd:YAG laser at λ = 532 nm, was not sufficient to compensate the overall loss and drive the system over the stimulated emission threshold, it was high enough to observe a change in the transmission spectra under pumping. Lasing would require much higher levels of gain, which were proposed to be achieved by using nanosecond optical pump pulses or by lowering the sample temperature.

5.4.2 Enhancement of Localized Surface Plasmons in Fractal Aggregate of Silver Nanoparticles

In Reference [50], optical loss in fractal aggregates of silver nanoparticles was compensated by gain in surrounding Rhodamine 6G (R6G) laser dye. Following Reference [48], an enhancement of Rayleigh scattering by nanoparticles was used as evidence of increased quality factor of surface plasmon resonance at sufficiently large values of gain.

Experimentally, poly(vinylpyrrolidone)-passivated silver aggregate suspended in ethanol was prepared according to the procedure described in Reference [86]. The concentrations of R6G molecules and Ag nanoparticles in the mixture were equal to $\approx 2.1 \times 10^{-5}$ M and $\approx 8.7 \times 10^{13}$ cm^{-3}, respectively. The absorption spectrum of Ag aggregate has one structureless band covering the whole visible range and extending to near-infrared; Figure 5.11a. The major feature in the absorption spectrum of R6G is the band peaking at ≈ 528 nm. The emission spectrum of R6G is dominated by the band with the maximum at 558 nm; Figure 5.11b.

In the pump-probe scattering experiment, R6G-Ag aggregate mixtures were pumped with frequency-doubled Q-switched Nd:YAG laser ($\lambda = 532$ nm, $t_{pump} \approx 10$ ns). A fraction of the pumping beam was split off and used to pump a simple laser consisting of the cuvette with R6G dye placed between two mirrors (Figure 5.12a). Its emission wavelength (~558 nm) corresponded to the maximum of the gain spectrum of R6G dye in the mixtures studied. The beam of the R6G laser, which was used as a probe in the Rayleigh

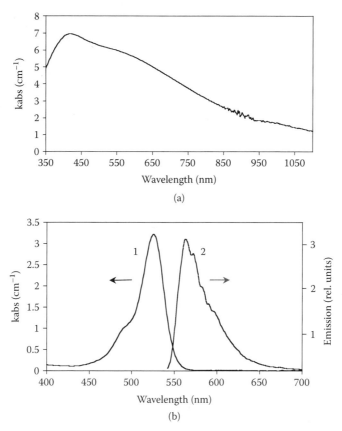

FIGURE 5.11   (a) Absorption spectrum of the suspension of Ag aggregate and (b) absorption (1) and emission (2) spectra of methanol solution of Rhodamine 6G laser dye.

scattering experiment, was aligned with the pumping beam in the beamsplitter and sent to the sample through a small (0.5 mm) pinhole. The pump and probe beams were collinear, and their diameters at the pinhole were larger than 0.5 mm.

The scattered probe light, along with scattered pumping light and spontaneous emission of dye, was collected by an optical fiber that was placed within several millimeters from the cuvette at angles ranging from ~45° to ~135° relative to the direction of the beam propagation. To detect scattered probe light, the emission spectrum was recorded (using a monochromator) between 540 and 650 nm. The scattered probe light was seen in the spectrum as a relatively narrow (~5 nm) line on the top of a much broader spontaneous emission band (Figure 5.12b). Thus, it could be easily separated from spontaneous emission.

Experimentally, the energy of the probe light was kept constant, and the intensity of the *scattered* probe light was measured as a function of the pumping light energy. The sixfold increase of Rayleigh scattering observed in the dye–Ag aggregate mixture with the increase of pumping energy (Figure 5.12c) unambiguously proves the compensation of loss in metal and enhancement of the quality factor of SP resonance by optical gain in the surrounding dielectric.

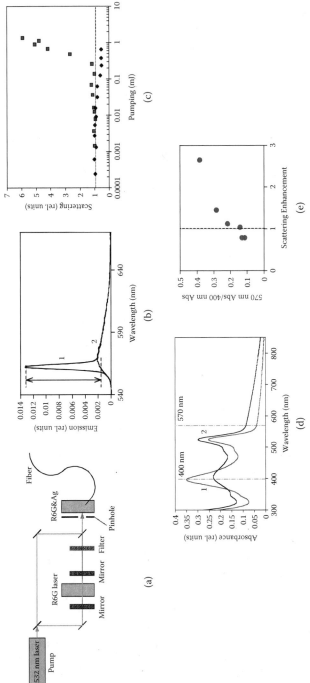

FIGURE 5.12 *See color insert.* (a) Pump-probe experimental setup for Rayleigh scattering measurements. (b) Spectra of scattered light and spontaneous emission (1) and spontaneous emission only (2). (c) Intensity of Rayleigh scattering as a function of pumping energy in two different dye–Ag aggregate mixtures. (d) Absorption spectra of dye–Ag aggregate mixtures; R6G—2.1×10^{-5} M, Ag aggregate—8.7×10^{13} cm^{-3}. Squares in (c) and trace 1 in (d) correspond to one particular mixture, and diamonds in (c) and trace 2 in (d) correspond to another mixture. (e) The ratio of absorption coefficients of dye–Ag aggregate mixtures at 570 nm and 400 nm plotted versus enhancement of the Rayleigh scattering measured at 0.46 mJ. (Adapted from Noginov, M. A. 2008. *J. Nanophotonics* 2: DOI:10.1117/1.3073670.)

Although physical properties of the mixtures were primarily determined by concentrations of ingredients, such factors as the regime of steering, steps by which one ingredient was added to another, time, etc., could make the absorption spectra of two nominally identical mixtures different from each other. Depending on the shape of the absorption spectrum of the Ag aggregate–dye mixture (Figure 5.12d), the gain-induced increase of the Rayleigh scattering could be large (as in Figure 5.12c, squares), small, or even negative (Figure 5.12c, diamonds). Figure 12e shows a monotonic dependence of the scattering enhancement measured at 0.46 mJ pumping energy versus the ratio of absorption coefficients in the mixtures at 570 and 400 nm. One can see that the relatively strong absorption of the mixture at 570 nm, which is a signature of aggregated Ag nanoparticles, helps to observe enhanced Rayleigh scattering. Even if the relationship between the absorption one spectra and the physical properties of the mixtures that govern the results of the scattering experiments is not clearly understood, the fact that the correlation exists is beyond doubt.

A plausible explanation for different scattering properties of different mixtures could be in line with the theoretical model developed in Reference [49], which predicts that depending on the thickness of an amplifying shell, the absorption of the complex can increase or decrease with increase of gain in dye [49]. Similarly, one can speculate that in the experiments described earlier, different mixtures had different numbers of adsorbed molecules per metallic nanoparticle, which determined the enhancement or reduction of the Rayleigh scattering.

5.4.3 Experimental Demonstration of Spaser and Nanolaser

A spaser—a plasmonic analog of a laser—should have a medium with optical gain in the close vicinity of a metallic nanostructure that supports surface plasmon oscillations [53]. In Reference [55], a spaser was fabricated by employing a modified synthesis technique for *Cornell dots* (C dots), high-brightness luminescent core-shell silica nanoparticles [64,87]. The produced spaser nanoparticles were composed of a gold core (providing for plasmon modes), surrounded by a silica shell containing organic dye Oregon Green 488 (OG-488); Figure 5.5e.

The diameter of the Au core and the thickness of the silica shell were equal to ~14 nm and ~15 nm, respectively (Figure 5.13a,b). The number of dye molecules per nanoparticle was estimated to be 2.7×10^3, and the concentration of nanoparticles in a water suspension was equal to 3×10^{11} cm^{-3}.

The extinction spectrum of the suspension of nanoparticles is dominated by the SP resonance band at ~0.52 μm and the broad short-wavelength band corresponding to interstate transitions between d states and hybridized s-p states of Au (Figure 5.14). The quality factor of the SP resonance, estimated from the width of its spectral band, is equal to Q = 13.2, in good agreement with the theoretical predictions. The SP band overlaps with both the emission and excitation bands of the dye (Figure 5.14).

The emission decay kinetics detected at 480 nm were nonexponential (Figure 5.15). Their fit with the sum of two exponentials resulted in two characteristic decay times, 1.6 and 4.1 ns. As expected of dyes, the absorption and emission spectra in OG-488 are nearly symmetrical to each other (Figure 5.14). This allows one to assume that the peak emission cross section, σ_{em}, is equal to the experimentally measured peak absorption

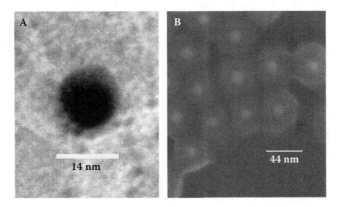

FIGURE 5.13 TEM image of (a) Au core and (b) SEM image of Au/silica/dye core-shell nanoparticles. (Adapted from Noginov, M. A. et al. 2009. *Nature* 460:1110–12.)

cross section, $\sigma_{abs} = 2.55 \times 10^{-16}$ cm^2. Using the known formula relating the strength and the width of the emission band with the radiative life-time τ [88], the latter was estimated to be equal to 4.3 ns, very close to the experimentally measured longer component of the emission kinetics. The decay-time shortening (down to 1.6 ns) of dye molecules in an effective plasmonic nanocavity was explained by the Purcell effect [55,89].

When the emission was detected in the spectral band 520 ± 20 nm, corresponding to the maximum of the emission and gain, its kinetics had a second maximum (Figure 5.15) characteristic of the development of the stimulated emission pulse [90]. This observation is consistent with the spaser effect described later. In fact, the delay of the stimulated emission pulse relative to the pumping pulse as well as the oscillating

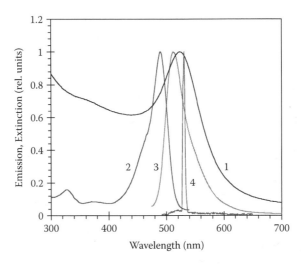

FIGURE 5.14 (1) Normalized extinction, (2) excitation, (3) spontaneous emission, and (4) stimulated emission spectra of Au/silica/dye nanoparticles. (Adapted from Noginov, M. A. et al. 2009. *Nature* 460:1110–12.)

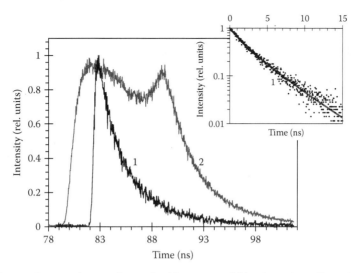

FIGURE 5.15 Emission kinetics detected at (1) 480 nm and (2) 520 nm. Inset: Trace 1 plotted in semilogarithmic coordinates (dots) and the corresponding fitting curve. (Adapted from Noginov, M. A. et al. 2009. *Nature* 460:1110–12.)

behavior of stimulated emission (relaxation oscillations) are known in lasers [90,91] and are expected in the spaser.

In the stimulated emission studies, the samples were loaded in 2 mm cuvettes and pumped at λ = 488 nm with ~5 ns pulses from an Optical Parametric Oscillator (OPO) lightly focused into a ~2.4 mm spot. At weak pumping, the emission spectra resembled those measured in the spectrofluorometer (Figure 5.14). When the pumping energy exceeded a critical threshold value, a narrow peak appeared at λ = 531 nm in the emission spectrum (Figure 5.16a). The intensity of this peak plotted versus pumping energy

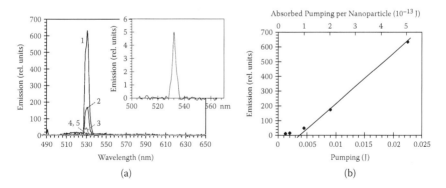

FIGURE 5.16 (a) Stimulated emission spectra of the nanoparticle sample pumped with (1) 22.5 mJ, (2) 9 mJ, (3) 4.5 mJ, (4) 2 mJ, and (5) 1.25 mJ OPO pulses at λ = 488 nm. (b) Corresponding input–output curve (lower axis—total launched pumping energy, upper axis—absorbed pumping energy per nanoparticle). Inset of (a): Stimulated emission spectrum at more than hundredfold dilution of the sample. (Adapted from Noginov, M. A. et al. 2009. *Nature* 460:1110–12.)

results in an input–output curve with a pronounced threshold characteristic of lasers and spacers [60] (Figure 5.16b). The ratio of the laser peak intensity to the spontaneous emission background increased with an increase of the pumping energy (as can be seen if the maximum intensities in Figure 5.16a are normalized to unity). As expected, the laser-like emission occurred at the wavelength at which the dye absorption, as evidenced by the excitation spectrum, is practically absent and the SP resonance is strong (Figure 5.14).

At the dilution of the sample, the emission intensity decreased, but the character of the spectral line remained unchanged (inset of Figure 5.16a) and the ratio of the stimulated emission intensity to the spontaneous emission background did not diminish. This proves that the observed stimulated emission was produced by *single nanoparticles* as opposed to a collective stimulated emission effect in a volume of gain medium with the feedback supported by the cuvette walls.

An aqueous solution of OG-488 dye at a concentration of 0.235 mM had spontaneous emission intensity approximately 1000 times stronger than that in the lasing nanoparticle sample. However, under pumping, the dye solution did not show any spectral narrowing or superlinear dependence of the emission intensity on pumping power. In contrast, the dependence of the emission intensity on pumping power was sublinear, which could be a result of dye photo-bleaching. This is further evidence of the stimulated emission occurring in individual hybrid Au/silica/dye nanoparticles rather than in the macroscopic volume of the cuvette. Lastly, the stimulated emission kinetics in Figure 5.15 (trance 2) was measured in a microscope-based setup, which sampling volume contained, on average, less than one nanoparticle. This is auther proof at the stimulated emission being generated by individual nanoparticles.

The diameter of the hybrid C dots, 44 nm, was too small to support visible stimulated emission in a purely photonic mode. Thus, the observed stimulated emission could originate only from a spaser. According to the model, the critical threshold gain in spaser nanoparticles can be achieved if the number of excited dye molecules per nanoparticle exceeds 2.0×10^3, which is smaller than the number of OG-488 molecules available per nanoparticle in the experimental sample, $\sim 2.7 \times 10^3$ [55]. The outcoupling of spaser oscillations to photonic modes constitutes a nanolaser.

5.4.4 Negative Index Materials with Gain

Multiple theoretical and experimental studies have been aimed at understanding and experimental realization of a new class of low-loss and active metamaterials enabled by optical gain. In 2003, it was understood that optical gain can significantly improve the performance of a superlens based on a slab of a negative index material, which otherwise would strongly suffer from loss [92]. The theoretical concepts of the metamaterials with gain and the computational methods were further developed in References [57,59,69,93–95].

The compensation of loss by gain in a fishnet NIM structure was theoretically studied in Reference [53]. In particular, it has been shown (1) that optical loss can be compensated and even overcompensated by gain at realistic concentrations of dye molecules ($\sim 10^{19}$ cm^{-3}) embedded in a dielectric medium and (2) that it is very critical to place optical gain *inside* the fishnet structure where it has good spatial and spectral overlap with hot spots of electric field.

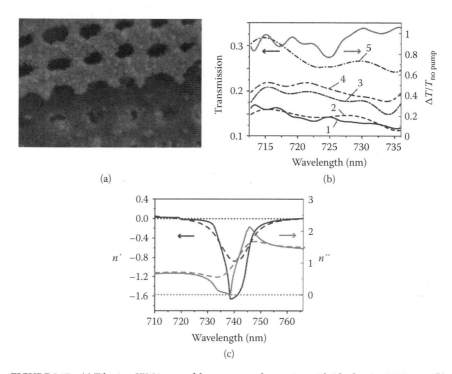

FIGURE 5.17 (a) Tilt-view SEM image of the structure after coating with Rhodamine 800/epoxy. (b) Sample's transmission (line 1) without pumping, (line 5) with the optimized delay between pump and probe (probe pulse is 54 ps later than the pump) and 1-mW pumping power, (line 3) with the optimized delay and 0.12 mW pumping power, (line 4) with the optimized delay and 0.16 mW pumping power, and (line 2) with the pump preceding the probe by 6 ps and 1 mW pumping power. The wavelength-dependent relative transmission change from the pump–probe experiment is shown by the red solid line. (c) Retrieved values of real (n′) and imaginary (n″) parts of the refraction index in (solid lines) pumped (dashed lines) and not pumped samples. (Adapted from. Xiao S. et al. 2010. *Nature* 466: 735–38.)

The experimental demonstration of a lossless NIM was achieved in a recent breakthrough study [96]. The NIM structure in this work was a well-known fishnet (see Chapter 2 of this book), and the great experimental challenge was to substitute alumina spacer separating metallic layers with the gain medium, epoxy-doped with Rhodamine 800 dye. The scanning electron microscope (SEM) image of the structure (Figure 5.17a) indicates that no damage has occurred during fabrication.

The samples' transmission spectra were measured with and without pumping dye molecules with 690 nm picosecond laser pulses. The transmission and reflection were probed with a supercontinuum white-light source generated by pumping water with an amplified 800 nm Ti-sapphire laser. It has been shown that the transmission of the sample is increased by more than 100% at the optimized delay (much smaller than the emission lifetime of the dye molecules) between the pump and probe beams (Figure 5.17b).

The transmittance and reflectance data were used to retrieve the spectra of electric permittivity, magnetic permeability, and refraction index. One can see that in the vicinity of 740 nm, optical pumping and gain (i) help to make the real part of the refraction index (n') more negative, (ii) make the metamaterial essentially lossless, and (iii) significantly improve the metamaterial's quality characterized by the figure of merit $F = n'/n''$ (where n'' is the imaginary part of the refraction index).

The experimental demonstration of gain-assisted lossless negative index metamaterial solves the inherent problem of loss in metal–dielectric structures and promotes a family of new applications for optical metamaterials.

5.5 Summary

To summarize, we have reviewed the major concepts, underlying theory, and experimental demonstrations of metamaterials with gain, which address two major challenges: compensation of optical loss and a need for active materials and devices.

We show how gain can offset loss in simple plasmonic structures, such as fractal aggregates of silver nanoparticles and thin silver films, as well as in much more complex systems, such as negative index fishnet metamaterials. These demonstrations pave the way for many metamaterials' applications, including high-resolution imaging, subdiffraction focusing, and optical cloaking, which are hindered by high loss.

When gain in a plasmonic mode exceeds loss, it causes amplification and stimulated emission of surface plasmons. Thus, stimulated emission of surface plasmon polaritons in open paths has been demonstrated in simple metallic films with adjacent dye-doped dielectric. The presence of a feedback, for example, in microring cavities, leads to a stimulated emission characterized by well-defined spectrally narrow modes. Surface plasmon polaritons provide for the stimulated emission mode in a variety of semiconductor lasers operating at infrared and visible wavelengths, including a nanolaser, which mode is substantially subwavelength in two dimensions.

Finally, stimulated emission of surface plasmons in metallic nanostructures with adjacent gain medium enables a spaser and a nanolaser. The developed class of stimulated emission sources of coherent surface plasmons and photons has the potential to revolutionize future nanophotonic, nanoelectronic, and information technology.

Acknowledgments

The work was partly supported by the NSF PREM grant # DMR 0611430, NSF NCN grant # EEC-0228390, AFOSR grant # FA9550-09-1-0456, and NSF IGERT grant DGE 0966188. The author cordially thanks all his co-authors and collaborators, in particular Yu. A. Barnakov, V. P. Drachev, E. E. Narimanov, V. A. Podolskiy, V. M. Shalaev, U. B. Wiesner, and G. Zhu, for contributions, stimulating discussions, and continuing support.

References

1. Ritchie, R. H. 1973. Surface plasmons in solids. *Surf. Sci.* 34:1–19.
2. Fleischmann, M., P. J. Hendra, and A. J. McQuillan. 1974. Raman spectra of pyridine adsorbed at a silver electrode. *Chem. Phys. Lett.* 26:163–66.

3. Moskovits, M. 1985. Surface-enhanced spectroscopy. *Rev. Mod. Phys.* 57:783–826.

4. Kreibig, U. and M. Vollmer. 1995. *Optical Properties of Metal Clusters.* Springer: New York.

5. Su, K.-H., Q.-H. Wei, X. Zhang, J. J. Mock, D. R. Smith, and S. Schultz. 2003. Interparticle coupling effects on plasmon resonances of nanogold particles. *Nano Lett.* 3:1087–90.

6. Quinten, M. 1999. Optical effects associated with aggregates of clusters. *J. Cluster Sci.* 10:319–58.

7. Quinten, M., A. Leitner, J. R. Krenn, and F. R. Aussenegg. 1998. Electromagnetic energy transport via linear chains of silver nanoparticles. *Opt. Lett.* 23:1331–33.

8. Averitt, R. D., S. L. Westcott, and N. J. Halas. 1999. Linear optical properties of gold nanoshells. *J. Opt. Soc. Am. B* 16:1824–32.

9. Brongersma, M. L., J. W. Hartman, and H. A. Atwater. 2000. Electromagnetic energy transfer and switching in nanoparticle chain arrays below the diffraction limit. *Phys. Rev. B* 62:R16356–59.

10. Mock, J. J., M. Barbic, D. R. Smith, D. A. Schultz, and S. Schultz. 2002. Shape effects in plasmon resonance of individual colloidal silver nanoparticles. *Chem. Phys.* 116:6755–59.

11. Kneipp, K., H. Kneipp, I. Itzkan, R. R. Dasari, and M. S. Feld. 2002. Topical review: Surface-enhanced Raman scattering and biophysics. *J. Phys.* 14:R597–R624.

12. Kneipp, K., Y. Wang, H. Kneipp et al. 1997. Single molecule detection using surface-enhanced Raman scattering (SERS). *Phys. Rev. Lett.* 78:1667–70.

13. Nie, S. and S. R. Emory. 1997. Probing single molecules and single nanoparticles by surface-enhanced Raman scattering. *Science* 275:1102–04.

14. Raether, Heinz. 1988. *Surface Plasmons on Smooth and Rough Surfaces and on Gratings*: Springer-Verlag: Berlin.

15. Maier, Stefan A. 2007. *Plasmonics: Fundamentals and Applications.* Springer: New York.

16. Bozhevolnyi, Sergey I. (Editor). 2009. *Plasmonic Nanoguides and Circuits.* Pan Stanford Publishing: Singapore.

17. Bozhevolnyi, S. I., V. S. Volkov, and K. Leosson. 2002. Localization and waveguiding of surface plasmon polaritons in random nanostructures. *Phys. Rev. Lett.* 89:186801.

18. Boltasseva, A., S. I. Bozhevolnyi, T. Søndergaard, T. Nikolajsen, and L. Kristjan. 2005. Compact Z-add-drop wavelength filters for long-range surface plasmon polaritons. *Opt. Express* 13:4237–43.

19. Maier, S. A., P. G. Kik, H. A. Atwater et al. 2003. Local detection of electromagnetic energy transport below the diffraction limit in metal nanoparticle plasmon waveguides. *Nat. Mater.* 2:229–32.

20. Karalis, A., E. Lidorikis, M. Ibanescu, J. D. Joannopoulos, and S. Marin. 2005. Surface-plasmon-assisted guiding of broadband slow and subwavelength light in air. *Phys. Rev. Lett.* 95:063901.

21. Stockman, M. 2004. Nanofocusing of optical energy in tapered plasmonic waveguides. *Phys. Rev. Lett.* 93:137404.

22. Sirtori, C., C. Gmachl, F. Capasso et al. 1998. Long-wavelength ($\lambda \approx 8\text{-}11.5$ μm) semiconductor lasers with waveguides based on surface plasmons. *Opt. Lett.* 23:1366–68.

23. Tredicucci, A., C. Gmachl, F. Capasso, A. L. Hutchinson, D. L. Sivco, and A. Y. Cho. 2000. Single-mode surface-plasmon laser. *Appl. Phys. Lett.* 76:2164.

24. Tredicucci, A., C. Gmachl, M. C. Wanke et al. 2000. Surface plasmon quantum cascade lasers at λ~19 μm. *Appl. Phys. Lett.* 77:2286.

25. Ferrell, T. L. 1994. Thin-foil surface-plasmon modification in scanning-probe microscopy. *Phys. Rev. B* 50:14738–41.

26. Sánchez, E. J., L. Novotny, and X. S. Xie. 1999. Near-field fluorescence microscopy based on two-photon excitation with metal tips. *Phys. Rev. Lett.* 82:4014–17.

27. M. I. Stockman, 1989. Possibility of laser nanomodification of surfaces with the use of the scanning tunneling microscope. *Optoelectronics, Instrumentation and Data Processing* 3:27–37.

28. Ghaemi, H. F., T. Thio, D. E. Grupp, T. W. Ebbesen, and H. J. Lezec. 1998. Surface plasmons enhance optical transmission through subwavelength holes. *Phys. Rev. B* 58:6779.

29. Schaadt, D. M., B. Feng, and E. T. Yu. 2005. Enhanced semiconductor optical absorption via surface plasmon excitation in metal nanoparticles. *Appl. Phys. Lett.* 86:063106.

30. Pendry, J. B. 2000. Negative refraction makes a perfect lens. *Phys. Rev. Lett.* 85:3966–69.

31. Shalaev, V. M., W. Cai, U. Chettiar et al. 2005. Negative index of refraction in optical metamaterials. *Opt. Lett.* 30:3356–58.

32. Podolskiy, V. A. and E. E. Narimanov. 2005. Near-sighted superlens. *Opt. Lett.* 30:75–77.

33. Jacob, Z., L. V. Alekseyev, and E. E. Narimanov. 2006. Optical hyperlens: Far-field imaging beyond the diffraction limit. *Opt. Express* 14:8247–56.

34. Salandrino, A. and N. Engheta. 2006. Far-field subdiffraction optical microscopy using metamaterial crystals: Theory and simulations. *Phys. Rev. B* 74:075103.

35. Liu, Z., H. Lee, Y. Xiong, C. Sun, and X. Zhang. 2007. Far-field optical hyperlens magnifying sub-diffraction-limited objects. *Science* 315:1686.

36. Smolyaninov, I. I., Y.-J. Hung, and C. C. Davis. 2007. Magnifying superlens in the visible frequency range. *Science* 315:1699–701.

37. Milton, G.W. and N. P. Nicorovici. 2006. On the cloaking effects associated with anomalous localized resonance. *Proc. Royal Soc. A: Mathematical, Physical Eng. Sci.* 462:3027–59.

38. Cai, W., U. K. Chettiar, A. V. Kildishev, and V. M. Shalaev. 2007. Optical cloaking with metamaterials. *Nat. Photonics* 1:224–27.

39. Valentine, J., J. Li, T. Zentgraf, G. Bartal, and X. Zhang. 2009. An optical cloak made of dielectrics. *Nat. Mater.* 8:568–71.

40. Sudarkin, A. N. and P. A. Demkovich. 1989. Excitation of surface electromagnetic waves on the boundary of a metal with an amplifying medium. *Sov. Phys. Tech. Phys.* 34:764–66.

41. Nezhad, M. P., K. Tetz, and Y. Fainman. 2004. Gain assisted propagation of surface plasmon polaritons on planar metallic waveguides. *Opt. Express* 12:4072–79.

42. Avrutsky, I. 2004. Surface plasmons at nanoscale relief gratings between a metal and a dielectric medium with optical gain. *Phys. Rev. B* 70:155416.

43. De Leon, I. and P. Berini. 2008. Theory of surface plasmon-polariton amplification in planar structures incorporating dipolar gain media. *Phys. Rev. B* 78:161401.

44. Noginov, M. A., V. A. Podolskiy, G. Zhu et al. 2008. Compensation of loss in propagating surface plasmon polariton by gain in adjacent dielectric medium. *Opt. Express* 16:1385–92.

45. Noginov, M. A. 2008. Compensation of surface plasmon loss by gain in dielectric medium. *J. Nanophotonics* 2: DOI:10.1117/1.3073670.

46. Plotz, G. A., H. J. Simon, and J. M. Tucciarone. 1979. Enhanced total reflection with surface plasmons. *JOSA* 69:419–21.

47. Ma, X. and C. Soukoulis. 2001. Schrödinger equation with imaginary potential. *Physica B* 296:107–11.

48. Lawandy, N. M. 2004. Localized surface plasmon singularities in amplifying media. *Appl. Phys. Lett.* 85:5040–42.

49. Lawandy, N. M. 2005. Nano-particle plasmonics in active media. *Proceedings of SPIE Complex Mediums VI: Light and Complexity*, ed. M. W. McCall, G. Dewar, and M. A. Noginov (SPIE, Bellingham, WA). 5924:59240G.

50. Noginov, M. A., G. Zhu, M. Bahoura et al. 2006. Enhancement of surface plasmons in an Ag aggregate by optical gain in a dielectric medium. *Opt. Lett.* 31:3022–24.

51. Drachev, V. P., A. K. Buin, H. Nakotte, and V. M. Shalaev. 2004. Size dependent $\chi(3)$ for conduction electrons in Ag nanoparticles. *Nano Lett.* 4:1535–39.

52. Johnson, P. B. and R. W. Christy. 1972. Optical constants of the noble metals. *Phys. Rev. B* 6:4370–79.

53. Bergman, D. and M. Stockman. 2003. Surface plasmon amplification by stimulated emission of radiation: Quantum generation of coherent surface plasmons in nanosystems. *Phys. Rev. Lett.* 90:027402.

54. Averitt, R. D., D. Sarkar, and N. J. Halas. 1997. Plasmon resonance shifts of Au coated Au_2S nanoshells: Insight into multicomponent nanoparticle growth. *Phys. Rev. Lett.* 78:4217–20.

55. Noginov, M. A., G. Zhu, A. M. Belgrave et al. 2009. Demonstration of a spaser-based nanolaser. *Nature* 460:1110–12.

56. Gordon, J. A. and R. W. Ziolkowski. 2007. The design and simulated performance of a coated nano-particle laser. *Opt. Express* 15:2622–53.

57. Gordon, J. A. and R. W. Ziolkowski. 2008. CNP optical metamaterials. *Opt. Express* 16:6692–716.

58. Stockman, M. I. 2008. Spasers explained. *Nat. Photonics* 2:327–29.

59. Sarychev, A. K. and G. Tartakovsky. 2007. Magnetic plasmonic metamaterials in actively pumped host medium and plasmonic nanolaser. *Phys. Rev. B* 75:085436.

60. Stockman, M. I. 2010. The spaser as a nanoscale quantum generator and ultrafast amplifier. *J. Opt.* 12: 024004.

61. Zheludev, N. I., S. L. Prosvirnin, N. Papasimakis, and V. A. Fedotov. 2008. Lasing spaser. *Nat. Photonics* 2:351–54.

62. Stockman, M. I., S.V. Faleev, and D. J. Bergman. 2001. Localization versus delocalization of surface plasmons in nanosystems: Can one state have both characteristics? *Phys. Rev. Lett.* 87:167401.

63. Gavrilenko, V. I. and M. A. Noginov. 2006. *Ab initio* study of optical properties of rhodamine 6G molecular dimmers. *J. Chem. Phys.* 124:044301.

64. Ow, H., D. R. Larson, M. Srivastava, B. A. Baird, W. W. Webb, and U. Wiesner. 2005. Bright and stable core-shell fluorescent silica nanoparticles. *Nano Lett.* 5:113–17.

65. Asada, M., A. Kameyama, and Y. Suematsu. 1984. Gain and intervalence band absorption in quantum-well lasers. *IEEE J. Quantum Electron.* QE-20:745–54.

66. Meuer, C., J. Kim, M. Laemmlin et al. 2008. Static gain saturation in quantum dot semiconductor optical amplifiers. *Opt. Express* 16:8269–79.

67. Lahoza, F., N. Capujb, C. J. Otonc, and S. Cheyland. 2008. Optical gain in conjugated polymer hybrid structures based on porous silicon waveguides. *Chem. Phys. Lett.* 463:387–90.

68. Ambati, M., S. H. Nam, E. Ulin-Avila, D. A. Genov, G. Bartal, and X. Zhang. 2008. Observation of stimulated emission of surface plasmon polaritons. *Nano Lett.* 8:3998–01.

69. Wegener, M., J. L. García-Pomar, C. M. Soukoulis, N. Meinzer, M. Ruther, and S. Linden. 2008. Toy model for plasmonic metamaterial resonances coupled to two-level system gain. *Opt. Express* 16: 19785–98.

70. Fang, A., T. Koschny, M. Wegener, and C. M. Soukoulis. 2009. Self-consistent calculation of metamaterials with gain. *Phys. Rev. B* 79:241104(R).

71. Seidel, J., S. Grafstroem, and L. Eng. 2005. Stimulated emission of surface plasmons at the interface between a silver film and an optically pumped dye solution. *Phys. Rev. Lett.* 94:177401.

72. Noginov, M. A., G. Zhu, M. Mayy, B. A. Ritzo, N. Noginova, and V. A. Podolskiy. 2008. Stimulated emission of surface plasmon polaritons. *Phys. Rev. Lett.* 101:226806.

73. Weber, W. H. and C. F. Eagen. 1979. Energy transfer from an excited dye molecule to the surface plasmons of an adjacent metal. *Opt. Lett.* 4:236–38.

74. Ray, K., H. Szmacinski, J. Enderlein, and J. R. Lakowicz. 2007. Distance dependence of surface plasmon-coupled emission observed using Langmuir-Blodgett films. *Appl. Phys. Lett.* 90:251116.

75. Selanger, K., A. J. Falnes, and T. Sikkeland. 1977. Fluorescence lifetime studies of Rhodamine 6G in methanol. 1977. *J. Phys. Chem.* 81:1960–63.

76. Ling, Y., H. Cao, A. L. Burin, M. A. Ratner, X. Liu, and R. P. H. Chang. 2001. Investigation of random lasers with resonant feedback. *Phys. Rev. A* 64:063808.

77. Noginov, M. A. 2005. *Solid-State Random Lasers.* Springer: New York.

78. Frolov, S. V., Z. V. Vardeny, and K. Yoshino. 1998. Plastic microring lasers on fibers and wires. *Appl. Phys. Lett.* 72:1802–04.

79. Polson, R. C., G. Levina, and Z. V. Vardeny. 2000. Spectral analysis of polymer microring lasers. *Appl. Phys. Lett.* 76:3858–60.

80. Kitur, J. K., V. A. Podolskiy and M. A. Noginov. 2011. Stimulated emission of surface plasmon polaritions in microcylinder cavity. *Phys. Rev. Lett.* 106: 183903 (2011)

81. Zhu, G., M. Mayy, M. Bahoura, B. A. Ritzo, H. V. Gavrilenko, V. I. Gavrilenko, and M. A. Noginov. 2008. Elongation of surface plasmon polariton propagation length without gain. *Opt. Express* 16:15576–83.

82. Tredicucci, A., C. Gmachl, M. C. Wanke et al. 2000. Surface plasmon quantum cascade lasers at lambda ~ 19 μm. *Appl. Phys. Lett.* 77:2286–88.

83. Hill, M. T., M. Marell, E. S. P. Leong et al. 2009. Lasing in metal-insulator-metal subwavelength plasmonic waveguides. *Opt. Express* 17:11107–112.

84. Oulton, R. F., V. J. Sorger, T. Zentgraf et al. 2009. Plasmon lasers at deep subwavelength scale. *Nature* 461:629–32.
85. Plum, E., V. A. Fedotov, P. Kuo, D. P. Tsai, and N. I. Zheludev. 2009. Towards the lasing spaser: Controlling metamaterial optical response with semiconductor quantum dots. *Opt. Express* 17:8548–51.
86. Noginov, M. A., G. Zhu, C. Davison et al. 2005. Effect of Ag aggregate on spectroscopic properties of Eu:Y_2O_3 nanoparticles. *J. Modern Opt.* 52:2331–41.
87. Enüstün, B.V. and J. Turkevich. 1963. Coagulation of colloidal gold. *JACS: Physical and Inorganic Chemistry* 85:3317–28.
88. Noginov, M. A., G. B. Loutts, and C. E. Bonner. 2000. Crystal growth and characterization of a new laser material, $Nd:Ba_5(PO_4)_3Cl$. *JOSA B* 17:1329–34.
89. Purcell, E. M. 1946. Spontaneous emission probabilities at radio frequencies. *Phys. Rev.* 69:681.
90. Noginov, M. A., I. Fowlkes, G. Zhu, and J. Novak. 2004. Neodymium random lasers operating in different pumping regimes. *J. Modern Opt.* 51:2543–53.
91. Svelto, O. 1998. *Principles of Lasers* 4th edition. Plenum Press: New York.
92. Ramakrishna, S. A. and J. B. Pendry. 2003. Removal of absorption and increase in resolution in a near-field lens via optical gain. *Phys. Rev. B* 67:201101(R).
93. Sivan, Y., S. Xiao, U. K. Chettiar, A. V. Kildishev, and V. M. Shalaev. 2009. Frequency-domain simulations of a negative-index material with embedded gain. *Opt. Express* 17:24060–74.
94. Klar, T. A., A. V. Kildishev, V. P. Drachev, and V. M. Shalaev. 2006. Negative-index materials: Going optical. *IEEE J. Sel. Top. Quantum Electron.* 12:1106–15.
95. Fang, A., T. Koschny, M. Wegener, and C. M. Soukoulis. 2009. Self-consistent calculation of metamaterials with gain. *Phys. Rev. B* 79:241104(R).
96. Xiao S., V. P. Drachev, A. V. Kildishev, X. Ni, U. K. Chettiar, H.-K. Yuan, and V. M. Shalaev. 2010. Loss-free and active optical negative-index metamaterials. *Nature* 466: 735–38.

6

Anisotropic and Hyperbolic Metamaterials

Viktor A. Podolskiy

*Physics Department,
Oregon State University,
Corvallis, Oregon and
Department of Physics
and Applied Physics,
University of
Massachusetts Lowell
Lowell, Massachusetts*

6.1 Introduction

For years, optics has been developed with the implicit assumption that underlying materials relatively weakly interact with optical radiation. Thus, optical media have been implicitly assumed to be nonmagnetic. Optical magnetism [1], experimentally realized [2–4] during the last decade, is fueling revolutionary growth in the area of metamaterials, nanostructured composites with tailored optical properties.

On the other hand, while fabrication of magnetic media at GHz frequencies is relatively straightforward [5–7,], fabrication of magnetic systems in the visible part of the spectrum typically involves three-dimensional patterning with 10 nm resolution [2–4, 8, 9], and is therefore extremely challenging and expensive.

However, as we will show in this chapter, most of the benefits offered by magnetic media can be mimicked in nonmagnetic structures under some restrictions. Moreover, the benefits offered by nonmagnetic media may surpass the over of conventional, magnetic, materials. In these nonmagnetic systems, an alternative (to magnetism) mechanism is required to provide additional (with respect to isotropic permittivity) control of optical radiation. Such an additional control can be achieved with either chirality (see Chapter 6) or anisotropy. Here, we focus on anisotropy. Thus, for the remainder of this chapter, we assume that the relative magnetic permeability is equal to one ($\mu = 1$), while the relative dielectric permittivity is described by a tensor $\hat{\epsilon}$.

The rest of this chapter is organized as follows. Section 6.2 presents several examples of metamaterials with extreme optical anisotropy (where components of $\hat{\epsilon}$ have different signs). As will be seen, these metamaterials can be successfully fabricated with existing technology. In Section 6.3, we consider propagation of light in anisotropic systems, as well as the reflection, refraction, and diffraction in systems involving anisotropic components. Section 6.4 describes the applications of anisotropic media for confined propagation of light, including subdiffraction light propagation and compression, negative refraction of waveguide modes, and mode profile engineering in waveguides. Section 6.5 describes deviation of optical properties of anisotropic composites from effective-medium predictions due to nonlocalities.

6.2 Effective-Medium Response of Anisotropic Metamaterials

In contrast to conventional homogenous media where dielectric response is primarily due to polarization of atoms [10,11], the optical properties of metamaterials are defined by the interaction of light with nano-structured components of these composites (meta-atoms). In a sense, metamaterial design introduces an additional length scale into consideration: the scale of the metamaterial elements. As result, dielectric permittivity describing the optical properties of the system depends on the particular length scale.

On the microscopic scale, electric and magnetic fields are strongly inhomogeneous and have to be determined by solving for the interaction of electromagnetic fields with each individual atom. Obviously, such a level of detail is impractical in most applications and can be rarely achieved in realistic systems.

Macroscopic electric field ($\vec{E}$) and polarization ($\vec{P}$) can be introduced as result of averaging of the microscopic electric field and the local dipole moment over the groups of atoms (inside particular component of metamaterial composite). Thus, the defined averaged fields are relatively homogeneous inside the components of metamateiral, but are still highly inhomogeneous on the scale of metamaterial as a whole (meta-scale). It is often the case that the size of the inclusion in metamaterial is much smaller than the wavelength (some examples of such metamaterials are presented later). In this case, the fields can be averaged again, over a scale that includes several inclusions but is still smaller than the wavelength. This time, the averaged fields are relatively smooth, and the ratio between average polarization and the average electric field can be used to introduce the effective (metamaterial) polarizability, which in turn can be used to calculate effective permittivity of the composite. The same effective permittivity results from the ratio of average displacement field $<\vec{D}>$ and average electric field $<\vec{E}>$. Note that in general the effective permittivity is a tensor quantity:

$$<D>_{\alpha}=\epsilon_{\alpha\beta}<E>_{\beta} \qquad (6.1)$$

where the brackets ($< >$) denote the averaging over the microscopically large (multi-inclusion), macroscopically small (subwavelength) spatial area; Greek indices denote (Cartesian) components; and summation over repeated indices is assumed. The relationship between the relative scale of the inclusion and the scale of field variation (internal wavelength λ_{in}) is of principal interest for a meaningful definition of the tensor of effective permittivity. When the size of an inclusion a is small in comparison with λ_{in}, the averaging in Equation 6.1 is quasi-static, and the values of effective permittivity are robust with respect to variations of sizes or positions of individual inclusions. In this *effective-medium regime*, the value of $\epsilon_{\alpha\beta}$ is completely determined by the averaged characteristics of the metamaterial (average inclusion shape, average inclusion concentration, etc.) [11].

In the limit $a \sim \lambda_{in}$, the optical properties of composites are affected by interference between the electromagnetic waves scattered by individual inclusions. Optical properties of composites with periodically arranged inclusions (photonic crystals [12]) are in fact dominated by these interference effects.

We now present two metamaterial designs of the strongly anisotropic composite for optical and infrared spectrum ranges that are successfully described by the EMTs.

6.2.1 Nanolayered Structures

We first consider the permittivity of a multilayer metamaterial, schematically shown in Figure 6.1. We assume that the layers are aligned parallel to the (y, z) plane. We further assume that the stack is composed from the (on average) periodically repeated layers, each layer having a set of components with permittivity ϵ_j and thickness d_j.

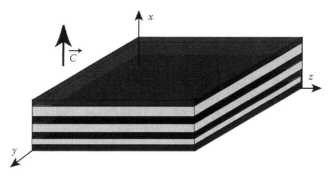

FIGURE 6.1 The schematics of the layered structure described in the text; $\vec{C}$ represents the direction of the optical axis.

In general, wave propagation in the layered materials strongly depends on polarization, the ratio of the typical layer thickness d to the wavelength λ_0, and the microgeometry of the system, and may become extremely complicated due to excitation of coupled surface plasmon polariton modes, or 1D photonic crystal-related effects (see e.g., References [12–17] and references therein for a more comprehensive analysis of the wave dynamics and underlying mode structure in layered composites).

Here, we focus on the case of ultrathin layers ($d \ll \lambda_0$), where some of the propagating modes have plane-wave-like structure, and can be successfully described by quasi-static effective permittivity derived via effective medium approximation.

To compute the effective permittivity* ϵ, we note that the E_y, E_z, and D_x have to be continuous throughout the system [11]. Equation 6.1 yields the uniaxial anisotropic effective permittivity with diagonal components of the permittivity tensor being

$$\epsilon_{yy} = \epsilon_{zz} = \epsilon_{yz} = \frac{\sum_j d_j \epsilon_j}{\sum_j d_j}$$

$$\epsilon_{xx} = \frac{\sum_j d_j}{\sum_j d_j/\epsilon_j}.$$

(6.2)

For two-component multilayers, Equation 6.2 reduce to

$$\epsilon_{yz} = \frac{d_1\epsilon_1 + d_2\epsilon_2}{d_1 + d_2}$$

$$\epsilon_{xx} = \frac{\epsilon_1\epsilon_2(d_1+d_2)}{d_1\epsilon_2 + d_2\epsilon_1}.$$

(6.3)

Not surprisingly, the same equations can be obtained as a quasi-static limit ($|k_{\{x,y,z\}}d| \ll 1$) of 1D photonic crystal formed by a periodic layered system [15]. Note that in this limit, the properties of the layered composites are controlled not by individual sizes of inclusions (d_1, d_2) but by their relative concentration $N = d_1/(d_1 + d_2)$.

Layered structures can be fabricated using subsequent deposition of different materials. Here, we are primarily interested in materials with extreme anisotropy of the effective permittivity tensor ($\epsilon_{xx} \cdot \epsilon_{yz} < 0$). To achieve such a strong anisotropy, it becomes necessary to combine layers of materials with negative permittivity ($\epsilon_1 < 0$) with layers of materials with positive permittivity ($\epsilon_2 > 0$). Negative permittivity can be achieved in plasmonic structures (noble metals, doped semiconductors) or in polar crystals (SiC) [18]; positive permittivity is routinely obtained in nonresonant dielectrics or in undoped semiconductors.

The effective permittivities of several two-component layered composites are shown in Figure 6.2. As seen, strong anisotropy can be achieved throughout the UV, visible, and IR frequency ranges. In the UV and visible frequencies, strong anisotropy requires the use of

---

* From this point onward, we assume that the tensor of dielectric permittivity is diagonal. The notation ϵ_{yz} refers to the values of ϵ_{yy} and ϵ_{yz} components of this tensor as defined by Eq (6.2). All off-diagonal components of the permittivity tensor are assumed to be zero.

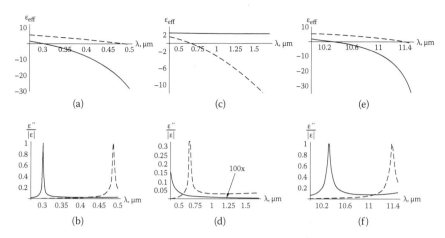

FIGURE 6.2 The real part (a,c,e) and absorption (b,d,f) of effective ϵ_{xx} (solid lines) and ϵ_{yz} (dashed lines) for layered systems; (a,b): Ag–Si stack; metal concentration $N = 0.6$; (c,d): Ag–SiO$_2$ stack; $N = 0.1$; Note the small absorption of this system; (e,f) SiC–Si stack; SiC concentration $N = 0.1$. (From R. Wangberg et al., *J. Opt. Soc. Am B* 23, 498, 2006.)

noble metals, and the tuning of metamaterial parameters is primarily achieved by controlling the concentration of plasmonic components in the overall volume of the metamaterial.

At lower frequencies, extreme anisotropy can be achieved in all-semiconductor systems (such as AlInAs-InGaAs metamaterials [20]). In these systems, the doping concentration becomes an additional (with respect to layer thickness) control parameter. Current technology allows manufacture of anisotropic nanolayered semiconductor composites with operating frequencies throughout the mid-IR and long-wave IC frequency ranges.

6.2.2 Nanowire Metamaterials

Another class of metamaterials with extreme anisotropy is based on arrays of aligned plasmonic nanowires deposited in dielectric substrate (Figure 6.3a). These composites can be successfully fabricated with electrochemistry [21,22]. The majority of nanowire composites are based on anodized alumina substrates, where the preetched holes are filled with noble metals (typically Au). The radius of individual gold nanowire r can be as small as 10 nm, and the separation between the wires d is on the order of 50 nm. The length of the wire (and thus the thickness of the composite) can range from ~100 nm to ~50 μm [21,22]. In the postprocessing stages, it is possible to remove alumina host matrix (either partially or completely) and fill the free-standing wire composite with a different dielectric core, adding tunability to the metamaterial response.

Similar to nanolayered composites, the cross section of nanowires and the interwire separation are much smaller than the operating wavelength $\lambda_0 \sim 1$ μm. Moreover, the surface concentration of plasmonic media in the composite $N = \pi r^2/d^2$ is usually small: $N \ll 1$. Under these assumptions, the optical properties of nanowire materials are adequately described by Maxwell-Garnett (MG) effective medium theory [23]. Practical limitations for MG applicability are discussed later.

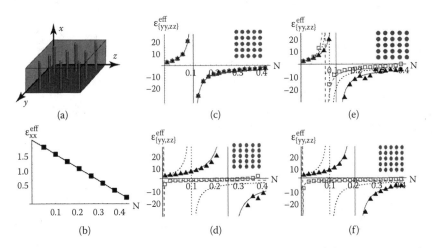

FIGURE 6.3 (a) Schematic geometry of a nanowire composite. (b) ϵ_{xx}^{eff} for Ag nanowires $\epsilon^{in} = -2.5$ in a polymer $\epsilon^{out} = 2$ for $\lambda_0 \approx 360nm$. (c–e) ϵ_{yy}^{eff} (solid triangles and lines) and ϵ_{zz}^{eff} (empty rectangles, dashed lines) for the composite in (a), with $\Lambda_y = \Omega_y = 0$ (c), $\Lambda_y = 0.2$; $\Omega_y = -0.2$ (d), $\Lambda_y = 0.2$; $\Omega_y = 0$ (e), $\Lambda_y = 0$; $\Omega_y = -0.2$ (f); quasi-static numerical calculations (symbols); and GMG (lines); dotted lines in (d–f) are identical to lines in (c). Insets show cross sections of composites for $N = 0.35$; the breakdown of GMG occurs when the local field becomes inhomogeneous on the scale of r_α. (From J. Elser et al., *Appl. Phys. Lett.* 89, 261102, 2006.)

Note that stretching or compression of the nanorod composite along the y or z axes will introduce anisotropy in the yz plane of the metamaterial. Depending on the composition of the material and on the applied stress, the stretching can lead to the deformation of cross sections of nanowires (making these elliptical, with semiradii r_y, r_z, rather than circular) or to dislocation of nanowires with respect to each other (effectively stretching or compressing the "lattice" of the metamaterial). When the deformation is relatively small, the Maxwell-Garnett approach can be extended to describe the response of fully bi-anisotropic nanowire metamaterial [24]. Here, we present the derivation of the generalized Maxwell-Garnett formalism.

In the quasi-static limit, the boundary conditions of Maxwell equations require the E_x component to be constant across the unit cell. Therefore, ϵ_{xx} is still given by

$$\epsilon_{xx} = N\epsilon^{in} + (1-N)\epsilon^{out} \tag{6.4}$$

and the in-plane (yz) components of the permittivity are given by Equation 6.1:

$$\epsilon_{\alpha\alpha}^{eff} = \frac{N\epsilon^{in}E_\alpha^{in} + (1-N)\epsilon^{out}E_\alpha^{out}}{NE_\alpha^{in} + (1-N)E_\alpha^{out}}, \tag{6.5}$$

where ϵ^{in} and ϵ^{out} are the permittivities of nanowires and of the materials and substrate E^{in} and E^{out} are the average (across the typical nanowire cell) fields inside and in-between the nanowires. Note that these average fields need to be distinguished from the excitation field E^0 acting on the system.

The total field across the nanowire cell is composed of the external field E^0, the field scattered by the nanowire E^{sc}, and the field scattered by the nanowires in the neighboring cells $\hat{\chi}E^{sc}$. If nanowires are isotropically distributed across the composite, the latter component is zero [10,23]. In the case of anisotropic distribution of nanowires, the feedback field depends on the microarrangement of nanowires in the composite. If nanowires are positioned at the vertices of the rectangular lattice with lattice constants l_y, l_z, the feedback tensor $\chi_{\alpha\beta}$ becomes diagonal, and the field acting on a nanowire can be estimated by $E^0 + \sum_j \hat{\chi}^j E^0 = [1 - \chi_{\alpha\alpha}]^{-1} E^0_\alpha$.

To calculate an explicit expression for the feedback tensor in nanowire composites with nanowires distributed over the rectangular grid, we introduce a two-dimensional lattice distortion vector $\{\Lambda_y, \Lambda_z\} = \{l_y/l_z - 1, l_z/l_y - 1\}$, and a dimensionless function

$$S(\xi) = \sum_{i,j\neq 0} \frac{i^2}{i^2 + \xi j^2},$$ (6.6)

where the summation is extended over all sites of rectangular lattice except the origin. The total feedback field tensor can then be expressed as

$$\chi_{\alpha\alpha} = \frac{(\epsilon_{in} - \epsilon_{out})r_x r_y P_\alpha}{4 l_x l_y} \left[(\Lambda_\alpha + 1) S(\Lambda_\alpha + 1) - \frac{1}{\Lambda_\alpha + 1} S\left(\frac{1}{\Lambda_\alpha + 1}\right) \right]$$ (6.7)

$$\simeq -0.16 N \Lambda_\alpha P_\alpha (\epsilon^{in} - \epsilon^{out}),$$

$$P_\alpha = \frac{1}{\epsilon^{out} + n_\alpha(\epsilon^{in} - \epsilon^{out})},$$ (6.8)

with $\{n_y, n_z\} = \{r_z/(r_y+r_z), r_y/(r_y+r_z)\}$ being the depolarization factors [11,25].

In the limit $N \ll 1$ considered here, the total field acting on the individual nanowire can be considered homogeneous. Therefore, the field inside the nanowire will also be homogeneous and will be given by

$$E^{in}_\alpha = \frac{\epsilon^{out} P_\alpha}{1 - \chi_{\alpha\alpha}} E^0_\alpha.$$ (6.9)

Each nanowire will add a dipolar contribution to the total field in the composite. Direct calculation of the field across the typical metamaterial cell yields

$$E^{out}_\alpha \simeq \left[1 + \frac{N P_\alpha (\epsilon^{in} - \epsilon^{out})(Q(N) \cdot (\Omega_\alpha + \Lambda_\alpha) - \pi\Omega_\alpha)}{2\pi(1-N)(1-\chi_{\alpha\alpha})} \right] E^0_\alpha,$$ (6.10)

with $Q(N) = \pi - 1 - N(\pi - 2)$ and shape vector $\{\Omega_y, \Omega_z\} = \{r_y/r_z - 1, r_z/r_y - 1\}$.

Naturally, in the absence of lattice- or nanowire-shape anisotropy, the presented Generalized Maxwell-Garnett formalism (GMG) [24] yields the expected expression [23]

$$\epsilon_{yz} = \frac{2N\epsilon_{in}\epsilon_{out} + (1-N)\epsilon_{out}(\epsilon_{out}+\epsilon_{in})}{2N\epsilon_{out} + (1-N)(\epsilon_{out}+\epsilon_{in})}$$

(6.11)

$$\epsilon_{xx} = N\epsilon_{in} + (1-N)\epsilon_{out}.$$

To test the GMG, we numerically generated a set of nanowire composites with different lattice and shape parameters, solved for microscopic field distribution in these composites with a commercial finite-element PDE solver [26], and used the microscopic field distribution to extract components of effective permittivity tensors. The agreement between analytical and numerical results is shown in Figure 6.3. It is seen that GMG adequately describes the properties of nanowire composites in the limit of small-to-moderate concentrations and deformations $(N, \Lambda, \Omega \gtrsim 0.3)$ [24] .

Our numerical simulations also demonstrate that (as predicted by GMG) the properties of nanowire metamaterials are governed by averaged parameters (concentration, average lattice constants, and averaged radii); the randomization of positions/radii of individual nanowires results in a relatively minor modification of metamaterial properties provided that the randomization does not affect the averaged parameters of metamaterials.

6.3 Optical Response of Bulk Anisotropic Metamaterials

6.3.1 Dispersion Relations

We first consider the properties of the electromangnetic waves supported by anisotropic composites. Assuming the time- and space-harmonic, plane wave propagation with

$$\vec{E} = \vec{E}^0 \exp(-i\omega t + i\vec{k}\cdot\vec{r})$$

$$\vec{H} = \vec{H}^0 \exp(-i\omega t + i\vec{k}\cdot\vec{r}),$$

(6.12)

Maxwell equations in an anisotropic nonmagnetic medium are reduced to

$$\begin{cases} \vec{k}\times\vec{H} = -\frac{\omega}{c}\hat{\epsilon}\vec{E} = -\frac{\omega}{c}\vec{D} \\ \vec{k}\times\vec{E} = \frac{\omega}{c}\vec{H}. \end{cases}$$

(6.13)

Note that $\vec{D}\perp\vec{k}, \vec{D}\perp\vec{H}, \vec{H}\perp\vec{E}, \vec{H}\perp\vec{k}$. The vectors $\vec{D}$ and $\vec{E}$ are not necessarily co-aligned in anisotropic media, so that the vectors $\vec{E}$ and $\vec{k}$ are not necessarily orthogonal to each other (see Figure 6.4).

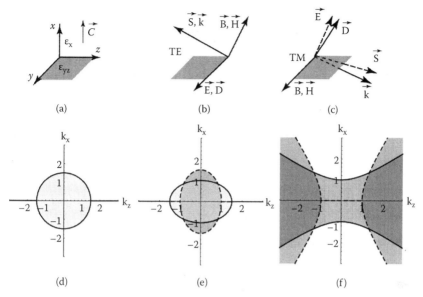

FIGURE 6.4 (a) Tensor of dielectric permittivity of uniaxial material ($\vec{C}$ denotes optical axis) and relationship between the fields in (b) ordinary (TE-polarized) and (c) extraordinary (TM-polarized) plane waves propagating in this material. Panels (d–f) illustrate the dispersion in isotropic and in anisotropic media (each point of the line corresponds to allowed $\{k_x, k_z\}$ combination; $k_y = 0$ is assumed): (d) dispersion in isotropic materials and dispersion of TE wave in anisotropic material ($\epsilon_{yz} = 2.25$), (e) dispersion of TM wave in anisotropic material with $\epsilon_{xx} = 2.25$, $\epsilon_{yz} = 1$ (solid line) and with $\epsilon_{xx} = 1$, $\epsilon_{yz} = 2.25$ (dashed line), (f) dispersion of TM wave in hyperbolic material with $\epsilon_{xx} = -2.25$, $\epsilon_{yz} = 1$ (solid line) and in material with $\epsilon_{xx} = 1$, $\epsilon_{yz} = -2.25$ (dashed line); all components of $\vec{k}$ are normalized to w/c.

Substituting the magnetic field from the second of Equation 6.13 to the first one, we see that any plane wave that satisfies Maxwell equations in anisotropic material has to satisfy the relationship

$$k^2 \vec{E}^0 - \vec{k}(\vec{k} \cdot \vec{E}^0) - \hat{\epsilon} \frac{\omega^2}{c^2} \vec{E}^0 = 0, \tag{6.14}$$

which in component form becomes

$$\left(\delta_{\alpha\beta} k^2 - k_\alpha k_\beta - \epsilon_{\alpha\beta} \frac{\omega^2}{c^2} \right) E_\alpha^0 = 0. \tag{6.15}$$

As before, here the Greek indices stand for Cartesian components, summation over the repeated indices is assumed, $\delta_{\alpha\beta}$ is Kroenecker delta symbol, and $k^2 = \vec{k} \cdot \vec{k}$.

The requirement for the Equation 6.15 to have nonzero amplitude of electric field results in the dispersion relation for the wave propagating in anisotropic material:

$$\det\left|\delta_{\alpha\beta}k^2 - k_\alpha k_\beta - \epsilon_{\alpha\beta}\frac{\omega^2}{c^2}\right| = 0.\tag{6.16}$$

For uniaxial materials with

$$\hat{\epsilon} = \begin{pmatrix} \epsilon_{xx} & 0 & 0 \\ 0 & \epsilon_{yz} & 0 \\ 0 & 0 & \epsilon_{yz} \end{pmatrix},\tag{6.17}$$

Equations 6.15 and 6.16 yield two solutions (modes). The propagation and structure of the first mode, also known as *transverse-electric* (TE) wave, or *ordinary wave*, is not affected by material anisotropy:

$$\begin{cases} k^2 = \epsilon_{yz}\frac{\omega^2}{c^2} \\ \vec{E}^0 = A\left\{0, \frac{k_z}{\sqrt{k_y^2+k_z^2}}, -\frac{k_y}{\sqrt{k_y^2+k_z^2}}\right\} \\ \vec{H}^0 = A\frac{c}{\omega}\sqrt{k_y^2+k_z^2}\left\{-1, \frac{k_x k_y}{k_y^2+k_z^2}, \frac{k_x k_z}{k_y^2+k_z^2}\right\}, \end{cases}\tag{6.18}$$

with A being a constant describing the amplitude of the electromagnetic field, and c being the speed of light in vacuum. Note that the electric field of the TE-polarized wave lies in the yz plane, and that dispersion of this mode is affected only by ϵ_{yz}. As seen from Equation 6.18, the properties of this wave are completely identical to the properties of a plane wave propagating in isotropic homogeneous dielectric with $\epsilon = \epsilon_{yz}$ [10]. Most notably, in an ordinary wave, $\vec{D}\|\vec{E}$, so that the direction of the wavevector $\vec{k}$ coincides with the direction of the Poynting vector, $\vec{S} = c\vec{E}\times\vec{H}/4\pi$. Furthermore, the components of the wavevector of the ordinary wave lie on a sphere of the radius $\sqrt{\epsilon_{yz}}\omega/c$.

The latter property can be easily related to the appearance of the diffraction limit, both in isotropic dielectrics, and in TE-polarized imaging in anisotropic materials. Indeed, the field scattered by the object can be represented as a set of plane waves propagating away from this object. Assuming that the propagation takes place in the $+z$ direction and that the object is located at $z = 0$, we can represent the field at any $z > 0$ location as

$$\vec{E}(\vec{r}) = \int_{-\infty}^{\infty} \vec{A}(k_x, k_y)exp(i\vec{k}\cdot\vec{r})dk_x\,dk_y,\tag{6.19}$$

where $\vec{A}(k_x, k_y)$ is given by the Fourier transform of the original scattered field $\vec{E}(x, y, z = 0)$. As is known from Fourier transform theory [10,27], the features of the size Δ of the object are transformed into $k_x^2 + k_y^2 \sim 4\pi^2/\Delta^2$. Hence, the information about subwavelength features of the object is encoded into plane waves with high-transverse wavenumber components ($|k_x, k_y| > \sqrt{\epsilon_{yz}}\,\omega/c$). As is evident from the dispersion relation

(Equations 6.12 and 6.18), these waves correspond to imaginary values of k_z, and thus they exponentially decay away from the scattering object. Since no far-field instrument can reliably restore the exponentially suppressed information[10,28,29], the subdiffraction features are lost in the far-field imaging process.

The situation is different for the second type of plane wave supported by anisotropic media, often called *transverse-magnetic* (TM) waves, or *extraordinary waves*, whose properties are given by[*1]

$$
\begin{cases}
\dfrac{k_x^2}{\epsilon_{yz}} + \dfrac{k_y^2 + k_z^2}{\epsilon_{xx}} = \dfrac{\omega^2}{c^2} \\[2mm]
\vec{E}^0 = A \left\{ 1, -\dfrac{k_x k_y}{k_y^2 + k_z^2} \dfrac{\epsilon_{xx}}{\epsilon_{yz}}, -\dfrac{k_x k_z}{k_y^2 + k_z^2} \dfrac{\epsilon_{xx}}{\epsilon_{yz}} \right\} \\[2mm]
\vec{H}^0 = A \dfrac{\omega}{c} \epsilon_{xx} \left\{ 0, \dfrac{k_z}{k_y^2 + k_z^2}, -\dfrac{k_y}{k_y^2 + k_z^2} \right\}.
\end{cases}
\tag{6.20}
$$

The dispersion of the extraordinary plane waves is either elliptical (if both $\epsilon_{xx} > 0$ and $\epsilon_{yz} > 0$) or hyperbolic (if ϵ_{xx} and ϵ_{yz} have different signs); the materials where all components of the permittivity tensor are negative do not support propagating modes. The possibility of hyperbolic dispersion opens new ways of managing light in metamaterials, and enables such unique applications as negative refraction, subwavelength light confinement, and subwavelength imaging, described in detail in the following sections.

As mentioned earlier, any subwavelength optics has to utilize high-transverse wavevector waves. In contrast to TE waves and to isotropic media, high-wavenumber TM-polarized modes propagate in materials with hyperbolic dispersion. This simple fact, seen in Equation 6.20 and illustrated in Figure 6.4, underlies the majority of fascinating applications of hyperbolic metamaterials. In some sense, a hyperbolic medium can postpone the onset of diffraction limit due to an extremely large refractive index of a particular wave (see later text).

In the rest of this chapter, we primarily focus on the propagation of TM-polarized light in anisotropic metamaterials, although our numerical simulations do include TE-polarized modes to satisfy the boundary conditions, as mentioned later.

6.3.2 Reflection and Refraction in Bulk Metamaterials

We now consider the wave dynamics at the interface between an isotropic medium (we use vacuum as the example) and an anisotropic uniaxial material. For simplicity, we consider the case when the interface lies in the yz plane, so that the optical axis of the anisotropic medium is perpendicular to the interface; the case when the interface lies in the xy plane so that the optical axis is parallel to the interface is considered later (Section 6.4.1)

[*1] Note that the case $k_y = k_z = 0$ requires special treatment. In this particular case, both waves behave as ordinary modes, and both electric and magnetic fields lie in the yz plane, with $H^0 = c\, \vec{K} \times E^0 / w$.

on example of refraction in planar waveguides. The problem of light refraction when the optical axis is oriented obliquely to the interface can be solved in a similar fashion [11,35].

The geometry of the problem is schematically shown in Figure 6.5. The interface between the vacuum (medium 1; $x < 0$) and anisotropic material (medium 2; $x > 0$) lies in the yz plane. In the remainder of this section, we will denote the components of the fields in medium 1 with superscript 1, and components of the fields in medium 2 with superscript 2; furthermore, we will use the $+$ or $-$ sign to define the direction of the energy transferred by the waves (upward or downward, respectively).

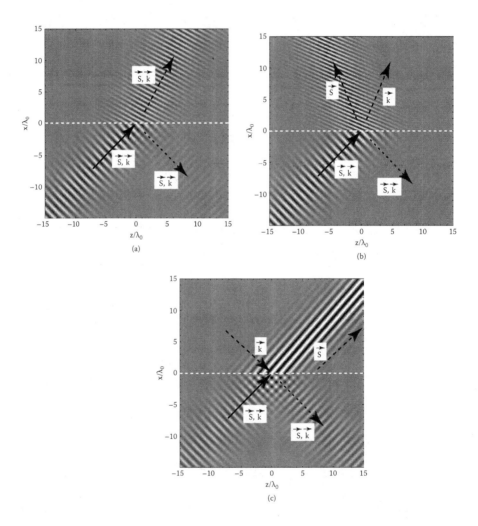

FIGURE 6.5 Reflection and refraction with isotropic (a) and hyperbolic (b,c) media (E_z field shown); The half-space $x < 0$ is vacuum; the half-space $x > 0$ is filled with (a) isotropic material, $\epsilon = 2.25$; (b) anisotropic material, $\epsilon_{xx} = -2.25$, $\epsilon_{yz} = 2.25$; (c) anisotropic material, $\epsilon_{xx} = 0.1$, $\epsilon_{yz} = -0.1$; angle of incidence is 45°; arrows show the directions of the Poynting flux $\vec{S}$ and wavevector $\vec{k}$.

Note that direction of the energy, given by the Poynting vector, can be different from the direction of the wavevector. In particular, in our case, the components of the Poynting vector of the TM wave are given by

$$\vec{S} = \frac{c^2}{4\pi\omega}(\vec{E}\cdot\vec{D})\left\{\frac{k_x}{\epsilon_{yz}}, \frac{k_y}{\epsilon_{xx}}, \frac{k_z}{\epsilon_{xx}}\right\}. \tag{6.21}$$

By definition, $S_x^{(1,+)} > 0$ and $S_x^{(2,+)} > 0$.

In this particular situation, it is convenient to choose the orientation of y and z axes in such a way that the wavevector of the incident wave lies in the xz plane. It is clear [10] that the wavevectors of reflected and refracted waves will also lie in the same plane.

In the selected coordinate frame, TE waves have nonzero H_x, H_z, and E_y components, while TM waves have nonzero H_y, E_x, and E_z fields. Hence, the two polarizations are completely orthogonal to each other, and can be considered separately. Here, we focus on TM waves.

By design, the k_y of all waves is equal to 0. Moreover, boundary conditions require all three waves to have the same k_z component,

$$k_z^{(1,-)} = k_z^{(1,+)} = k_z^{(2,-)} = k_z^{(2,+)}. \tag{6.22}$$

Since $k_z = nw/c \sin\theta$, the latter requirement can be used to derive Snell's law in isotropic optics.

Note that in anisotropic composites, it is possible to introduce two different refractive indices. The first of these, phase index n_p (given by Equation 6.22), relates the direction of the wavevector of the incident wave to the direction of the wavevector of the refracted wave. The second one, group index n_g, describes the relationship between the directions of the power fluxes, given by the Poynting vectors. In experiments, n_p governs the phase advance/delay of the wavepacket, while n_g determines the direction and amplitude of the beamshift. Both these depend on the direction of the light propagation, which can be related to the angle of incidence θ_i. Assuming that material (1) is vacuum, the phase and group indices of anisotropic medium are given by

$$n_p = \sqrt{\epsilon_{yz}^{(2)} + \left(1 - \frac{\epsilon_{yz}^{(2)}}{\epsilon_{xx}^{(2)}}\right)\sin^2\theta_i}, \tag{6.23}$$

$$n_g = \frac{\epsilon_{xx}^{(2)}}{\epsilon_{yz}^{(2)}}\sqrt{\epsilon_{yz}^{(2)} - \frac{\epsilon_{yz}^{(2)}}{\epsilon_{xx}^{(2)}}\left(1 - \frac{\epsilon_{yz}^{(2)}}{\epsilon_{xx}^{(2)}}\right)\sin^2\theta_i}. \tag{6.24}$$

Note that when $\epsilon_{xx}^{(2)} < 0, \epsilon_{yz}^{(2)} > 0$, the phase index is positive, while the group index is negative. Thus, the beam of light incident on the planar slab of anisotropic metamaterial will undergo negative refraction (Figure 6.5b), which will result in the negative lateral shift of the beam, experimentally measured in References [9,20]. Furthermore, the slab of uniaxial media may behave as a planar lens [1,30–32], although the resolution of such a lens will be limited by the diffraction limit of the surrounding material, and may also be affected by aberrations [30].

The opposite situation, when $\epsilon_{xx}^{(2)} > 0, \epsilon_{yz}^{(2)} < 0$, leads to negative phase velocity, but positive refraction of the beam. The light propagating in these materials will experience a negative phase shift. Note, however, that the angle of incidence should be sufficiently large to excite the propagating wave inside these hyperbolic media; for small angles of incidence, light exhibits total reflection from materials with $\epsilon_{yz} < 0$.

Instead of considering the cases of light incidence from vacuum and from anisotropic material separately, we will combine these two cases together and present the results of transfer and scattering matrix formalisms, powerful techniques that are often used in modal-expansion methods. The transfer matrix relates the amplitudes of the incoming and reflected fields in one layer to the amplitudes of the incoming and reflected fields in the second layer. The scattering matrix relates the amplitudes of two incoming waves to the amplitudes of two reflected (scattered) waves [33–35].

To derive expressions for components' transfer matrix, we start from boundary conditions. Using Equation 6.20, the continuity of D_x and E_z components reduces to[*2]

$$\begin{cases} \epsilon_{xx}^{(1)}[A^{(1,+)} + A^{(1,-)}] = \epsilon_{xx}^{(2)}[A^{(2,+)} + A^{(2,-)}] \\ k_x^{(1,+)}[A^{(1,+)} - A^{(1,-)}] = k_x^{(2,+)}[A^{(2,+)} - A^{(2,-)}], \end{cases} \tag{6.25}$$

leading to

$$\begin{pmatrix} A^{(2,-)} \\ A^{(2,+)} \end{pmatrix} = \hat{T} \begin{pmatrix} A^{(1,-)} \\ A^{(1,+)} \end{pmatrix}, \tag{6.26}$$

$$\hat{T} = \frac{1}{2} \frac{\epsilon_{xx}^{(1)}}{\epsilon_{xx}^{(2)}} \begin{pmatrix} 1 + \frac{k_x^{(1,+)} \epsilon_{yz}^{(2)}}{k_x^{(2,+)} \epsilon_{yz}^{(1)}} & 1 - \frac{k_x^{(1,+)} \epsilon_{yz}^{(2)}}{k_x^{(2,+)} \epsilon_{yz}^{(1)}} \\ 1 - \frac{k_x^{(1,+)} \epsilon_{yz}^{(2)}}{k_x^{(2,+)} \epsilon_{yz}^{(1)}} & 1 + \frac{k_x^{(1,+)} \epsilon_{yz}^{(2)}}{k_x^{(2,+)} \epsilon_{yz}^{(1)}} \end{pmatrix}. \tag{6.27}$$

The elements of the scattering matrix of the system can be expressed in terms of the elements of the transfer matrix:

$$\begin{pmatrix} A^{(1,-)} \\ A^{(2,+)} \end{pmatrix} = \hat{S} \begin{pmatrix} A^{(2,-)} \\ A^{(1,+)} \end{pmatrix}, \tag{6.28}$$

$$\hat{S} = \frac{1}{T_{11}} \begin{pmatrix} 1 & -T_{12} \\ T_{21} & T_{11}T_{22} - T_{12}T_{21} \end{pmatrix}. \tag{6.29}$$

[*2]Here, we present the expressions for TM-polarized light. Transfer matrix for TE-polarized waves can be derived in a similar way [33,34].

In multilayered composites, the transfer matrices can be calculated for each interface (in this case, the transfer matrix must incorporate the position of the interface) [10,36]. The relationship between the fields in any two layers of the multilayer stack is then given by the product of the transfer matrices for the interfaces separating the two layers. The scattering matrix for the system can still be expressed in terms of the combined transfer matrix.

When designing transfer matrices, it is important to remember that when the optical axis is not perpendicular to the interface, the two polarizations may get coupled, so that the incident TM beam may excite the refracted TE wave, etc. In these cases, both the *T* and *S* matrices are 4 × 4 matrices.

6.3.3 Hypergratings: Diffraction and Subwavelength Focusing with Hyperbolic Metamaterials

The possibility of focusing light on subwavelength areas is among the most interesting applications of metamaterials. Subwavelength focusing and imaging are of the great importance for nanofabrication, nanosensing, and nanoimaging. The separation between the object and a focal spot (focal distance) and the size of the focal spot are the main figures of merit for focusing and imaging devices.

Several techniques have been suggested to achieve subwavelength imaging. Some of these techniques—scanning near-field optical microscopy (SNOM) [37], superlens [28,38–40], and near-field plates [41–43] rely on exponentially decaying (evanescent) fields with $|k_z| > \omega/c$ to surpass the diffraction limit. Unfortunately, the realistic applications of these techniques are limited to near-field proximity of the imaging system [29].

Another class of structures either uses transformation-optics techniques (light compressors) [44–48], or make use of ultra-high-index modes in plasmonic [49,50] or hyperbolic [51–54] media. While these systems are able to achieve subwavelength light manipulation in the far-field, their fabrication requires three-dimensional patterning. Moreover, the devices themselves are often nonplanar, which further restricts their integration into complex photonic units.

Here, we discuss a planar system capable of far-field subwavelength light manipulation. The system comprises a (planar) slab of strongly anisotropic metamaterial covered with a diffraction grating. As explained earlier, the diffraction grating is responsible for generating the high-wavevector components of electromagnetic fields, and the slab is used for the routing of the resulting subwavelength signals.

We begin by comparing the diffraction through a thin slit (size *d*) in isotropic and hyperbolic materials (Figure 6.6). We assume for simplicity that the slit is parallel to the *y* coordinate axis and the external field is normally incident at the slit from the $x < 0$ region. To calculate the field behind the slit, we note that the magnetic field just behind the slit is given by

$$H_y(x=0,z) = \begin{cases} 1, & |z| \le d/2 \\ 0, & |z| > d/2 \end{cases} = -\frac{1}{\pi} \int_{-\infty}^{\infty} \frac{\sin(k_z d/2)}{k_z} e^{ik_z z} dk_z. \qquad (6.30)$$

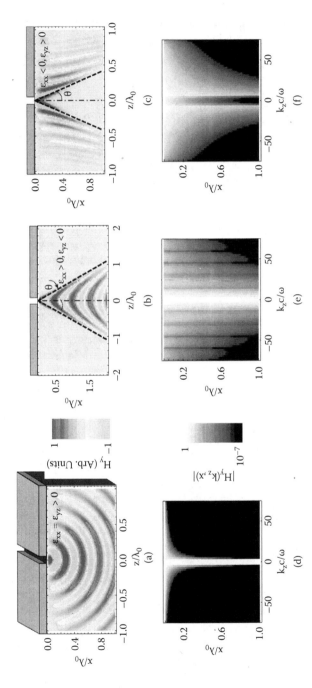

FIGURE 6.6 (a) Diffraction-limited propagation in Si; $\lambda_0 = 15\ \mu m$; $d = 200\ nm$; $\epsilon_{xx} = \epsilon_{yz} = 16$; (b,c) subdiffraction propagation in nanowire (b) and nanolayer (c) anisotropic metamaterials; $d = 100\ nm$ (b) and $d = 200\ nm$ (c); dashed lines correspond to critical cones, reflecting the direction of $|k_x c/\omega| \gg 1$ modes (Equation 6.32); panels (d–f) show spectra of the systems in (a–c), respectively. (From S. Thongrattanasiri and V.A. Podolskiy, *Opt. Lett.* 34, 890, 2009.)

Hence, the field behind the grating ($x > 0$) is given by

$$H_y(x,z) = -\frac{1}{\pi} \int_{-\infty}^{\infty} \frac{\sin(k_z d/2)}{k_z} e^{ik_x x + ik_z z} dk_z. \tag{6.31}$$

The field distributions behind the slit in isotropic and hyperbolic materials are compared in Figure 6.6a–c. Note that in strongly anisotropic media, the phase fronts of the diffracted light form hyperbolas, and thus the radiation lies completely inside or outside the so-called critical cones. To find the angle of the critical cone, we use Equations 6.20 and 6.21:

$$\tan\theta = \lim_{k_z \to \infty} \frac{S_z}{S_x} = \sqrt{-\frac{\epsilon_{yz}}{\epsilon_{xx}}}. \tag{6.32}$$

Figure 6.6d–f shows the wavevector-space evolution of the diffracted wave behind the slit. It is clearly seen that in an isotropic medium, the wavepacket immediately collapses onto a relatively narrow region of propagating waves; all information about subwavelength features of the slit is lost, and the wave cannot be refocused onto the tight spot.

The situation in hyperbolic materials is drastically different from the one in isotropic media. Regardless of the sign of ϵ_{yz}, strongly anisotropic materials support the propagation of the high-wavevector components (although the extremely high-k_z components relatively quickly attenuate due to absorption). Thus, it becomes possible to refocus light to a subwavelength spot inside the hyperbolic medium.

To achieve this goal, we use an approach that is often used in Fresnel optics [10]. We first select a focal point inside the material. We then design a *hypergrating* on top of the medium in such a way that light from open regions (zones) of the grating comes to the focal spot in phase [55].

For the focal spot at $\{x, z\} = \{f, 0\}$, the coordinates of boundaries of Fresnel zones in anisotropic media are given by

$$\sqrt{\epsilon_{yz} f^2 + \epsilon_{xx} z_m^2} - \sqrt{\epsilon_{yz} f^2 + \epsilon_{xx} z_0^2} = \pm \frac{m\pi c}{\omega}, \tag{6.33}$$

with z_0 corresponding to the horizontal displacement of the first Fresnel zone with respect to the horizontal position of the focal point; for materials with $\epsilon_{yz} > 0$, $z_0 \geq 0$. The sign in Equation 6.33 corresponds to the sign of ϵ_{xx}. Note that a Fresnel lens with $\epsilon_{yz} > 0$, $\epsilon_{xx} < 0$ is "left-handed": for oblique incidence, its focal point stays on the same side of the normal as the incident beam.

Examples of light focusing in several hypergratings are shown in Figure 6.7. In these simulations we have used dielectric permittivity given by effective medium theories (Equations 6.2 and 6.11) for an Au-alumina nanowire system [9,22,24,57] and AlInAs-InGaAs nanolayer [20] structures operating at 1.5 μm and 15 μm, respectively. The particular material parameters that we use in our simulations are $\epsilon_{yz} = 3.6 + 0.005i$;

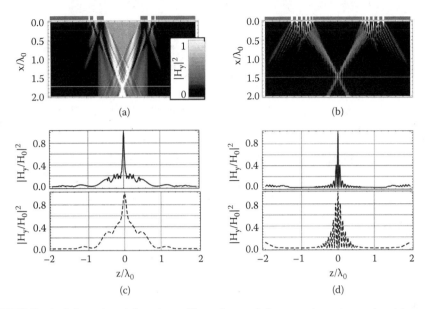

FIGURE 6.7 Subwavelength focusing and loss tolerance in hypergratings; sizes and positions of the slits are given by Equation 6.33; panels (a,b) correspond to low-loss nanowire (a) and nanolayer (b) structures (ϵ'' is reduced 10 times with respect to values shown in the text); panels (c,d) show field profiles in focal planes; solid lines: in low-loss systems, dashed lines: full loss structures. (From S. Thongrattanasiri and V.A. Podolskiy, *Opt. Lett.* 34, 890, 2009.)

$\epsilon_{xx} = -12.2 + 1.36i$ for a nanowire system and $\epsilon_{yz} = -6.4 + 1.4i$; $\epsilon_{xx} = 36 + 3.4i$ for a multi-layered structure. Note that in both cases hypergratings produce focal spots on the order of $\lambda_0/20$ in the far field of the grating.

In the limit of far-field operation ($f \gg \lambda_0$) and vanishingly small loss, hypergratings with $\epsilon_{yz} > 0$ behave as thin lenses. The imaging by such a lens is governed by the equation

$$\frac{1}{s_i} + \frac{\epsilon_{xx}}{\sqrt{\epsilon_{yz}}} \frac{1}{s_o} = -\frac{1}{f}, \tag{6.34}$$

with s_i and s_o being the distance from the grating to the image and object, respectively, and where the relationship between the focal distance and the open Fresnel zones is given by Equation 6.33 [56].

The device may magnify the object located inside the hyperbolic medium with magnification given by

$$M = -\frac{\epsilon_{xx}}{\epsilon_{yz}} \frac{s_i}{s_o}. \tag{6.35}$$

In the regime $M \gg 1$, the hypergrating may form a magnified image of a subwavelength object in free space, where it may be imaged with conventional microscopes.

6.4 Waveguiding with Anisotropic Metamaterials

In this section, we consider the applications of anisotropic metamaterials for confined propagation of light—waveguide photonics. We will first discuss optics in planar waveguides, and will address the benefits of uniaxial media for subwavelength light propagation, waveguide-based sensing, imaging, and mode profile engineering. We will then consider the one-dimensional (fiberlike) waveguides where hyperbolic materials enable straighforward coupling between micro- and nanoscale.

6.4.1 Waveguide Photonics

We start from analyzing light propagation in the planar waveguide, schematically shown in Figure 6.8. We assume that the waveguide core (thickness d) contains an anisotropic metamaterial with the optical axis directed perpendicular to the waveguide walls.

6.4.1.1 General Properties of Waveguide Modes

We first for simplicity assume that the waveguide has perfect electric conducting (PEC) material and discuss the generalization of the presented technique to different waveguide cladding material later.

The total field inside the waveguide can be represented as a linear combination of waveguide modes. The dispersion of the mode in the planar waveguide can be deduced from Equation 6.20 and from boundary conditions, requiring [10,11]

$$E_y(x = \pm d/2) = E_z(x = \pm d/2) = 0. \tag{6.36}$$

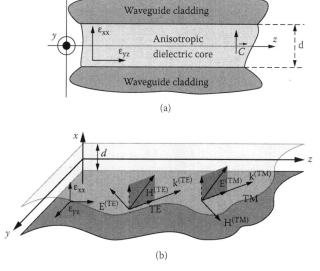

(a)

(b)

FIGURE 6.8 The geometry of planar guiding structure (a) and field structure of planar waveguide modes (b). (From V.A. Podolskiy and E.E. Narimanov, *Phys. Rev. B* 71, 201101(R), 2005.)

The modes of the planar waveguide therefore can be parameterized by their polarization (TE or TM), the direction of their propagation (given by the in-plane components of the modal wavevector $\{k_y, k_z\}$), and by the modal parameter κ, describing the profile of the mode in the x direction. Explicitly, the fields of the waveguide modes can be represented as [19,58]

$$
\begin{cases}
\vec{E}^{(\text{TM})} = \left\{ A(x), i\frac{k_y}{k_y^2+k_z^2}\frac{\epsilon_{xx}}{\epsilon_{yz}}A'(x), i\frac{k_z}{k_y^2+k_z^2}\frac{\epsilon_{xx}}{\epsilon_{yz}}A'(x) \right\} \exp(-i\omega t + ik_y y + ik_z z) \\[3mm]
\vec{H}^{(\text{TM})} = \left\{ 0, \frac{k_z}{k_y^2+k_z^2}\frac{\omega}{c}\epsilon_{xx}A(x), -\frac{k_y}{k_y^2+k_z^2}\frac{\omega}{c}\epsilon_{xx}A(x) \right\} \exp(-i\omega t + ik_y y + ik_z z)
\end{cases}
$$

(6.37)

$$
\begin{cases}
\vec{E}^{(\text{TE})} = \left\{ 0, \frac{k_z}{\sqrt{k_y^2+k_z^2}}A(x), -\frac{k_y}{\sqrt{k_y^2+k_z^2}}A(x) \right\} \exp(-i\omega t + ik_y y + ik_z z) \\[3mm]
\vec{H}^{(\text{TE})} = -\frac{c\sqrt{k_y^2+k_z^2}}{\omega}\left\{ A(x), \frac{ik_y}{k_y^2+k_z^2}A'(x), \frac{ik_z}{k_y^2+k_z^2}A'(x) \right\} \exp(-i\omega t + ik_y y + ik_z z),
\end{cases}
$$

(6.38)

with $A(x)$ being either $\cos(\kappa x)$ or $\sin(\kappa x)$ depending on mode symmetry, A' being the derivative of A with respect to x, and the mode parameter κ defined by the boundary conditions at the core-cladding interface. For PEC boundaries:

$$
\kappa^{(\text{TM})} = \begin{cases}
\frac{2m\pi}{d}, & A(x) = a\cos(\kappa x) \\[2mm]
\frac{(2m+1)\pi}{d}, & A(x) = a\sin(\kappa x)
\end{cases}
$$

(6.39)

$$
\kappa^{(\text{TE})} = \begin{cases}
\frac{(2m+1)\pi}{d}, & A(x) = a\cos(\kappa x) \\[2mm]
\frac{2m\pi}{d}, & A(x) = a\sin(\kappa x)
\end{cases}
$$

(6.40)

with a being (constant) field amplitude.

In both cases, propagation of a mode is given by the following dispersion relation [58]:

$$
k_y^2 + k_z^2 = \epsilon v \frac{\omega^2}{c^2}
$$

(6.41)

with

$$
v = 1 - \frac{\kappa^2}{\epsilon_{yz}\frac{\omega^2}{c^2}}
$$

(6.42)

$$
\epsilon = \begin{cases}
\epsilon_{xx}, & \text{TM modes} \\[2mm]
\epsilon_{yz}, & \text{TE modes}
\end{cases}
$$

(6.43)

In the form of Equation 6.41, dispersion of the mode in the planar waveguide is identical to the dispersion of the plane wave in isotropic metamaterial. The effective refractive index of the mode $n = \sqrt{\epsilon \nu}$ is given by the product of two scalar parameters. The propagation is possible only when both parameters have the same sign. Furthermore, when both ϵ and n are positive, the mode is "right-handed," meaning that the components $\vec{E}$, $\vec{H}$, and $\vec{k}_{yz} = \{k_y, k_z\}$ form a right-handed trio, and the propagation of the energy is parallel to the in-plane wavevector, while in the case when both ϵ and n are negative, the mode is left-handed so that the energy propagates in the direction opposite to $\vec{k}_{yz}$. The explicit relationship between the averaged Poynting vector of the mode $\vec{S}_{av} = \frac{c}{4\pi d}\int_{-d/2}^{d/2}\vec{E} \times \vec{H}^* dx$ and the material parameters of the core is given by

$$\vec{S}_{av}^{(TE)} = \frac{c^2|a|^2}{8\pi\omega}\vec{k}_{yz},$$ (6.44)

$$\vec{S}_{av}^{(TM)} = \frac{\omega\epsilon_{xx}|a|^2}{8\pi(k_y^2 + k_z^2)}\vec{k}_{yz}.$$ (6.45)

In both cases, it is possible to relate the average Poynting vector to the average energy density stored in the mode, U_{av} [19]:

$$U_{av} = \frac{1}{8\pi d}\int_{-d/2}^{d/2}(\vec{D}\cdot\vec{E}^* + \vec{H}\cdot\vec{H}^*)dx,$$

$$\vec{S}_{av} = \frac{c^2 U_{av}}{\epsilon\omega}\vec{k}_{yz}.$$ (6.46)

Note, however, that the foregoing expressions are valid only in the limit of small losses and of relatively weak material dispersion ($|\epsilon/\omega| \gg |d\epsilon/d\omega|$) [11,19].

Since the optical axis of the core is perpendicular to the direction of mode propagation, the photonics of waveguide modes is isotropic in the waveguide plane. Its energy propagation is parallel (or antiparallel) to the wavevector, and its refractive index is given by the product of two scalar numbers. From this point of view, the two-dimensional photonics of modes in planar guides is completely identical to the three-dimensional photonics of the plane waves in isotropic bulk materials with nontrivial permittivity and permeability [58].

Therefore, planar waveguide photonics provides full benefits of magnetic metamaterials, including planar lensing, inversed Doppler effect, and Cherenkov radiation, etc., with no requirements of actually having magnetism. An additional benefit is that the planar optics brings about superb compactness of light propagation ($n < 0$ assumes subwavelength waveguide size or excitation of high-order mode).

6.4.1.2 Refraction and Imaging in Planar Waveguides

We further now consider the propagation of waveguide modes through the interface between two planar waveguides. We assume that the two waveguides have the same size d and both have PEC cladding. In this case, both waveguides have the same mode spectrum.

We further assume that the fundamental TM_0 mode is incident from a waveguide with isotropic cladding ($\epsilon = 1/2 + 0.002i$) into the waveguide with hyperbolic cladding

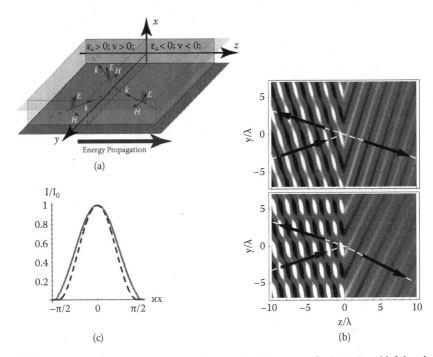

(a)

(c)

(b)

FIGURE 6.9 Reflection and refraction at the boundary between right-(RHM) and left-handed waveguides (LHM). (a) Schematic illustration of refraction of a TM wave at the RHM–LHM interface (TE waves not shown). (b) (top) Results of exact numerical calculations of refraction of the mode in planar waveguide with perfectly conducting walls; $\chi = \omega/2c$; RHM parameters ($z < 0$): $\epsilon = \nu = 1/2 + 0.002i$; LHM ($z > 0$): $\epsilon = \nu = -1/2 + 0.003i$; angle of incidence: $\pi/10$, normalized real part of E_x shown; (bottom): same as (top), but with finite-conductive waveguide (silver; $\lambda = 0.75$ μm; $\epsilon_w = -25 + 0.3i$). Arrows show the direction of energy flux in incident, reflected, and refracted waves, calculated using Snell's law. (c) Intensity profile of E_z for the systems in (b); dashed and solid lines correspond to perfect metal (b, top) and silver (b, bottom) waveguide walls. (From V.A. Podolskiy and E.E. Narimanov, *Phys. Rev. B* 71, 201101(R), 2005.)

($\epsilon_{xx} = -1/2 + 0.003i$, $\epsilon_{yz} = 1/6 + 0.0003i$); in these simulations, we keep the losses artificially small to demonstrate the physics behind mode propagation in hyperbolic waveguides.

The field distribution across the guide is shown in Figure 6.9. It is clearly seen that the beam undergoes negative refraction as the incident TM_0 wave is refracted into the TM_0 mode of the anisotropic waveguide. Moreover, it is seen that the angle of refraction perfectly agrees with Snell's law predictions.

However, Snell's law does not provide complete information about refracting beams as it cannot predict the amplitude of the transmitted wave. The full description of the refraction of the incident TM wave requires consideration of both TM and TE waves, since under oblique incidence ($k_y \neq 0$), the H_z and E_z components of the incident beam have the potential to excite TE-polarized beams. While TE waves exponentially decay inside hyperbolic media, they do get excited at the interface, and they do propagate in an isotropic part of the system. In fact, in the specific example, 85% of the incident energy is converted into a negatively refracted TM_0 wave, 13% is reflected in a TE_0 wave, and 2% is reflected in TM_0 mode.

Due to their blockage of TE radiation, waveguides with $\nu < 0$ act as efficient filters that may help limit interpolarization cross-talk in propagating beams. In particular, polarization mixing does not affect planar lensing in waveguides with hyperbolic cores.

The resolution of the planar lenses based on the hyperbolic waveguides, as with the resolution of any far-field imaging system, is limited by the wavelength inside the device. For hyperbolic waveguides, the resolution Δ is estimated by $\Delta \simeq \lambda_0/(2\sqrt{\epsilon\nu})$. However, since hyperbolic waveguides support propagating waves when $d \ll \lambda_0$, the effective modal index $n = \sqrt{\epsilon\nu}$ can be much larger than one ($n \sim \sqrt{-\frac{\epsilon\kappa^2 c^2}{\epsilon_{yz}\omega^2}} \gg 1$), so that the achievable resolution inside the waveguide can be much better than the one achievable in free space. Of course, the waveguide imaging does not directly provide information about the x structure of the object, which represents a trade-off of planar photonics.

The perspectives of subwavelength imaging with hyperbolic materials are illustrated in Figure 6.10, where planar lens imaging is clearly seen. In these simulations, we assume that free-space wavelength of incident light is $\lambda_0 = 1.5\ \mu$m, and the waveguide thickness $d = 0.3\ \mu$m. The isotropic parts of the system ($z < 2.5\ \mu$m and $z > 5\ \mu$m) are filled with Si

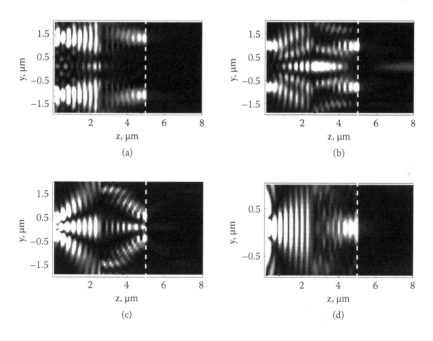

FIGURE 6.10 Imaging by a planar NIM-waveguide-based lens. (a) The intensity distribution in the system with $d = 0.3\ \mu$m; LHM region is between $z = 2.5\ \mu$m and $z = 5.0\ \mu$m; the focal plane corresponds to $z = 5.0\ \mu$m (white dashed line); slit size $w = 0.8\ \mu$m; (b) same as (a) but with $w = 0.6\ \mu$m; (c) same as (a), but $w = 0.3\ \mu$m, corresponding to the far-field resolution limit of the system; (d) same as (a) but $w = 0.15\ \mu$m, below the resolution limit of the system. (From V.A. Podolskiy et al., *Proc. SPIE* 6002, 600205, 2005.)

($\epsilon = 13$), while the anisotropic part of the waveguide is filled with Ag nanowire composite, resulting in ($\epsilon_{xx} \simeq -3.91$; $\epsilon_{yz} = 2.33$) [19].

To determine the resolution of the planar lens, we analyze the evolution of the image of the double-slit source (slit thickness w) as we reduce the size of the slits and the distance between them. The calculations employ the mode matching technique.

In particular, first the field of the double-slit source (each slit size w, positioned at $w \leq |y| \leq 2w$) is represented as a set of waveguide modes:

$$\vec{E}(x,y,z) = \int_{-\infty}^{\infty} \vec{E}^{(TM)}(x,k_y,z)dk_y, \tag{6.47}$$

where $\vec{E}^{TM}(x,k_y,z)$ is given by Equation 6.37 with modal amplitude $A(x, k_y)$:

$$A(x,k_y) = \frac{2\sin(\pi x/d)}{k_y}[\sin(2wk_y) - \sin(wk_y)]. \tag{6.48}$$

The propagation of each component of the above wavepacket corresponding to a particular value of k_y is then considered individually, and the reflected and transmitted waves excited by this component are determined; both the *TE* and *TM* parts of the spectrum are considered in these simulations. Finally, the response of the system to the wavepacket is calculated as a sum of the responses to the individual wavepacket components.

The resolution limit, calculated from these numerical solutions of Maxwell equations, is $\Delta \simeq \lambda_0/5$, which is consistent with the far-field estimates, taking into account the modal index of $|n| \simeq 2.5$ [19].

Note that in contrast to the superlens [28] that resonantly enhances subwavelength information and is severely affected by material losses [29], planar imaging in hyperbolic materials is almost unaffected by material absorption. In fact, our simulations show that incorporation of losses leads to suppression of image intensity without loss of resolution [19].

6.4.1.3 Waveguides with Non-PEC Claddings

The foregoing formalism can be straightforwardly extended to multilayered planar waveguides with non-PEC claddings. As before, the field inside the system can be represented as a linear combination of modes. The full spectrum consists of two fundamentally distinct groups of modes.

The modes of the first group, known as *guided waves*, represent the modes that exponentially decay inside the cladding. The field profile of these waves is given by transfer matrix formalism, described earlier, while their dispersion (dependence of in-plane wavevector on frequency) is given by

$$T_{11}^{1N}(\vec{k}_{yz}, \omega) = 0, \tag{6.49}$$

where T_{11}^{1N} is the element of the transfer matrix relating the field in the bottom cladding layer (layer 1) to the fields in the top cladding layer (layer N).

The waves of the second group, also known as *open-waveguide modes*, represent the response of the planar system to the excitation by a plane wave with *real k_x* from either the top or the bottom-most layers.

In an isolated planar guide, the components of the modal spectrum are all orthogonal to each other, and therefore can be considered independently. The handedness of the mode will be given by the relationship between the Poynting flux (integrated over the entire x axis) and wavevector of this mode. Note that the Poynting flux in hyperbolic materials with $\epsilon_{xx} < 0$ is opposite to $\vec{k}_{yz}$ while the Poynting flux in isotropic dielectrics is parallel to the wavevector. Therefore, hyperbolic waveguides with dielectric claddings open new perspectives for building highly compact systems where the energy is "circulated" between the core and cladding regions [30]. This energy circulation drastically reduces the group velocity of the waveguide modes, making hyperbolic waveguides one of the promising candidates for ultra-slow-light structures.

The slow-light modes are of prime importance for sensing applications since energy compression typically yields higher local fields and better sensitivity. In addition, the composite nature of hyperbolic metamaterials provides additional benefits for the higher surface area of micro- and nanosensors and for easy-to-implement size selectivity given by metamaterial geometry.

The hyperbolic-metamaterial biosensor, based on plasmonic nanowire composites, is illustrated in Figure 6.11. The system comprises a planar waveguide with plasmonic nanowire core, grown on top of a dielectric prism and submerged into a flow-cell. From the optical standpoint, the device is a metamaterial analog of the surface plasmon polariton (SPP) biosensor [60,61]. The fluid to be sensed is dispersed through the flow-cell, while the system is interrogated by an incident light from the prism side. The changes in the reflectivity profile provide the measurement of changes in the refractive index of the fluid.

The numerical simulations and experiments [62] clearly demonstrate that metamaterial-based sensors are drastically more sensitive than their SPP counterparts. The superior sensitivity of metamaterial structure is combined with an increased figure of merit, defined as $\frac{\Delta\lambda}{\Delta n \Delta w}$ (where Δw is the full width of the resonant dip at half maximum and $\Delta\lambda$ is the resonance shift for Δn refractive index change), which takes into account the sharpness of the resonance. This parameter reaches 23 for SPP sensors, but is 330 and higher for hyperbolic-metamaterial guides [61,62].

As mentioned earlier, in addition to drastically increased sensitivity, metamaterial-based sensors offer the benefits of tunable spectral response and size selectivity for particular applications

6.4.1.4 Coupled Planar Guides

Planar guides offer the benefits of realizing the subwavelength optics in the plane of the optical chip. It should be noted, however, that manipulation of the propagation parameters of the mode (such as effective modal index) that is necessary to steer the wavepacket is typically accompanied by changes in the modal profile. The sensitivity of the modal profile to variation of modal index is usually higher in waveguides with dielectric cores. While this sensitivity is beneficial for sensing applications, it may severely complicate the applications of planar optics.

Thus, in waveguides with dielectric cores, every reflection and refraction event that is accompanied by mode reshaping will necessarily lead to modal cross-talk: a single mode, incident at the boundary will excite a (large) number of reflected and refracted modes. Since some radiation will potentially be scattered into bulk modes (in realistic guiding

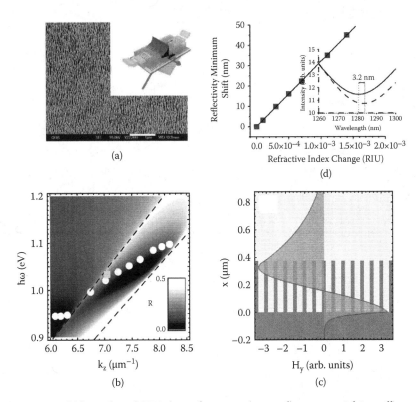

FIGURE 6.11 (a) [main figure] SEM photo of nanowire (nanorod) metamaterial; inset illustrates the principle of operation of a flow-cell metamaterial-based sensor; (b) calculated reflectivity profile (plot) and measured dispersion of the mode in the sensor (symbols); field distribution in the mode corresponding to the parameters marked by a square in (b) is shown in (c); note the strong overlap between the electric field and the dielectric; (d) experimentally measured dependence of the shift in the absorption profile on the change in refractive index of the dielectric. (From A. Kabashin et al., _Nature Materials_, 8, 867–871, 2009.)

systems, this amount can be as high as 30%) [63–65], both guided and open-waveguide modes need to be included in calculations. The technique for numerical computation of light propagation in coupled planar guides is presented in Reference [36].

Fortunately, anisotropic materials allow independent control of the refractive index of the mode and its profile. Therefore, a carefully designed optical circuit may completely eliminate the problems of modal cross-talk and of out-of-plane scattering of radiation, raising the possibility of _truly planar optics_, which would be completely analogous to volume optics of plane waves: every reflection/refraction would preserve the polarization of the incident beam, every incident mode will excite exactly one reflected mode and one refracted mode, and most importantly, the amplitudes of incident, reflected, and refracted waves would be related to each other though the analogs of Fresnel equations.

Here we consider the development of truly planar optics on the example of one inter-face between two planar guides. In order to realize complete mode matching, the two layered structures must (i) have the same number of guided modes, and (ii) the modal index of the guided modes must not depend on the modal profile. Naturally, the layers on both sides of the interface should be aligned with each other.

As first shown in Reference [64], these conditions can be satisfied when

$$\epsilon_{yz}^{(L_j)} = \epsilon_{yz}^{(R_j)}$$
$$n^2 \epsilon_{xx}^{(L_j)} = \epsilon_{xx}^{(R_j)}, \qquad (6.50)$$

with L and R corresponding to "left" ($z < 0$) and "right" ($z > 0$) regions of the guide, and n being the constant number that does not depend on layer number j and corresponds to the change of the modal index.

As can be explicitly verified, when Equation 6.50 is satisfied, the intermode cou-pling is absent across the interface. The interface remains completely transparent to TE-polarized waves, while reflection and refraction of TM-polarized modes are con-trolled by the ratio of out-of-plane permittivities. The direction and amplitudes of the reflected and refracted modes are related to the direction and amplitude of the incident modes via Snell's law:

$$n_i \sin(\theta_i) = n_i \sin(\theta_r) = n_t \sin(\theta_t) \qquad (6.51)$$

and Fresnel coefficients:

$$\frac{A_r^{(j)}}{A_i^{(j)}} = \frac{k_{i_z} - k_{t_z}}{k_{i_z} + k_{t_z}}, \quad \frac{A_t^{(j)}}{A_i^{(j)}} = \frac{2k_{i_z}}{k_{i_z} + k_{t_z}}. \qquad (6.52)$$

In the foregoing equations, the subscripts i, r, and t correspond to direction (θ), ampli-tudes (A), and wavevectors ($\vec{k}$) of incident, reflected, and transmitted modes.

The concept of truly 2D optics is illustrated in Figure 6.12 with the example of an air–Si–air system coupled to a metamaterials waveguide. As expected, the reflection of a single mode in a conventional planar system is accompanied not only by significant radiation scattering and modal cross-talk, but also by cross-polarization coupling. In contrast, for metamaterial guides the single incident mode excites one reflected wave and one transmitted wave.

Planar optics can be realized for both volume and surface guides [36,64]. In the latter case, it becomes very promising to achieve almost-ideal conditions with an electrooptical system. In this case, the surface plasmon polariton guide is formed by a metallic substrate with electrooptical dielectric deposited on top, covered with control circuitry, capable of modifying the potential distribution above the metal plane. The voltage, applied perpen-dicular to the metal surface, would modulate the *xx* component of dielectric permittivity of the electrooptical layer, resulting in steering of plasmonic beams. The schematic of such an *electroplasmonic* device is shown in Figure 6.13.

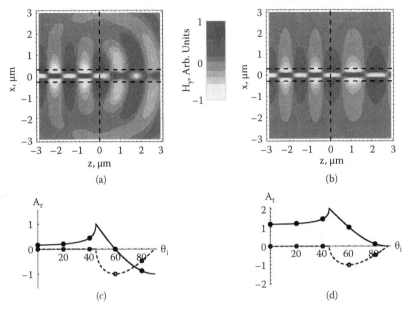

FIGURE 6.12 (a) An interface between the air–Si–air waveguide and isotropic air ($\epsilon = 6.06$)–air guide leads to substantial modal cross-talk, polarization mixing, and out-of-plane scattering (evident from "curvature" of phase fronts), while the interface between the air-Si-air waveguide and its anisotropic truly planar optics analog allows for ideal mode matching with light steering capabilities; (b–d); dashed lines in (a,b) correspond to material interfaces; the system is excited by a TM_2 mode propagating at the angle 40° to the $z = 0$ interface; panels (c,d) show the amplitudes of reflected (c) and transmitted (d) waves, normalized by the amplitude of incident TM_2 mode; solid and dashed lines correspond to real and imaginary parts of the amplitudes, respectively; lines are due to analytical Equation 6.52, dots represent results of numerical simulations according to Reference [36]. (From S. Thongrattanasiri et al., *J. Opt. Soc. Am. B* 26, B102, 2009.)

6.4.2 Photonic Funnels: Merging Micro- and Nanophotonics

Guided propagation of optical signals in isotropic dielectric fibers is possible only when the light-transmitting region of a fiber is at least on the order of the free-space wavelength [11]. This fact strongly limits resolution of modern optical microscopy and spectroscopy, and prevents the construction of compact all-optical processing units and further development in other areas of photonics [37,67–71]. Although it is possible to propagate gigahertz radiation in deep subwavelength areas in coaxial cables or anisotropic magnetic systems [72–77], the direct scale-down of these techniques to optical or IR domains is problematic [78].

Until recently, all designs involving optical light transport in subwavelength areas relied on the excitation of a surface wave: a special kind of electromagnetic wave propagating at the boundary between the materials with positive and negative dielectric constants [49,79–82]. The spatial structure of surface waves, however, is fundamentally different from that of "volume" fiber modes or free-space radiation. While it is possible

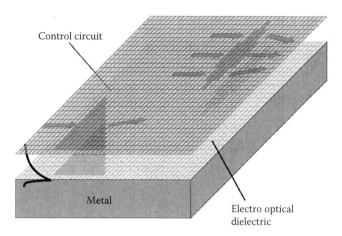

FIGURE 6.13 Schematics of dynamically reconfigurable electroplasmonic system. (From V.A. Podolskiy and J. Elser, Electroplasmonics: Dynamical plasmonic circuits with minimized parasitic scattering, in *CLEO/QELS 2008*, OSA, Washington DC, 2008, QTuJ2.)

to couple the radiation between the volume and surface modes, such a coupling is typically associated with substantial scattering losses and involves materials and devices of substantial (in comparison with optical wavelength) size [11,83].

In this section, we consider the perspectives of subwavelength light compression and propagation in waveguides with anisotropic cores. The geometry of the system is shown in Figure 6.14. Note that in contrast to the previous discussions, here we use a cylindrical reference frame. The optical axis of the core material is aligned with the direction of mode propagation (z). The dielectric permittivity is of the core uniaxial, with two distinct components being $\epsilon_{r\varphi}$ and ϵ_{zz}.

6.4.2.1 Light Propagation in Cylindrical Waveguides with PEC Cladding

As before, we first consider the case of perfectly conducting waveguide cladding. In this particular case, the modes of a cylindrical waveguide have well-defined polarization, *TE*

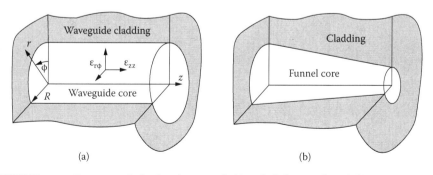

FIGURE 6.14 Geometry of cylindrical waveguide (a) and of photonic funnel (b).

$(E_z = 0)$ or *TM* $(H_z = 0)$. The field profile of the modes can be related to the profile of the nonzero z-components via

$$
\begin{cases}
E_r^{\mathrm{TE}} = \pm \dfrac{i\omega}{r\kappa_{TE}^2 c} \dfrac{\partial H_z}{\partial \phi} \\[2mm]
E_\phi^{\mathrm{TE}} = \mp \dfrac{i\omega}{\kappa_{TE}^2 c} \dfrac{\partial H_z}{\partial r} \\[2mm]
H_r^{\mathrm{TE}} = \pm \dfrac{ik_z}{\kappa_{TE}^2} \dfrac{\partial H_z}{\partial r} \\[2mm]
H_\phi^{\mathrm{TE}} = \pm \dfrac{ik_z}{r\kappa_{TE}^2} \dfrac{\partial H_z}{\partial \phi}
\end{cases}
$$

$$
\begin{cases}
E_r^{\mathrm{TM}} = \pm \dfrac{ik_z}{\kappa_{TM}^2} \dfrac{\epsilon_z}{\epsilon_{r\phi}} \dfrac{\partial E_z}{\partial r} \\[2mm]
E_\phi^{\mathrm{TM}} = \pm \dfrac{ik_z}{r\kappa_{TM}^2} \dfrac{\epsilon_z}{\epsilon_{r\phi}} \dfrac{\partial E_z}{\partial \phi} \\[2mm]
H_r^{\mathrm{TM}} = \mp \dfrac{i\epsilon_z \omega}{r\kappa_{TM}^2 c} \dfrac{\partial E_z}{\partial \phi} \\[2mm]
H_\phi^{\mathrm{TM}} = \pm \dfrac{i\epsilon_z \omega}{\kappa_{TM}^2 c} \dfrac{\partial E_z}{\partial r},
\end{cases}
$$

$$(6.53)$$

where c is the speed of light in vacuum, and the parameters $\kappa_{\mathrm{TE,TM}}$ describe wave structure in the radial direction.

The z-component of cylindrical waves must satisfy the following relations:

$$
\begin{aligned}
\Delta_2 E_z \pm \kappa_{\mathrm{TM}}^2 E_z &= 0, \\
\Delta_2 H_z \pm \kappa_{\mathrm{TE}}^2 H_z &= 0,
\end{aligned}
$$

$$(6.54)$$

with $\Delta_2 = \Delta - \frac{\partial^2}{\partial z^2}$ being the 2D Laplacian operator, and the additional condition that the electric field components E_ϕ, E_z must be continuous along the core-cladding interface [11]. The solutions corresponding to the top signs in the foregoing equations represent the field profile of the volume waveguide modes. The solutions corresponding to the bottom signs represent the fields that exponentially decay or increase in a radial direction and are often used to describe the fields of the surface modes and the fields in the non-PEC waveguide claddings, considered in the next section.

The volume-mode solution to Equation 6.54, yields

$$
\begin{aligned}
E_z(r,\phi,z) &= aJ_m(\kappa_{\mathrm{TM}}r)e^{im\phi + ik_z z} \\
H_z(r,\phi,z) &= aJ_m(\kappa_{\mathrm{TE}}r)e^{im\phi + ik_z z}
\end{aligned}
$$

$$(6.55)$$

with

$$
\kappa_{(\mathrm{TM|TE})} = \frac{X_{(\mathrm{TM|TE})}}{R},
$$

$$(6.56)$$

where R is a waveguide radius and X is given by $J_m(X_{TM}) = 0$ for *TM* waves and $J'_m(X_{TE}) = 0$ for *TE* waves, respectively.[*3]

Similar to the case of planar waveguides, the propagation parameter of the mode k_z can be related to the frequency and mode profile parameter via the dispersion equation:

$$k_z^2 = \epsilon_{r\phi} v \frac{\omega^2}{c^2}$$

(6.57)

with the polarization-dependent parameter n:

$$v_{TE} = 1 - \frac{\kappa_{TE}^2 c^2}{\epsilon_{r\phi} \omega^2},$$

$$v_{TM} = 1 - \frac{\kappa_{TM}^2 c^2}{\epsilon_{zz} \omega^2}.$$

(6.58)

Equation 6.57 is once again fundamentally similar to the dispersion of a plane wave in isotropic material with the combination $n = k_z c/\omega = \pm\sqrt{\epsilon_{r\phi} v}$ playing the role of the effective index of refraction that is combined from two (mode-dependent) scalar quantities. The mode propagation requires both propagation constants $\epsilon_{r\phi}$ and n to be of the same sign.

While the permittivity depends solely on the dielectric properties of the core material, the propagation parameter n can be controlled (through κ) by changing the waveguide (or mode) geometry. Since κ is inversely proportional to the waveguide size (see Equation 6.56) [11], for every bulk mode propagating in the waveguide with isotropic dielectric core there exists a *cut-off radius* $R_{cr} \sim \lambda/2$ corresponding to $n = 0$. The modes propagate in structures with $R > R_{cr}$, and are reflected from thinner systems, as illustrated in Figure 6.15a,d. This appearance of the cut-off radius in all dielectric waveguides can be considered a manifestation of a diffraction limit: it is impossible to localize a combination of plane waves to a region much smaller than the wavelength inside the material $\lambda = \lambda_0/n$.

Material anisotropy in a sense opens another direction in controlling light propagation in waveguides. Indeed, anisotropy of the dielectric constant makes the *TM*-mode parameters $\epsilon_{r\phi}$ and n completely independent of each other (*TE* waves are not affected by material anisotropy [58,84]). Hyperbolic optical materials may be used to achieve volume-mode propagation in deep subwavelength waveguides.

As directly follows from Equations 6.57 and 6.58, when $\epsilon_{r\phi} > 0$, $\epsilon_{zz} < 0$, the parameter n is positive regardless of the size of the system. Thus, the cut-off radius does not exist at all: the decrease of the waveguide radius is accompanied by a decrease of the internal wavelength of the mode $2\pi/k_z \propto R$, in a sense "postponing" the diffraction limit in the system.

[*3] Here we use the notation $J'_m(\xi) = \frac{dJ_m(\xi)}{d\xi}$.

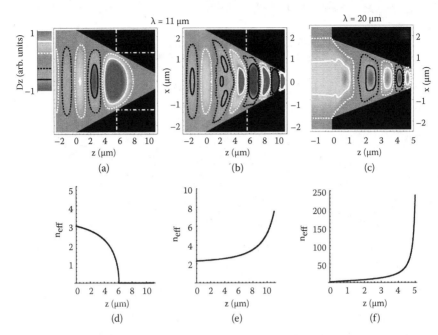

FIGURE 6.15 (a–b) TM_{01} mode propagation ($\lambda = 11\ \mu m$) from cylindrical Si waveguide ($z < 0$) to a conical one with circular core; D_z component shown; (a) Si core structure. Note the reflection from the point where radius reaches the cut-off value $R_0 \simeq 1.2\ \mu m$ (dash-dotted lines). Only 10^{-10} of the energy is transmitted from $R = 2.3\ \mu m$ to $R = 0.35\ \mu m \sim \lambda/31$. This behavior is similar to that in tips of near-field microscopes. [36,37] (b) Field concentration in photonic funnel made from 100-nm-thick Si–SiC nanolayer composite: 13% of energy is transmitted to $R = 0.35\ \mu m$, 16% is reflected back to Si ($z < 0$) waveguide. (c) Energy transfer from the AlInAs waveguide ($z < 0$) to a photonic funnel with an InGaAs-AlInAs metamaterial core. Panels (d–f) show the dependence of the modal index on the coordinate within the conical waveguide section. The modal index inside the waveguide with isotropic core decreases with decrease of waveguide radius, reaching zero at the cut-off radius (a,d). In contrast to this behavior, the modal index in waveguides with hyperbolic cores is inversely proportional to the radius, postponing the onset of the diffraction limit. (From A.A. Govyadinov and V.A. Podolskiy, *Phys. Rev. B* 73(15), 155108, 2006.)

The case of opposite anisotropy ($\epsilon_{r\varphi} < 0$, $\epsilon_{zz} > 0$) is of a special interest. The mode propagation is now possible only when $n < 0$, which in turn *requires* the waveguide cross section to be extremely small. Furthermore, the requirement of the positive energy flux now requires the phase velocity of a propagating mode to be *negative* [58,84]. In a sense, such a waveguide is the complete antipode of a conventional dielectric fiber, in terms of phase propagation, as well as in terms of cut-off radius.

Figure 6.15e,f shows the dependence of the refractive index on the waveguide radius. It is clearly seen that the "subdiffractional" light propagation indeed follows from the reduction of the internal wavelength (increase of effective n) in thinner waveguides.

The existence of the propagating modes in highly confined geometries allows for unique applications of fibers with hyperbolic cores: in the structure with a gradually tapered waveguide core, the mode will self-adjust to the new confinement scale, leading to the possibility of "funneling" the propagating radiation from the micro- to nanoscale. Such photonic funnels, illustrated in Figure 6.15, are in a sense ideal systems for the new generations of near-field microscopy tips. Indeed, photonic funnels combine the benefits of spectral tunability offered by metamaterials with the benefits of direct coupling between the guided modes of optical fibers and nanoscale emitters or absorbers of radiation.

In designing the tapering profile of a photonic funnel, it is important to balance the reflection losses (caused by reflection from nonadiabatic mode compression) with absorption losses (caused by light propagation through absorbing metamaterials) [49,84]. However, despite these significant loss channels, our numerical simulations suggest that it is possible to transfer ~20% of incident radiation to a scale as small as $\lambda_0/20$.

6.4.2.2 Photonic Funnels with Non-PEC Cladding

Similar to the case of perfectly conducting walls described earlier, light propagation in fibers with any isotropic cladding can be related to propagating waves with *TE* and *TM* polarizations. In this approach, the field is represented as a series of *TM* and *TE* waves with the same propagating constant k_z and frequency w, and the boundary conditions are used to find the effective refractive index $n = k_z c/w$. Note that the *TE* and *TM* components of the mode have similar but not identical spatial structure inside the anisotropic core, given by Equations 6.53 and 6.54.

Explicitly, solutions to Equation 6.54 result in the z-components of fields in the waveguide core to be proportional to $J_m(k_{(TE|TM)}r)\exp(im\varphi)$ with $\kappa_{TE}^2 = \epsilon_{r\phi}\omega^2/c^2 - k_z^2$, and $\kappa_{TM}^2 = \epsilon_{zz}(\omega^2/c^2 - k_z^2/\epsilon_{r\phi})$. On the other hand, the components of the fields in the cladding material are proportional to $K_m(k_{cl}r)\exp(im\varphi)$ with $\kappa_{cl}^2 = k_z^2 - \epsilon_{cl}\omega^2/c^2$, and ϵ_{cl} being the permittivity of the cladding, with all other field components given by Equation 6.53.

The boundary conditions yield the following dispersion relation for a propagation constant of a mode in waveguide with anisotropic core:

$$[J_m^+(\kappa_{TE}R) + K_m^+(\kappa_{cl}R)][\epsilon_{zz}J_m^+(\kappa_{TM}R) + \epsilon_{cl}K_m^+(\kappa_{cl}R)] =$$

$$\frac{1}{k_z^2}\frac{m^2\omega^2}{R^2c^2}\left(\frac{\epsilon_{r\phi}}{\kappa_{TE}^2} + \frac{\epsilon_{cl}}{\kappa_{cl}^2}\right)\left(\frac{\epsilon_{zz}}{\kappa_{TM}^2} + \frac{\epsilon_{cl}}{\kappa_{cl}^2}\right), \tag{6.59}$$

where $L_m^+(\kappa R) = L_m'(\kappa R)/[\kappa L_m(\kappa R)]$. The two terms in the left-hand side of the equation correspond to the contributions from *TE* and *TM* modes, respectively. As follows from Equation 6.59, the "pure" *TM* and *TE* modes are only possible when (1) $m = 0$, or (2) $\epsilon_{cl} \rightarrow -\infty$. The latter case corresponds to the perfectly conducting metallic walls described earlier. Solutions of Equation 6.59 can be separated into two fundamentally different groups: the ones with $\kappa_{(TE|TM)}^2 > 0$ describe volume modes, while the ones with $\kappa_{(TE|TM)}^2 < 0$ correspond to surface waves.

It can be shown [85] that in case of highly conductive cladding, cross-polarization mixing of the volume modes mode leads to only quantitative corrections with respect to the foregoing results.

6.5 Beyond Effective Medium Response of Anisotropic Composites

The majority of exciting applications of metamaterials, most notably those associated with strong confinement of electromagnetic radiation, rely on the validity of effective-medium description of their optical properties. However, substantial compression of electromagnetic energy in hyperbolic materials is usually related to substantial decrease of the scale of typical field variation—internal wavelength. As the scale of field variation becomes smaller, the validity of effective medium description of metamaterial properties (that rely on quasi-static averaging of fields) breaks down, and the composite nature of the structure starts to dominate the electromagnetic properties of the system.

The relatively weak deviation of optical response of the composite from the EMT predictions can be adequately described by spatially dispersive permittivity [11,86],

$$\hat{\epsilon} = \hat{\epsilon}(\omega, \vec{k}). \tag{6.60}$$

We note that, in principle, the spatially dispersive permittivity can describe the properties of any composite structure. In the majority of the cases, the permittivity of individual components of the metamaterial remains local, and the nonlocality appears only as a result of the homogenization process. One should be careful to analyze the limits of the applicability of such nonlocal EMTs: they may perfectly describe the direction of the propagation of the waves inside the material, but would be less accurate in predicting the amplitudes of transmission or reflection. Moreover, these EMTs will as a rule completely neglect the scattering of light by the composite surface. In general, the condition that the typical inclusion inside the composite must be smaller than the free-space wavelength is required (but may be not sufficient) [86] in order to meaningfully introduce the (nonlocal) permittivity.

Here, we consider the case of weak spatial dispersion, where the dependence of permittivity on the components of the wavevector is much weaker than the dependence of permittivity on frequency. Under these conditions, the tensor of permittivity can be represented as

$$\epsilon_{ij}(\omega, \vec{k}) = \epsilon_{ij}^0(\omega) + \alpha_{ij} \frac{k_i k_j c^2}{\omega^2}. \tag{6.61}$$

For simplicity, here we consider the case when α_{ij} is a diagonal tensor with $|\alpha_{ii}| \ll 1$, and $\alpha_{yy} = \alpha_{zz} =$ denotes as α_{yz}, and when $k_y = 0$; more complicated cases can be considered similarly [87].

While the properties of plane waves propagating in spatially dispersive media are still given by Equation 6.15, the equation now has not two, but three linearly independent solutions. One of these corresponds to *TE*-polarized wave, while two others correspond to the *TM*-polarized mode. Dispersion of these waves is given by [87,90]:

$$k_x^2 = \epsilon_{yz}^0 \frac{\omega^2}{c^2} - k_z^2(1+\alpha_{yz}), \text{ TE modes}$$

(6.62)

$$k_x^2 = \frac{1}{2\alpha_{xx}} \left[\left(\epsilon_{xx}^0 + \xi\right)\frac{\omega^2}{c^2} \pm \frac{\omega}{c} \sqrt{\left(\epsilon_{xx}^0 - \xi\right)^2 \frac{\omega^2}{c^2} + 4\xi k_z^2} \right], \text{ TM modes}$$

with $\xi = \alpha_{xx}\left(\epsilon_{yz}^0 - \alpha_{yz}k_z^2 c^2/\omega^2\right)$.

The existence of the third (additional) wave in the bulk of (meta-) material requires the introduction of additional boundary conditions in order to adequately describe the micro- and even nanostructure of the interface. These conditions require knowledge of the material composition, and are therefore not universal.

In the limit $\alpha_{ii} \to 0$, two of the three waves described by Equation 6.62 recover the results of local equations (Equations 6.18 and 6.20). The remaining (additional) wave has $k_x^2 \to \infty$, so that it cannot be excited.

Naturally, when $|\alpha| \ll |\epsilon^0|$, coupling to an additional wave is relatively weak, and spatial dispersion represents *quantitative* corrections to the results of local electromagnetism. In nanolayered composites, the nonlocal corrections lead to corrections to effective modal indices and additional absorption of highly confined modes (in comparison with predictions of local EMT). The corrections to permittivities are given by [88]*4

$$\epsilon_x^{eff} = \frac{\epsilon_x^{(0)}}{1 - \alpha_{xx}(k_x, \omega)}$$

(6.63)

$$\epsilon_{yz}^{eff} = \frac{\epsilon_{yz}^{(0)}}{1 - \alpha_{yz}(k_x, \omega)},$$

where the nonlocal corrections are given by

$$\alpha_{xx} = \frac{a_1^2 a_2^2 (\epsilon_1 - \epsilon_2)^2 \epsilon_x^{(0)2}}{12(a_1 + a_2)^2 \epsilon_1^2 \epsilon_2^2} \left(\epsilon_{yz}^{(0)} \frac{\omega^2}{c^2} - \frac{k_x^2 (\epsilon_1 + \epsilon_2)^2}{\epsilon_{yz}^{(0)2}} \right),$$

(6.64)

$$\alpha_{yz} = \frac{a_1^2 a_2^2 (\epsilon_1 - \epsilon_2)^2}{12(a_1 + a_2)^2 \epsilon_{yz}^{(0)}} \frac{\omega^2}{c^2}.$$

The validity of these expressions is illustrated in Figure 6.16.

Similar corrections have been derived for the low-frequency response of metallic nanowire composites [91–95].

---

*4 Note that since components of the wavevector are related to each other via Equation 6.20, the choice of k_x and ω/c as opposed to k_z or k_y (in Equation 6.64) is somewhat arbitrary and primarily depends on the geometry.

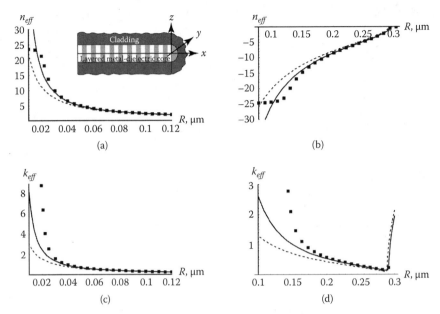

FIGURE 6.16 The comparison between effective refractive indices $n_{eff} = Re(k_x c/w)$ (a,b) and extinction coefficients $k_{eff} = Im(k_x c/w)$ (c,d) of TM_{01} modes in cylindrical waveguides with multilayer cores and perfectly conducting claddings (inset); dashed lines, solid lines, and dots correspond to results of local EMT, nonlocal EMT, and exact solution of the dispersion equation (from P. Yeh et al., *J. Opt. Soc. Am.* 67, 423, 1977), respectively. The multilayered core is composed from 15 nm layers with (a,c) $\epsilon_m = -1 + 0.1i$; $\epsilon_d = 1.444^2$, and (b,d) $\epsilon_m = -114.5 + 11.01i$; $\epsilon_d = 1.444^2$ (Au–SiO$_2$ structure); $\lambda_0 = 1.55$ μm; similar agreement between nonlocal EMT and exact dispersion equation is achieved for cylindrical systems with air claddings. (From J. Elser et al., *J. Nanomaterials*, 2007, 79469, 2007.)

The situation is dramatically different in the epsilon-near-zero (ENZ) regime when $|\epsilon_{xx}^0| \lesssim \alpha_{xx}$. In this case, the structure of the additional wave becomes comparable to the structure of the "main" *TM*-polarized mode. As result, both *TM* waves couple to the incident radiation, which drastically affects transmission through the system.

When the nonlocality is weak, the main mode represents a single channel of transmission. Hence, the amplitude of the transmitted light depends only on the losses of the main wave. In contrast to this behavior, in a regime of strong nonlocality, the transmission is dominated by the interference between the main and additional waves, so that the output is related to the amplitudes and phases of both modes.

Although nonlocality has been observed in homogenous media [96,97], these observations typically require ultra-high-purity samples, and can be only performed under liquid-He temperatures. Metamaterials bring nonlocal electrodynamics to the domain of room-temperature materials and applications.

The experimental study of excitation of additional waves in metamaterials is detailed in Reference [90]. The geometry of the experiment is shown in Figure 6.17. The *p*-polarized (*TM*-polarized) transmission through a planar slab of gold–alumina metamaterial

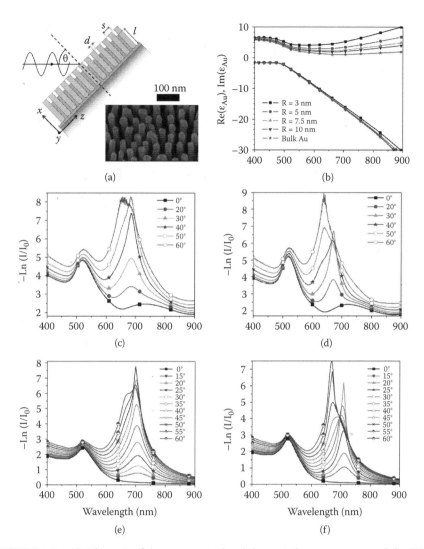

FIGURE 6.17 (a) Schematic of the metamaterial and the optical measurements and the SEM image of the nanorods after removal of the AAO matrix. (b) Spectra of Au permittivity calculated for different values of parameter R in Equation 6.65. (c,d) Extinction spectra measured with p-polarized incident light for different angles of incidence with the nanorod assemblies unannealed (c) and annealed at 300°C for 2 hours (d). (e,f) Extinction spectra derived from 3D FEM-based simulations for the ϵ_{Au} corresponding to $R = 5$ nm (e) and $R = 10$ nm (f). (From R.J. Pollard et al., *Phys. Rev. Lett.* 102, 127405, 2009.)

is analyzed as a function of angle and wavelength (an electron microscope image of the typical nanowire composite is shown in inset). When the wavelength of incident light was substantially different from the one corresponding to the $\epsilon_{xx} = 0$ condition, the properties of the composite were well-described by the Maxwell-Garnett theory. Note that in contrast to thin-film-based plasmonic metamaterials, the internal structure of solution-derived gold strongly limits the electron motion in the metal. Quantitatively, the effect of restricted mean free path of the electrons on the permittivity of gold is described by [98]

$$\epsilon_{Au}(\omega) = \epsilon_{bulk} + \frac{i\omega_p^2 \tau (R_b - R)}{\omega(\omega\tau + i)(\omega\tau R + iR_b)}, \quad R \leq R_b, \tag{6.65}$$

where ϵ_{bulk} is the permittivity of bulk metal [99], $R_b \simeq 35.7$ nm is the mean free path of the electrons in bulk Au, $\omega_p \simeq 13.7 \times 10^{15}$ Hz is the plasma frequency, $\tau \simeq 2.53 \times 10^{-14} s$ is the relaxation time for the free electrons in gold, and R is the effective (restricted) mean free path. For solution-based gold nanorods, it is not untypical to have $R \sim 3$ nm.

The unique advantage of nanorod composites over the typical metamaterial is that the electron confinement scale (and thus the absorption) may be drastically changed by annealing the samples. The internal structure of Au in nanorods changes during annealing so that the electron mean free path can be controlled. In our studies, the samples were annealed in different regimes (100°C –300°C for 1–3 h), resulting in the estimated increase of the restricted mean free path from 3 to 10 nm, and a corresponding change in the permittivity as shown in Figure 6.17 [90].

Spectroscopic measurements show that a change in the restricted mean free path causes a dramatic effect on the optical properties of metamaterials in the ENZ regime: namely, in samples with a short mean free path, the strong extinction resonance that corresponds to $\epsilon_{xx} = 0$ is almost independent of the wavelength, and has a homogeneous dependence on the incident angle. In contrast to this behavior, in systems with larger mean free path, the resonance breaks down into a series of interference-like maxima, with inhomogeneous dependence on angle and wavelength that indicate nonlocal origin of the observed phenomenon.

Numerical analysis of the eigenmodes of periodic nanowire composites, performed with finite-element-based (FEM) solver [26], reveals that the system supports three different waves: one transverse-electric (TE) wave and two transverse-magnetic (TM) waves. The ENZ frequency range coincides with the regime of strong interaction between the two *TM* waves, signified by the avoided crossing between their dispersion curves. This is shown in Figure 6.18b, where the dispersion of the two TM modes obtained from FEM-based solutions of Maxwell's equations is shown with dots, and the solution corresponding to Equation 6.62, with lines. Numerical fitting of the FEM data with Equation 6.62 results in $\alpha_{xx} \simeq 0.04 - 0.006i$, $\alpha_{yz} = 0$.

The two TM waves identified in our numerical simulations represent main and additional TM waves supported by the nanorod metamaterial. The main wave with relatively smooth E_x profile (Figure 6.18e,f) corresponds to the lower TM mode for wavelengths

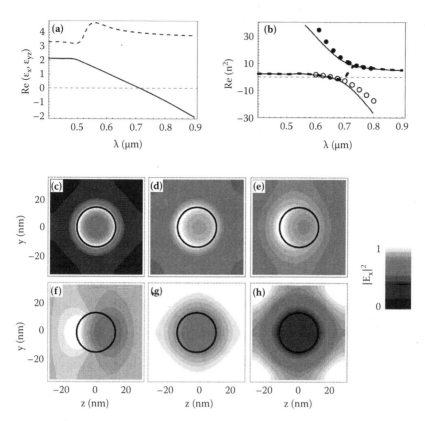

FIGURE 6.18 (a) Effective permittivities (real part) of nanorod metamaterial calculated with local EMT for $R = 10$ nm; solid and dashed lines correspond to ϵ_{xx} and ϵ_{yz}, respectively; nonlocal corrections are observed near $\epsilon_{xx} = 0$; (b) dispersion of the two TM-polarized modes calculated according to numerical simulations (dots) and after Equation 6.62 (solid lines), dashed black line corresponds to the main wave described by local EMT; (c–h) the profiles of electric field in the unit cell around the nanorod for the upper (c–e) and lower (f–h) branches for $\theta = 60°$; (c,f): $\lambda = 600$ nm, the "main" mode with uniform (almost constant) field (f) and additional wave with strongly nonuniform field (c) can be clearly distinguished; (d,g) regime of strong mode mixing; the profile of both modes is almost uniform, $\lambda = 700$ nm; (e,h) the main (e) and additional (h) waves can be distinguished again at $\lambda = 800$ nm. (From R.J. Pollard et al., *Phys. Rev. Lett.* 102, 127405, 2009.)

below the ENZ resonance and to the upper TM mode for larger wavelengths. Note that the behavior of this mode away from ENZ resonance is adequately described by local effective medium theory (dashed line in Figure 6.18b). At the same frequency range, the additional wave has a strongly nonuniform field distribution (Figure 6.18c,h) and can only weakly couple to incident plane waves. The strong interaction between these modes in proximity to ENZ resonance leads to substantial field mixing. As result of this process, the profiles of both main and additional waves become almost uniform, and both waves contribute to the optical response of the structure.

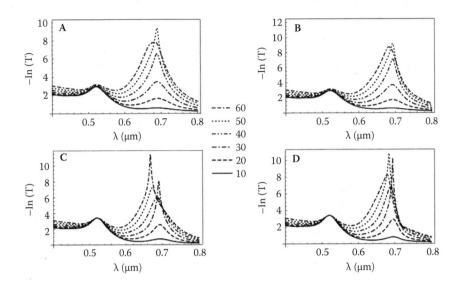

FIGURE 6.19 Extinction spectra calculated for different material parameters [$R = 5$ nm: (A,B); $R = 10$ nm: (C,D)] and different additional boundary conditions [$\vec{P}_{nonlocal} = 0$: (A,C); and $\partial \vec{P}_{nonlocal}/\partial \vec{n} = 0$: (b,d)]; Different lines correspond to different angles of incidence as defined in Figure 6.17; Panels (A,C) perfectly agree with experiment and 3D simulations.

Extinction of the nanowire composite taking into account nonlocal response of the metamaterial (Figure 6.19) is in perfect agreement with experimental results. Our simulations suggest that additional boundary conditions play only a quantitative role in the prediction of transmission through nanowire structure and that the condition $\vec{P}_{nonlocal} = 0$ should be used for nanowire systems. The latter assessment is consistent with the analytical analysis of Reference [100].

6.6 Summary

In this chapter, we have summarized the analytical techniques for description of optical properties of anisotropic (hyperbolic) metamaterials. Specific examples of hyperbolic metamaterials have been considered, and their unique optical properties have been discussed. Applications of anisotropic metamaterials for subwavelength light confinement, subwavelength propagation, and optical sensing were described. Finally, the nonlocal properties of metamaterials were presented.

Acknowledgment

This work has been supported in parts by the National Science Foundation (Grant # ECCS-0724763), Office of Naval Research via ONAMI-center initiative (Grant # N00014-07-1-0457), and Air Force Office of Scientific Research (Grant # FA9550-09-1-0029).

References

1. V.G. Veselago, The electrodynamics of substances with simultaneously negative values of ϵ and μ, *Soviet Phys. Uspekhi* 10, pp. 509–514, 1968.
2. V.M. Shalaev, W. Cai, U.K. Chettiar, H.-K. Yuan, A.K. Sarychev, V.P. Drachev, and A.V. Kildishev, Negative index of refraction in optical metamaterials, *Opt. Lett.*, 30, 3356, 2005.
3. G. Dolling, C. Enkrich, M. Wegener, C. M. Soukoulis, and S. Linden, Simultaneous negative phase and group velocity of light in a metamaterial, *Science* 312, 892–894, 2006.
4. J. Valentine, S. Zhang, T. Zentgraf, E. Ulin-Avila, D.A. Genov, G. Bartal, and X. Zhang, Three-dimensional optical metamaterial with a negative refractive index, *Nature* 455, 376, 2008.
5. J.B. Pendry and D.R. Smith, Reversing light with negative refraction, *Physics Today*, 57(6), pp. 37–43, 2004.
6. D.R. Smith, W.J. Padilla, D.C. Vier, S.C. Nemat-Nasser, and S. Shultz, Composite medium with simultaneously negative permeability and permittivity, *Phys. Rev. Lett.* 84, 4184–4187, 2000.
7. C. Parazzoli, R. Greegor, K. Li, B.E.C. Koltenbah, and M. Tanielian Experimental verification and simulation of negative index of refraction using Snells law, *Phys. Rev. Lett.* 90 107401-1–4, 2003.
8. V. M. Shalaev, Optical negative-index metamaterials, *Nat. Phot.* 1, 41–48, 2007.
9. J. Yao, Z. Liu, Y. Liu, Y. Wang, C. Sun, G. Bartal, A. M. Stacy, and X. Zhang, Optical negative refraction in bulk metamaterials of nanowires, *Science* 321, 930, 2008.
10. M. Born and E. Wolf, *Principles of Optics*, Cambridge U. Press, Cambridge, 1999.
11. L.D. Landau, E.M. Lifshitz, and L.P. Pitaevskii *Course of Theoretical Physics*, vol. 8, 2nd ed., Reed Ltd, Oxford, 1984.
12. J.D. Joannopoulos, S.G. Johnson, J. N. Winn, and R.D. Meade, *Photonic Crystals: Molding the Flow of Light*, Princeton U. Press,, Princeton, NJ, 2008.
13. A. Alú, and N. Engheta, in G. V. Eleftheriades and K. G. Balmain (eds.), *Negative Refraction Metamaterials: Fundamental Properties and Applications*, John Wiley & Sons, New York, 2005.
14. G. Shvets, Photonic approach to making a materials with a negative index of refraction, *Phys. Rev. B* 67, pp.035109-1–8, 2003.
15. L.M. Brekhovskikh, *Waves in Layered Media*, 2nd edition, Academic Press, New York, 1980.
16. M. Notomi, Theory of light propagation in strongly modulated photonic crystals: Refractionlike behavior in the vicinity of the photonic band gap, *Phys. Rev. B* 62 10696–10705, 2000.
17. A.A. Govyadinov and V.A. Podolskiy Metamaterials photonic funnels for sub-diffraction light compression and propagation, *Phys. Rev. B* 73, 155108, 2006.
18. E. Palik, ed., *The Handbook of Optical Constants of Solids*, Academic Press, New York, 1997.
19. R. Wangberg, J. Elser, E.E. Narimanov, and V.A. Podolskiy, Nonmagnetic nanocomposites for optical and infrared negative refraction index media, *J. Opt. Soc. Am. B* 23, 498, 2006.

20. A. J. Hoffman, L. Alekseyev, S.S. Howard, K.J. Franz, D. Wasserman, V.A. Podolskiy, E.E. Narimanov, D.L. Sivco, and C. Gmachl, Negative refraction in semiconductor metamaterials, *Nature Mat.* 6, 946, 2007.

21. P. Evans, W. R. Hendren, R. Atkinson, G. A. Wurtz, W. Dickson, A. V. Zayats, and R.J. Pollard, *Nanotechnology* 17, 5746–5753, 2006.

22. M.A. Noginov, Yu. A. Barnakov, G. Zhu, T. Tumkur, H. Li, and E.E. Narimanov, *Appl. Phys. Lett.* 94, 151105, 2009.

23. J.C.M. Garnett, *Philos. Trans. R. Soc. London*, Ser. B 203, 385, 1904.

24. J. Elser, R. Wangberg, V.A. Podolskiy, and E.E. Narimanov, Nanowire metamaterials with extreme optical anisotropy, *Appl. Phys. Lett.* 89, 261102, 2006.

25. G.W. Milton, *The Theory of Composites*, Cambridge U. Press, Cambridge, 2002.

26. For details see COMSOL Multiphysics User's Guide, AC/DC Module and RF Module User's Guide; COMSOL, 1994–2008); www.comsol.com.

27. K.F. Riley, M.P. Hobson, and S.J. Bence, *Mathematical Methods for Physics and Engineering*, 3rd ed., Cambridge U. Press, Cambridge, 1998.

28. J.B. Pendry, Negative refraction makes a perfect lens, *Phys. Rev. Lett.* 85, 3966–3969, 2000.

29. V.A. Podolskiy and E.E. Narimanov, Near-sighted superlens, *Opt. Lett.* 30, 75–77, 2005.

30. L.V. Alexeyev and E.E. Narimanov, Slow light and 3D imaging with nonmagnetic negative index systems, *Opt. Exp.* 14, 11184, 2006.

31. V.G. Veselago and E.E. Narimanov, The left hand of brightness: Past, present, and future of negative index materials, *Nat. Mat.* 5, 759, 2006.

32. P.A. Belov, Backward waves and negative refraction in uniaxial dielectrics with negative dielectric permittivity along the anisotropy axis, *Micr. Opt. Tech. Lett.* 37, 259, 2003.

33. S.M. Rytov, Electoromagnetic properties of a finely stratified medium, *Sov. Phys. JETP* 2, 466, 1956.

34. P. Yeh, A. Yariv, and C. Hong, Electromagnetic propagation in periodic stratified media. I. General theory, *J. Opt. Soc. Am.* 67, 423, 1977.

35. I. Avrutsky. *J. Opt. Soc. Am. B* 20 548, 2003.

36. S. Thongrattanasiri, J. Elser, and V.A. Podolskiy, Quasi-planar optics: computing light propagation and scattering in planar waveguide arrays, *J. Opt. Soc. Am. B* 26, B102, 2009.

37. A. Lewis, H. Taha, A. Strinkovski et.al. *Nat. Biotechnol.* 21 1378, 2003.

38. A. Grbic and G. Eleftheriades, *Phys. Rev. Lett.* 92, 117403, 2004.

39. N. Fung et.al., *Science* 308, 534, 2005.

40. R.J. Blaikie and D.O.S. Melville, *J. Opt. A* 7, S176, 2005.

41. R. Merlin, *Science* 317 927, 2007.

42. A. Grbic, L. Jiang, and R. Merlin, *Science* 320, 511, 2008.

43. L. Makley et.al. *Phys. Rev. Lett.* 101, 113901, 2008.

44. L.S. Dolin and Izv. VUZov, *Radiofizika*, 4, 964, 1961.

45. J. Pendry, D. Schurig, and D. Smith, *Science* 312, 1780, 2006.

46. A. Kildishev and V.M. Shalaev, *Opt. Lett.* 33, 43, 2008.

47. E.E. Narimanov and A.V. Kildishev, Optical black hole: Broadband omnidirectional light absorber, *Appl. Phys. Lett.* 95, 041106, 2009.
48. D.A. Genov, S. Zhang, and X. Zhang, and Mimicking celestial mechanics in metamaterials, *Nat. Phys.* v.5(a), p.687–692, 2009.
49. M.I. Stockman, *Phys. Rev. Lett.* 93, 137404, 2004.
50. C. Ropers, C.C. Neacsu, T. Elsaesser, M. Albrecht, M.B. Raschke, and C. Lienau, Grating-coupling of surface plasmons onto metallic tips: A nanoconfined light source, *Nano Lett.* 7, 2784, 2007.
51. Z. Jacob, L. Alekseyev, and E. Narimanov, *Opt. Exp.* 14, 8247, 2006.
52. A. Salandrino and N. Engheta, *Phys. Rev. B* 74, 075103, 2006.
53. Z. Liu et.al., *Science* 315, 1686, 2007.
54. I.I. Smolyaninov, Y.J. Hung, and C.C. Davis, *Science* 315, 1699, 2007.
55. S. Thongrattanasiri and V.A. Podolskiy, Hyper-gratings: Nanophotonics in planar anisotropic metamaterials, *Opt. Lett.* 34, 890, 2009.
56. V.A. Podolskiy and S. Thongrattanasiri, Hypergratings: Far-field subwavelength focusing in planar metamaterials, *Proc. SPIE*, vol. 7392, 73921A, eds. M.A. Noginov, N.I. Zheludev, A.D. Boardman, and N. Engheta, 2009.
57. C. Reinhardt et.al. *Appl. Phys. Lett.* 89, 231117, 2006.
58. V.A. Podolskiy and E.E. Narimanov, Strongly anisotropic waveguide as a nonmagnetic left-handed system, *Phys. Rev. B* 71, 201101(R), 2005.
59. V.A. Podolskiy, R. Wangberg, J. Elser, and E.E. Narimanov, Left-handed high energy density waveguides: Nano-light propagation and focusing, *Proc. SPIE* 6002, 600205, 2005.
60. H. Raether, *Surface Plasmons on Smooth and Rough Surfaces and on Gratings*, Springer-Verlag, Berlin, 1988.
61. J.N. Anker, W.P. Hall, O. Lyandres, N.C. Shah, J. Zhao, and R.P. Van Duyne, Biosensing with plasmonic nanosensors, *Nature Mat.* 7, 442, 2008.
62. A. Kabashin, P. Evans, S. Pastkovsky, W. Hendren, G. A. Wurtz, R. Atkinson, R. Pollard, V. Podolskiy, and A.V. Zayats, plasmonic nanorod metamaterials for biosensing accepted to *Nature Materials*, v. 8, p.867, 2009.
63. R.F. Oulton, D.F.P. Pile, Y. Liu, and X. Zhang, *Phys. Rev. B* 76, 035408, 2007.
64. J. Elser and V.A. Podolskiy, Scattering-free plasmonic optics with anisotropic metamaterials, *Phys. Rev. Lett.* 100, 066402, 2008.
65. T. Sondergaard and S.I. Bozhevolnyi, Out-of-plane scattering properties of long-range surface plasmon polariton gratings, *Phys. Stat. Sol. (b)*, 242, 3064, 2005.
66. V.A. Podolskiy and J. Elser, Electroplasmonics: Dynamical plasmonic circuits with minimized parasitic scattering, in *CLEO/QELS 2008*, OSA, Washington, DC, 2008, QTuJ2.
67. E. Betzig, J.K. Trautman, T.D. Harris, J.S. Weiner, and R.L. Kostelak, *Science* 251, 1468, 1991.
68. S.F. Mingaleev and Y.S. Kivshar, *J. Opt. Soc. Am. B* 19, 2241, 2002.
69. M.F. Yanik, S. Fan, M. Soljâcĉi, and J.D. Joannopoulos, *Opt. Lett.* 28, 2506, 2003.
70. D. Walba, *Science*, 270, 250, 1995.

71. Q. Xu, B. Schmidt, S. Pradhan, and M. Lipson, *Nature*, 435, 325, 2005.

72. A. Kramer, F. Keilmann, B. Knoll, and R. Guckenberger, *Micron*, 27, 413, 1996.

73. R. Marques, J. Martel, F. Mesa, and F. Medina, *Phys. Rev. Lett.* 89, 183901, 2002.

74. J. Baena, L. Jelinek, R. Marques, and F. Medina, *Phys. Rev. B*, bf 72, 075116, 2005.

75. P. Belov, and C. Simovski, *Phys. Rev. E*, 72, 036618, 2005.

76. S. Hrabar, *IEEE Trans. Ant. Prop.* 53, 110, 2005.

77. A. Alu, and N. Engheta, *IEEE Trans. Micr. Theory Tech.* 52, 199, 2004.

78. F. Demming, A. v-d Lieth, S. Klein, and K. Dickmann, *Adv. Func. Mat.* 11, 198, 2001.

79. S.A. Maier, P.G. Kik, H.A. Atwater, S.M. Meltzer, E. Harel, B.E. Koel, and A.G. Requicha, *Nature* 2, 229, 2003.

80. P.G. Kik, S.A. Maier, and H.A. Atwater, *Phys. Rev. B* 69, 045418, 2004.

81. A. Karalis, E. Lidorikis, M. Ibanescu, J.D. Joannopoulos, and M. Soljacic, *Phys. Rev. Lett.* 95, 063901, 2005.

82. S.I. Bozhevolnyi, V.S. Volkov, and K. Leosson, *Phys. Rev. Lett.* 89, 186801, 2002.

83. A. Bouhelier, J. Renger, M.R. Beversluis, and L. Novotny, *J. Microscopy*, 210, 220, 2002.

84. A.A. Govyadinov and V.A. Podolskiy, Metamaterials photonic funnels for sub-diffraction light compression and propagation, *Phys. Rev. B* 73(15), 155108, 2006.

85. A.A. Govyadinov and V.A. Podolskiy, Sub-diffraction light propagation in fibers with anisotropic metamaterial cores, *J. Mod. Opt.* 53, 2315, 2006.

86. Agranovich, V.M. and Ginzburg, V.L., *Crystal Optics with Spatial Dispersion and Excitons*, Springer Series in Solid-State Sciences, vol. 42, Springer-Verlag, Berlin, 1984.

87. Pekar, S.I., The theory of electromagnetic waves in a crystal in which excitons are produced, *Sov. Phys. JETP* 6, 785–796, 1958.

88. J. Elser, V.A. Podolskiy, I. Salakhutdinov, and I. Avrutsky, Non-local effects in effective-medium response of nano-layered meta-materials, *Appl. Phys. Lett.* 90, 191109, 2007.

89. J. Elser, A.A. Govyadinov, I. Avrustky, I. Salakhutdinov, and V.A. Podolskiy, Plasmonic nanolayer composites: Coupled plasmon polaritons, effective-medium response, and subdiffraction light manipulation, *J. Nanomaterials*, 2007, 79469, 2007.

90. R.J. Pollard, A. Murphy, W.R. Hendren, P.R. Evans, R. Atkinson, G.A. Wurtz, A.V. Zayats, and V.A. Podolskiy Optical nonlocalities and additional waves in epsilon-near-zero metamaterials, *Phys. Rev. Lett.* 102, 127405, 2009.

91. A.N. Lagarkov and A.K. Sarychev, Electromagnetic properties of composites containing elongated conducting inclusions, *Phys. Rev. B* 53, 6318–6336, 1996.

92. J.B. Pendry, A.J. Holden, W.J. Steward, and I. Youngs, Extremely low frequency plasmons in metallic mesostructures, *Phys. Rev. B* 76, 4773, 1996.

93. P.A. Belov, R. Marques, S.I. Maslovski, I.S. Nefedov, M. Silveirinha, C.R. Simovski, and S.A. Tretyakov, Strong spatial dispersion in wire media in the very large wavelength limit, *Phys. Rev. B* 67, 113103, 2003.

94. A.L. Pokrovsky and A.L. Efros, Nonlocal electrodynamics of two-dimensional wire mesh photonic crystals, *Phys. Rev. B* 65, 045110, 2002.

95. M. Silveirinha, Nonlocal homogenization model for a periodic array of epsilon-negative rods, *Phys. Rev. E* 73, 046612, 2006.

96. J.J. Hopfield and D.G. Thomas, Theoretical and experimental effects of spatial dispersion on the optical properties of crystals, *Phys. Rev.* 132, 563–572, 1963.

97. V.A. Kiselev, B.S. Razbirin, and I.N. Uraltsev, Additional waves and Fabri-Perot interference of photoexcitons (polaritons) in thin II-IV crystals, *Phys. Stat. Sol.* (b) 72, 161, 1975.

98. P.H. Lisseberger, and R.G. Nelson, Optical properties of think film Au-MgF$_2$ cements, *Thin Solid Films* 21, 159–172, 1974.

99. P.B. Johnson and R.W. Christy, Optical constants of noble metals, *Phys. Rev. B* 6, 4370–4379, 1972.

100. M.G. Silveirinha, Artificial plasma formed by connected metallic wires at infrared frequencies, *Phys. Rev. B*, 79, 035118, 2009.

7

Radiative Decay Engineering in Metamaterials

Leonid V. Alekseyev

Purdue University
West Lafayette Indiana

Evgenii Narimanov

School of Electrical and
Computer Engineering

7.1 Introduction

Interactions between emitters and nanophotonic structures are of central importance in contemporary optical science. Many devices in the fields of sensing, quantum information processing, and plasmonics, among others, may be modeled as dipoles interacting with some (possibly complex) medium. However, despite its fundamental importance, the problem of a dipole radiating in the presence of dielectrics is surprisingly nontrivial and has been the subject of a large body of research dating back to over a century. Even the most simple case—a dipole over a dielectric half-space—has required many decades to work out, with Sommerfeld, Weyl, Zenneck, and Norton among the many contributors [1–4].

Originally intended to study the propagation of Marconi's radio waves over a lossy Earth or sea, this problem is key to understanding the interaction of emitters with plasmonic structures, waveguides, and metamaterials. With this in mind, in this chapter we consider the problem of a classical dipole near a planar dielectric substrate. Although we only explicitly consider a half-infinite homogeneous substrate, we choose to use a method that readily generalizes to any stratified planar structure with a well-defined transmission function. Furthermore, we allow the substrate to be anisotropic, possibly admitting a dielectric tensor with differing signs of the principal components.

In addition to obtaining expressions for the electric fields, we study the emission characteristics of a dipole near a substrate, focusing on its radiative lifetime. As shown by Purcell, radiative properties of an emitter can be strongly influenced by its environment [5]. Accordingly, a variety of schemes for controlling radiative decay have been developed [6]. In recent years, many such schemes have exploited nanophotonic structures [7,8].

We start by solving the problem of a dipole in vacuum and derive its plane wave decomposition, which allows to treat the presence of planar structures via the usual Fresnel reflection and transmission coefficients. We proceed to show how to use the resultant solution in deriving the radiative lifetime of a dipole. When dealing with nonanalytic solutions, we take care to discuss potential difficulties in their numerical implementation.

7.2 Field of an Electric Dipole

7.2.1 Space Domain

The electric field obeys

$$\nabla \times \nabla \times E - k^2 E = i\omega\,\mu j. \tag{7.1}$$

On the other hand, the field (assuming the Lorenz gauge) also satisfies

$$E = i\omega\left[A + \frac{1}{k^2}\nabla(\nabla \cdot A) \right], \tag{7.2}$$

where $k \equiv \omega/c$ and the vector potential A satisfies

$$-(\nabla^2 + k^2)A = \mu j. \tag{7.3}$$

We define a dyadic Green's function G as

$$\nabla \times \nabla \times G - k^2 G = 1\delta(r - r'), \tag{7.4}$$

where the vector operators are understood to act on each column of G separately. Each ith column of G can be regarded as a field due to a point current source $j = (1/i\omega\mu)\,\hat{e}_i\delta(r - r')$. It follows from Equation 7.3 that the corresponding vector potential obeys

$$-(\nabla^2 + k^2)A = \frac{1}{i\omega}\,\hat{e}_i\delta(r - r'). \tag{7.5}$$

The solution of this equation is $A = \frac{1}{i\omega}g(r,r')\hat{e}_i$, where $g(r, r')$ is the well-known scalar Green's function for the Helmholtz equation:

$$-(\nabla^2 + k^2)g = \delta(r - r'); \tag{7.6}$$

$$g(r,r') = \frac{e^{ik|r-r'|}}{4\pi\,|r - r'|}. \tag{7.7}$$

Plugging this result into Equation 7.2, we obtain for the ith column of $\mathbf{G}$

$$G_i\left(\hat{e}_i + \frac{1}{k^2}\nabla\partial_i\right)g(\mathbf{r}-\mathbf{r}'),$$

which gives for the final form of the dyadic Green's function:

$$G = \left(1 + \frac{1}{k^2}\nabla\nabla\right)g(\mathbf{r},\mathbf{r}'). \tag{7.8}$$

The electric field can now be found from the given source current as

$$E(\mathbf{r}) = i\omega\mu\int G(\mathbf{r},\mathbf{r}')\cdot j(\mathbf{r}')d\mathbf{r}', \tag{7.9}$$

or (using Equation 7.7) as

$$E(\mathbf{r}) = i\omega\mu\left(1 + \frac{1}{k^2}\nabla\nabla\right)\int\frac{e^{ik|\mathbf{r}-\mathbf{r}'|}}{4\pi|\mathbf{r}-\mathbf{r}'|}j(\mathbf{r}')d\mathbf{r}'. \tag{7.10}$$

We can use this formula to find the fields due to a Hertzian dipole. To avoid introducing local field corrections [9,10], the surrounding medium is taken to be vacuum. Such dipole can be represented by a delta-function current source at a height h above the $z = 0$ plane, with $\mathbf{m}$ giving its strength and orientation:

$$j(\mathbf{r},t) = -i\omega\mathbf{m}\exp(-i\omega t)\delta(\mathbf{r} - h\hat{z}). \tag{7.11}$$

The electric field due to this dipole is given by

$$E(\mathbf{r}) = -i\omega\mu\left(1 + \frac{1}{k^2}\nabla\nabla\right)\left(-i\omega\mathbf{m}\frac{e^{ik|\mathbf{r}-h\hat{z}|}}{4\pi|\mathbf{r}-h\hat{z}|}\right). \tag{7.12}$$

It is straightforward to compute the actual expression. For example, in the case of a vertically oriented dipole ($\mathbf{m} = m\hat{z}$), the electric field becomes

$$E(\mathbf{r}) = \frac{\mu m\omega^2}{4\pi k^2}e^{ikR}\left[\left(\frac{k^2}{R} + \frac{ik}{R^2} - \frac{1}{R^3}\right)\hat{z} + \left(-\frac{k^2}{R^3} - \frac{3ik}{R^4} + \frac{3}{R^5}\right)(z-h)\mathbf{R}\right], \tag{7.13}$$

where $\mathbf{R} = x\hat{x} + y\hat{y} + (z-h)\hat{z}$ and $R \equiv |\mathbf{R}|$.

7.2.2 Spatial Frequency Domain

When considering a dipole in the presence of dielectric or conducting media, it is often necessary to compute the reflected and transmitted fields, or to study the effects of surface modes. Such calculations require decomposing the electric field into its spatial frequency spectrum. We will be using the following Fourier expansion:

$$E(r,\omega) = \int E(E,\omega)\exp(i k \cdot r)dk. \tag{7.14}$$

To get the dipole fields, we start with Equation 7.12, writing it as

$$E(r,\omega) = \frac{\omega^2 \mu_0}{(\omega/c)^2}[(\omega/c)^2 \mathbf{1} + \nabla\nabla]mg(r,r')\big|_{r'=hz}. \tag{7.15}$$

We can write

$$\nabla\nabla mg(r,r') = \nabla[\nabla \cdot mg(r,r')]$$

$$= \nabla\times\nabla\times[mg(r,r')] - m[\delta(r-r') + (\omega/c)^2 g(r,r')],$$

where we used Equation 7.6. With this substitution, Equation 7.15 becomes

$$E(r,\omega) = \frac{\omega^2\mu_0}{(\omega/c)^2}\{-m\delta(r-r') + \nabla\times\nabla\times[mg(r,r')]\}\big|_{r'=h\hat{z}}. \tag{7.16}$$

In our Fourier representation, we have

$$\delta(r-r') = \frac{1}{(2\pi)^3}\int e^{ik\cdot(r-r')}dk.$$

$$g(r-r') = \frac{1}{(2\pi)^3}\int \frac{e^{ik\cdot(r-r')}}{k^2-(\omega/c)^2}dk \tag{7.17}$$

$$= -\frac{1}{(2\pi)^3}\int dk_\perp e^{ik\perp\cdot r\perp}\int dk_z \frac{e^{i(k_z(z-h))}}{(\omega/c)^2-k_\perp^2-k_z^2}.$$

We can perform the integral

$$I_0 \equiv \int dk_z \frac{e^{ikz(z-h)}}{(q_z+k_z)(q_z-k_z)},$$

where we defined $q_z^2 := (\omega/c)^2 - k_\perp^2$. For $z > h$, we close the contour in the upper half-plane, in which case Cauchy's Residue Theorem gives

$$I_0 = -2\pi i \frac{e^{iq_z(z-h)}}{2q_z}$$

as the value of the integral. For $z < h$, we close the contour in the lower half-plane, in which case we get

$$I_0 = -2\pi i \frac{e^{iq_z(z-h)}}{2q_z}.$$

We therefore obtain

$$g(\mathbf{r},\mathbf{r}'=h\hat{z}) = \frac{i}{(2\pi)^2}\int d\mathbf{k}_\perp e^{i\mathbf{k}_\perp \cdot \mathbf{r}_\perp} \frac{1}{2q_z} e^{iq_z|z-h|}. \tag{7.18}$$

To make it more convenient to work with the $|z-h|$ term in our equations, we define
$\mathbf{k}_1 := k_x\hat{\mathbf{x}} + k_y\hat{\mathbf{y}} + q_z \operatorname{sign}(h-z)\hat{z}$.

We can finally express Equation 7.16 as an integral over the transverse wave vector $\mathbf{k}_\perp$:

$$\mathbf{E} = \frac{\omega^2\mu_0}{(\omega/c)^2(2\pi)^2}\int d\mathbf{k}_\perp e^{i\mathbf{k}_\perp \cdot \mathbf{r}_\perp}\left[-\mathbf{m}\delta(z-h) - i\mathbf{k}_1 \times \mathbf{k}_1 \times \mathbf{m}\frac{e^{iq_z|z-h|}}{2q_z}\right]. \tag{7.19}$$

In order to enable separate treatment of the two polarization components, we rewrite this equation as

$$\mathbf{E} = \frac{1}{\epsilon_0}\frac{1}{(2\pi)^2}\int d\mathbf{k}_\perp e^{i\mathbf{k}_\perp \cdot \mathbf{r}_\perp}\left\{-\mathbf{m}\delta(z-h) + i[(\mathbf{m}\cdot\mathbf{p})\mathbf{p} + (\mathbf{m}\cdot\mathbf{s})\mathbf{s}]\frac{e^{iq_z|z-h|}}{2q_z}\right\}, \tag{7.20}$$

where $\mathbf{p} = k_\perp\hat{z} + q_z\operatorname{sign}(h-z)\hat{\mathbf{k}}_\perp$ and $\mathbf{s} = (\hat{z}\times\hat{\mathbf{k}}_\perp)\sqrt{k_\perp^2 + q_z^2} = (\hat{z}\times\hat{\mathbf{k}}_\perp)\sqrt{\epsilon\mu}(\omega/c)$ (in the case of isotropic medium) are vectors indicating P- and S-polarization.

To make these equations more suitable for numerical computations, it is advantageous to rewrite the integral in polar coordinates and integrate over the angle variable. In particular, we make the substitutions

$$\mathbf{k}_\perp \cdot \mathbf{r}_\perp \to kr\cos(\theta),\ d\mathbf{k}_\perp \to kdkd\theta, \hat{\mathbf{k}}_\perp \to R_z(\theta).\hat{\mathbf{r}}_\perp, \tag{7.21}$$

with $R_z(\theta)$ being the usual matrix of rotations around the z axis. Performing the θ integral gives

$$E = \frac{1}{\epsilon_0}\frac{1}{(2\pi)^2}\int k\ dk\Bigg\{-J_0(kr)\mathbf{m}\delta(z-h)+$$

$$+i\begin{bmatrix}\left(-\dfrac{\tilde{q}_z^2 J_1(kr)(m_x(x^2-y^2)+2xym_y)}{kr^3} + \dfrac{x\tilde{q}_z^2 J_0(kr)(xm_x+ym_y)}{r^2} + \dfrac{ikxm_z\tilde{q}_z J_1(kr)}{r}\right)\\[1.2em] \left(\dfrac{\tilde{q}_z^2 J_1(kr)(m_y(x^2-y^2)+2xym_x)}{kr^3} + \dfrac{y\tilde{q}_z^2 J_0(kr)(xm_x+ym_y)}{r^2} + \dfrac{ikym_z\tilde{q}_z J_1(kr)}{r}\right)\\[1.2em] k^2m_z J_0(kr) + \dfrac{ik\tilde{q}_z J_1(kr)(xm_x+ym_y)}{r}\end{bmatrix}$$

$$+\begin{bmatrix}\left(\dfrac{J_1(kr)(m_x(x^2-y^2)+2xym_y)}{kr^3} + \dfrac{yJ_0(kr)(ym_x+ym_y)}{r^2}\right)\\[1.2em] \left(\dfrac{J_1(kr)(m_y(y^2-x^2)+2xym_x)}{kr^3} + \dfrac{xJ_0(kr)(xm_y+ym_x)}{r^2}\right)\end{bmatrix}\frac{e^{-i\tilde{q}_z(z-h)}}{2|\tilde{q}_z|}\Bigg\}, \tag{7.22}$$

where $k=\sqrt{k_x^2+k_y^2}$, $r=\sqrt{x^2+y^2}$, $\tilde{q}_z := sign(h-z)\sqrt{(\omega/c)^2-k^2}$, and the column elements are the Cartesian x, y, z components of the field. Note that the first term in the square brackets corresponds to P polarization, and the second term gives the S polarization.

For the case of a vertically oriented dipole ($m = m\hat{z}$), the S polarization component vanishes, and the P polarization component becomes substantially more tractable:

$$E=-\frac{1}{\epsilon_0}\frac{im}{(2\pi)^2}\int k^2\,dk\left[\frac{\tilde{q}_z}{r}(x\hat{x}+y\hat{y})iJ_1(kr)+kJ_0(kr)\hat{z}\right]\frac{e^{-i\tilde{q}_z(z-h)}}{2|\tilde{q}_z|}.\qquad(7.23)$$

(Note that from this point onwards we will drop the delta-function term at the location of the dipole, since we are interested in fields elsewhere.)

Although the resulting expression appears considerably simpler, there remain certain difficulties when treating this integral numerically. First, the integrand is a diverging oscillatory function in the plane of the dipole, $z = h$. It is therefore necessary to compute the integral for $z = h + \epsilon$ and follow a limiting procedure as $\epsilon \to 0$. Second, the integrand oscillates rapidly for $kr \gg 1$; as a result, standard numerical integration algorithms fail or suffer inaccuracies. This problem can be avoided with a change of variables, effectively integrating over the quantity kr. In particular, if we denote the integral in Equation 7.23 with a and b at the limits of integration as $I_1(a, b)$, we have

$$E=\frac{1}{\epsilon_0}\frac{im}{(2\pi)^2}[I_1(0,\omega/c)+I_1(,\omega/c,\infty)]$$

$$=\frac{1}{\epsilon_0}\frac{im}{(2\pi)^2}\left\{I_1(0,\omega/c)+\int_{\xi_0}^{\infty}\frac{\xi^2}{r^2}\,d\xi\times\right.\qquad(7.24)$$

$$\left.\times\left[sign(h-z)\frac{\sqrt{\xi_0^2-\xi^2}}{r}(x\hat{x}+y\hat{y})iJ_1(\xi)+\frac{\xi}{r}J_0(\xi)\hat{z}\right]\frac{e^{i\sqrt{\xi_0^2-\xi^2}|z-h|/r}}{\sqrt{\xi_0^2-\xi^2}}\right\},$$

where $\xi \equiv kr$, $\xi_0 = r\omega/c$.

7.3 Reflection and Transmission of Dipole Radiation

We now consider the problem of a dipole radiating above a homogeneous half-infinite medium with the boundary at $z = 0$, as illustrated in Figure 7.1. If the plane wave reflection and transmission coefficients at the boundary are known, with the help of Equation 7.20 we can compute the field everywhere in space. (This approach works not just for a half-infinite medium, but in fact for any stratified planar structure with a well-defined transfer function.) The exact expression is

$$E=\frac{i}{2(2\pi)^2}\int dk_\perp e^{ik_\perp\cdot r_\perp}\frac{1}{\epsilon_0}\frac{(m\cdot p_i)}{q_z^{(1)}}\left\{[p_ie^{iq_z^{(1)}|z-h|}+\right.$$

$$\left.+r_p p_r e^{iq_z^{(1)}(z+h)}]\theta(-z)+\epsilon^{-1}t_p p_t e^{-iq_z^{(2p)}z}e^{iq^{(1)}h}\theta(z)\right\},\qquad(7.25)$$

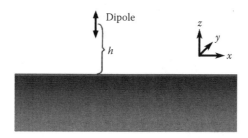

FIGURE 7.1 Geometry of the problem. The dipole is situated in vacuum in the vicinity of a planar substrate, with $z = 0$ as the boundary.

where $q_z^{(1)}$ and $q_z^{(2p)}$ denote the propagation vector in medium 1 and 2, respectively, ϵ (assumed to be diagonal) is the dielectric tensor in medium 2, and $\theta(z)$ is the step function. We have, furthermore,

$$\boldsymbol{p}_i = k_\perp \hat{z} + q_z^{(1)} \text{sign}(h-z)\hat{\boldsymbol{k}}_\perp$$

$$\boldsymbol{p}_r = k_\perp \hat{z} - q_z^{(1)}\hat{\boldsymbol{k}}_\perp \qquad (7.26)$$

$$\boldsymbol{p}_t = k_\perp \hat{z} + q_z^{(2p)}\hat{\boldsymbol{k}}_\perp \cdot$$

A very similar expression can be written for the S polarization:

$$E = \frac{i}{2(2\pi)^2} \int dk_\perp e^{ik_\perp \cdot r_\perp} \frac{1}{\epsilon_0} \frac{(\boldsymbol{m} \cdot \boldsymbol{s}_i)}{q_z^{(1)}} \left\{ [s_i e^{iq_z^{(1)}|z-h|} + \right.$$

$$\left. + r_s s_r e^{iq_z^{(1)}(z+h)}]\theta(-z) + \epsilon^{-1} t_p s_t e^{-iq_z^{(2s)}z} e^{iq^{(1)}h}\theta(z) \right\}, \qquad (7.27)$$

with $\boldsymbol{s}_i = \boldsymbol{s}_r = (\hat{z} \times \hat{\boldsymbol{k}}_\perp)\sqrt{k_\perp^2 + [q_z^{(1)}]^2}$; $\boldsymbol{s}_t = (z \times \hat{\boldsymbol{k}}_\perp)\sqrt{k_\perp^2 + [q_z^{(2s)}]^2}$.

Specifically, we would like to return to the case of a vertically oriented dipole and consider its field in the vicinity of a uniaxial anisotropic material. The dielectric tensor of this material is

$$\epsilon = \begin{pmatrix} \epsilon_x & 0 & 0 \\ 0 & \epsilon_x & 0 \\ 0 & 0 & \epsilon_x \end{pmatrix}, \qquad (7.28)$$

and the propagation vectors in this material are

$$q_z^{(2p)} = \sqrt{\epsilon_x \mu \left(\frac{\omega}{c}\right)^2 - \frac{\epsilon_x}{\epsilon_z} k_\perp^2} \qquad (7.29)$$

$$q_z^{(2s)} = \sqrt{\epsilon_x \mu \left(\frac{\omega}{c}\right)^2 - k_\perp^2}. \qquad (7.30)$$

The reflection and transmission coefficients are given by the standard formulas:

$$r_p = \frac{\epsilon_s^{(2)} q_z^{(1)} - \epsilon_x^{(1)} q_z^{(2)}}{\epsilon_x^{(2)} q_z^{(1)} + \epsilon_x^{(1)} q_z^{(2)}} \qquad t_p = \frac{2\epsilon_x^{(1)} q_z^{(1)}}{\epsilon_x^{(2)} q_z^{(1)} + \epsilon_x^{(1)} q_z^{(2)}} \tag{7.31}$$

As before, we can introduce polar coordinates and integrate over the angular dimension. With this approach, we can write the electric field of a z-oriented dipole over an anisotropic half-space as

$$
E = \frac{1}{\epsilon_0} \frac{im}{2(2\pi)^2} \int k^2 dk \frac{1}{q_z^{(1)}} \frac{1}{r} \left\{ \left[\begin{pmatrix} \tilde{q}_z^{(1)} x i J_1(kr) \\ \tilde{q}_z^{(1)} y i J_1(kr) \\ kr J_0(kr) \end{pmatrix} e^{-i\tilde{q}_z(z-h)} + \right. \right.
$$

$$
\left. + r_p \begin{pmatrix} -\tilde{q}_z^{(1)} x i J_1(kr) \\ -\tilde{q}_z^{(1)} y i J_1(kr) \\ kr J_0(kr) \end{pmatrix} e^{i\tilde{q}_z^{(1)}(z+h)} \right] \theta(-z) +
$$

$$
\left. + t_p \begin{pmatrix} q_z^{(2p)} x i J_1(kr)/\epsilon_x \\ q_z^{(2p)} y i J_1(kr)/\epsilon_x \\ kr J_0(kr)/\epsilon_z \end{pmatrix} e^{-i q_z^{(2p)} z} e^{i q^{(1)} h} \theta(z) \right\} \tag{7.32}
$$

(as before, column vectors correspond to the Cartesian components of the field).

This integral can be readily computed numerically using methods described above Equation 7.24. Figure 7.2 shows sample field intensity plots for the case of a strongly anisotropic "hyperbolic" metamaterial (i.e., $\epsilon_x > 0$, $\epsilon_z < 0$, leading to hyperbolic dispersion relation [11]) in the effective medium approximation [12].

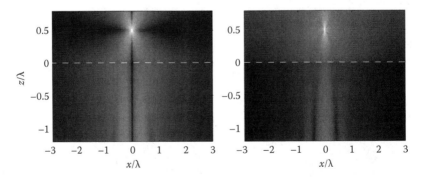

FIGURE 7.2 Electric field intensity for a vertical ($m = m\hat{z}$) electric dipole over a strongly anisotropic metamaterial substrate ($\epsilon_x = 2$, $\epsilon_z = -2$). Panels (a) and (b), respectively, show the x and z components of the electric field.

7.4 Radiative Lifetime and Spontaneous Decay Rates

An important characteristic of a radiating dipole is its rate of energy dissipation. It has long been known that this rate can be modified by the external environment. Much of the past work focused on studying dipole radiation in resonant cavities. In recent years, due to rising interest in waveguide and metamaterial structures, changes in dipole lifetime in close proximity to planar structures or stratified media are often helpful to consider. It turns out that these properties are readily derived from results obtained earlier.

The general expressions for computing power radiated by the dipole come from Poynting's theorem for harmonic fields:

$$\frac{dW}{dt} = -\frac{1}{2}\int \text{Re}[\boldsymbol{j}^* \cdot \boldsymbol{E}]d\boldsymbol{r} = \frac{\omega}{2}\text{Im}[\boldsymbol{m}^* \cdot \boldsymbol{E}(h\hat{z})], \quad (7.33)$$

where we used the dipole current defined in Equation 7.11.

We can now use Equation 7.20 to obtain the Larmor radiation formula for a dipole in vacuum. For this, we simply take the limit of the integrand as $r_\perp \to 0$, $z \to h$, rewrite the expression in polar coordinates, and take the angular integral. Note that the delta-function term in Equation 7.20 vanishes when taking the imaginary part in Equation 7.33. We are left with

$$\frac{dW}{dt} = \text{Im}\left\{ \frac{i\omega}{16\pi\epsilon_0} \int_0^\infty \left[\frac{k}{q_z}(k^2(|m_\perp|^2 + 2|m_z|^2) + 2q_z^2(|m_\perp|^2)) \right] dk \right\}, \quad (7.34)$$

where $|m_\perp|^2 \equiv |m_x|^2 + |m_y|^2$. Because we are taking the imaginary part, we require the integrand here to be purely real. Due to the q_z term, the integrand is in fact purely real for $k \in [0, \omega/c]$ and purely imaginary for $k > \omega/c$. Picking ω/c as the upper integration limit, we obtain the familiar Larmor formula:

$$\frac{dW}{dt} = \frac{|\boldsymbol{m}|^2 \omega^4}{12\pi\epsilon_0 c^3}. \quad (7.35)$$

The same procedure can be used to compute the power radiated by a dipole in the presence of a homogeneous or stratified medium. To do this, Equations 7.25 and 7.27 must be taken as the starting point. The integrals are only somewhat more involved. Because of the complex reflection coefficients, we can no longer decompose the integrand into a purely real and imaginary part, and hence the solution remains as an integral over all values of the transverse wave vector component.

We now proceed to compute a closely related quantity, and one that is often of greater practical importance: the decay rate γ. The radiative power dissipated by the dipole can be related to its decay rate via the semiclassical expression [13]

$$\gamma = P/\hbar\omega. \quad (7.36)$$

For any realistic emitter, particularly in the presence of other media, the total decay rate is a sum of radiative (γ_{rad}) and nonradiative (γ_{nr}) contributions. We can define the quantum efficiency η as

$$\eta = \frac{\gamma_{rad}}{\gamma_{rad} + \gamma_{nr}}. \tag{7.37}$$

We are interested in computing the decay rate as a function of the separation distance h between the dipole and the surface. We normalize the decay rate by that of a dipole in free space (meaning that it still admits nonradiative decay channels, but the radiative power is given precisely by Equations 7.35 and 7.36). We write this decay rate as

$$\tilde{\gamma}(h) = \frac{\gamma}{\gamma_0} \frac{\gamma_{nr} + \gamma_{rad}}{\gamma_{nr} + P_0/\hbar\omega}, \tag{7.38}$$

where P_0 is given by the Larmor formula, Equation 7.35. Because of the presence of scattered fields, the radiative decay rate term in the numerator can be written as $\gamma_{rad} = (P_0 + P_s)/\hbar\omega$, where P_s gives the radiated power computed via Equation 7.33 applied only to the reflected fields. Furthermore, assuming the quantum efficiency in Equation 7.37 is independent of h, we can express γ_{nr} in terms of η and P_0, resulting in a very simple expression for the reduced decay rate:

$$\tilde{\gamma}(h) = 1 + \eta \frac{P_s}{P_0}. \tag{7.39}$$

It is straightforward to compute P_s by starting with Equations 7.25 and 7.27, keeping only the terms with the reflection coefficients r_p and r_s. The result is

$$\tilde{\gamma}(h) = 1 + \eta \text{Re} \frac{3c^3}{4|m|^2 \omega^3} \int_0^\infty k\,dk[r_s|m_\perp|^2(k^2+q_z^2) - $$
$$-r_p(|m_\perp|^2 q_z^2 - 2|m_z|^2 k^2)]\frac{e^{2ihq_z}}{q_z}. \tag{7.40}$$

In Figure 7.3, we plot the normalized lifetime ($\tau \equiv \tilde{\gamma}^{-1}$) of a dipole as a function of its height above a substrate. The lifetime of the dipole in the vicinity of a medium (closer than about one vacuum wavelength) is seen to depend strongly on the nature of the material. Here we consider the case of a lossless dielectric, a metal (with loss), and a strongly anisotropic (hyperbolic) metamaterial (lossless). For the metallic substrate, the dipole lifetime approaches zero as the separation between metal and dipole vanishes: the emission is effectively quenched [14,15]. Similarly, the dipole lifetime tends to zero when metal is replaced with a hyperbolic medium. The dielectric substrate, in contrast, never completely suppresses the fluorescence.

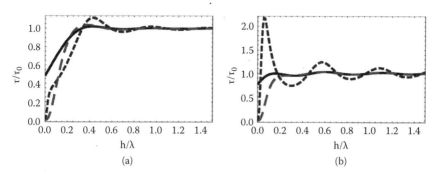

FIGURE 7.3 Normalized lifetime of a vertically [panel (a)] and horizontally [panel (b)] oriented dipole as dependent on distance h from a half-infinite material in the horizontal plane. Materials considered are dielectric, $\epsilon = 2$ (solid curves), strongly anisotropic metamaterial, $\{\epsilon_x, \epsilon_y\} = \{2, -5\}$ (dashed curves), and metal, $\epsilon = -30 + 8i$ (finely dashed curves).

We can understand the origin of this different behavior by examining the integrand of Equation 7.40 in the limit $h \ll \lambda$. In this limit, contributions to the integrand that arise from large values of the transverse wave vector, $k_x \gg \omega/c$, become important. With this in mind, we assume $k_x \in [\omega/c, \infty]$ and write $q_z \equiv i\kappa = i\sqrt{k_x^2 - (\omega/c)^2}$. For simplicity here (and for the rest of this chapter), we assume the dipole to be vertically oriented $(\boldsymbol{m} = \hat{z})$. We can then write Equation 7.40 as

$$\tilde{\gamma}(h) \approx 1 + \eta \frac{3}{2} \frac{1}{(\omega/c)^3} \int_{\omega/c}^{\infty} \mathrm{Im}(r_p) \frac{k_x^3}{\kappa} e^{-2h\kappa} dk_x$$

(7.41)

(we emphasize that the integration variable is the transverse wave vector coordinate k_x). Assuming that the leading contribution to the integral in Equation 7.40 comes from large wave vectors $(k_x \gg \omega/c)$, we can write $\kappa \approx k_x$. Furthermore, r_p (as given by Equation 7.31) becomes independent of k_x to leading order. We can, therefore, obtain a simple analytical solution for the decay rate. For the case of metallic substrate, it is

$$\tilde{\gamma}(h) \approx 1 + \eta \frac{3}{4} \frac{1}{(h\omega/c)^3} \mathrm{Im}\left(\frac{\epsilon - 1}{\epsilon + 1}\right)$$

$$= 1 + \eta \frac{3}{2} \frac{1}{(h\omega/c)^3} \frac{\epsilon''}{(1 + \epsilon')^2 + \epsilon''^2},$$

(7.42)

where we used the usual definition $\epsilon \equiv \epsilon' + i\epsilon''$.

We conclude that in the vicinity of a lossy substrate (be it metal or dielectric) the emitter lifetime vanishes as h^3 as the separation h between the emitter and the substrate goes to zero. The physical origin of this dependence is the nonradiative transfer of energy into leaky waves propagating along the surface of the substrate, which come to dominate over other decay channels at very close distances [16]. In light of this discussion, it is reasonable that the lifetime of an emitter plotted in Figure 7.3 tends to zero for a lossy metal, but approaches a finite limiting value in the case of a lossless dielectric. Furthermore, we can understand why a *lossless* hyperbolic metamaterial substrate induces a quenching effect similar to that observed for lossy materials. By examining Equations 7.29 and 7.31, we can see that, for the hyperbolic material with $\epsilon_x > 0$ and $\epsilon_z < 0$, $\mathrm{Im}(r_p) \neq 0$ even if the dielectric tensor is purely real; as a result, we get a similar h^3 dependence of dipole lifetime on height above the substrate. In this case, the energy is transferred to the bulk propagating waves inside the hyperbolic material. Indeed, one can view this transfer of energy as arising from the increased (formally infinite) density of photonic states that characterizes the hyperbolic medium [17].

In addition to the basic material parameters, device geometry plays a significant role in determining radiative decay characteristics. To provide a simple illustration of this, we turn out attention to slab waveguides (note that a filled half-space can be regarded as a limiting case of a waveguide with thickness $d \rightarrow \infty$). To determine radiative decay rate in the vicinity of a waveguide, we use Equations 7.40 and 7.41 with r_p given by

$$r_p = \frac{\frac{\epsilon_x^{(2)} q_z^{(1)}}{\epsilon_x^{(1)} q_z^{(2)}} - \frac{\epsilon_x^{(1)} q_z^{(2)}}{\epsilon_x^{(2)} q_z^{(1)}}}{\frac{\epsilon_x^{(2)} q_z^{(1)}}{\epsilon_x^{(1)} q_z^{(2)}} + \frac{\epsilon_x^{(1)} q_z^{(2)}}{\epsilon_x^{(2)} q_z^{(1)}} + 2i\cot(dq_z^{(2)})}, \tag{7.43}$$

where $q_z^{(2)}$ is given by Equation 7.29, $q_z^{(1)} = \sqrt{(\omega/c)^2 - k_x^2}$, and $\epsilon_x^{(1)} = 1$ for the case of a dipole in vacuum that we treat here.

In Figure 7.4, we plot the normalized lifetime t of the vertically oriented dipole as a function of its height above a metallic or hyperbolic waveguide (note the logarithmic scale on both axes). The different panels correspond to different waveguide thicknesses ($d = \infty$, $d = 0.01\lambda$, and $d = 0.001\lambda$), and metallic and hyperbolic curves are plotted on the same axes for ease of comparison; in addition, we plot the lifetime for several values of material losses ($\epsilon'' \in \{0, 0.001, 1\}$). Since the excitation of lossy waves provides an important decay channel (as shown earlier), it is instructive to consider the dipole lifetime in the limit of zero losses, gradually increasing them to more realistic values. It is interesting to note that, even though the zero-loss assumption substantially simplifies many problems in optics and electromagnetics, in this particular case it leads to increased

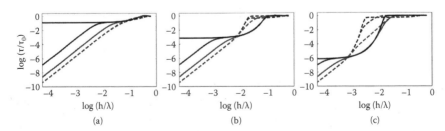

FIGURE 7.4 Normalized lifetime of a vertically oriented dipole versus distance from a metallic ($\epsilon' = -5$; solid curves) or hyperbolic ($\{\epsilon'_x, \epsilon'_y\} = \{5, -5\}$; dashed curves) waveguide; note the log–log scale. Different curves correspond to varying the amount of loss ($\epsilon'' \in \{0, 0.001, 1\}$), with higher losses resulting in lower lifetime. Panels (a), (b), and (c) correspond to waveguide thicknesses $d = \infty$, $d = 0.01\lambda$, and $d = 0.001\lambda$, respectively.

difficulties, since the denominator of Equation 7.43 acquires infinitely sharp resonances (two, corresponding to the excitation of surface plasmons in the case of the metallic slab, and infinitely many, corresponding to guided wave modes for a slab of hyperbolic material), causing simple numerical integration techniques to fail. However, the zero–loss limit admits a semi-analytic solution (see Appendix), simplifying its treatment.

The analysis of Figure 7.4 yields several interesting conclusions. Panel (a) allows for a clear comparison between half-infinite metallic and hyperbolic substrates as discussed earlier. The $t \sim h^3$ dependence for both types of materials shows clearly on the log–log scale. Furthermore, for the case of hyperbolic medium, the dipole lifetime is *independent* of the amount of losses. In contrast, for a metallic substrate, the dipole lifetime is a constant (for $h \ll \lambda$) in the zero-loss limit, and decreases with increasing loss (as expected from Equation 7.42), matching the lifetime of the hyperbolic medium only for a relatively high loss of $\epsilon'' \approx 5$. This suggests that excitation of propagating high-k_x waves in the hyperbolic material provides a decay channel for the emitter that is more efficient than coupling to lossy surface waves. By examining panels (b) and (d), we can identify two distinct regimes: $h \ll d \ll \lambda$ and $d \lesssim h \ll \lambda$. In the first regime, there is little difference between the waveguide systems and the half-infinite substrate; indeed, the only readily quantifiable effect appears in the lossless limit of a metallic substrate, where lifetime drops with decreasing d (log $t \sim$ log d). Ideal lossless hyperbolic waveguides, on the other hand, support an infinite number of modes for an arbitrary thickness; this singularity in the density of states implies that in the $h \ll d$ limit the waveguide thickness has no effect on decay rates even in the case of zero losses.

In the second regime, $d \lesssim h \ll \lambda$, we observe that for metallic waveguides dipole lifetime drops relative to that of a half-infinite substrate, while for hyperbolic systems, the lifetime *increases*. In fact, it can be shown (and is evident from the plots, in particular, Figure 7.4c) that the lifetime near a metal slab behaves as h^4 [18] (hyperbolic medium, in contrast, retains the h^3 dependence for sufficiently high losses). The reason for this is that for hyperbolic waveguides the coupling to high-k_x propagating modes is reduced for $d \lesssim h \ll \lambda$, while for the metals, coupling to surface plasmon modes becomes an important nonradiative decay channel, leading to lower lifetimes.

7.5 Conclusion

In this tutorial, we have presented the solution to the classic problem of dipole emission in the vicinity of a planar dielectric. We expressed the results in a way readily suitable for numerical integration via standard software packages. The method of decomposing dipole radiation into its plane wave spectrum readily generalizes to arbitrary stratified planar media, and is therefore highly relevant to studying the interaction of emitters with planar nanophotonic structures. One immediate application of the derived electric field is in computing the radiated power and the associated dipole lifetime. Such computations play a central role in the emerging field of radiative decay engineering in metamaterials, which promises new avenues for creating photonic devices with novel functionality.

The author acknowledges support under ARO MURI grant.

7.6 Appendix

In order to obtain a clear distinction between the behaviors of different systems we consider, and to fully understand the effect of losses, it is sometimes desirable to perform computations while setting losses to zero. This presents problems when equations of interest feature resonant denominators, as in the Equation 7.43.

The situation is remedied by the observation that, in the limit as losses approach zero, the resonant peaks resemble delta functions, which greatly simplifies the integration procedure. This result (sometimes called the "Dirac trick") is known as the Sokhotsky–Weierstrass theorem, which, in its simple form, states:

$$\lim_{\epsilon \to 0+} \int_a^b \frac{f(x)}{x \mp i\epsilon} dx = \mp i\pi f(0) + P \int_a^b \frac{f(x)}{x} dx. \tag{7.44}$$

Note that since all the quantities are assumed to be real, we trivially obtain the imaginary part of the integral via delta-function integration. For more complex denominators, such as the ones treated here, the theorem can be rewritten as

$$\lim_{\epsilon \to 0} \int_a^b \frac{g_1(x)}{g_2(x) \pm i\epsilon} dx = \mp i\pi \sum_{\{x_0\}} \frac{g_1(x_0)}{|g_2'(x_0)|} + P \int_a^b \frac{g_1(x)}{g_2(x)} dx, \tag{7.45}$$

where x_0 are the roots of $g_2(x)$. Once again, the imaginary part of the integral is obtained by an algebraic evaluation—provided the roots of $g_2(x)$ are known. For the systems described earlier, these roots had to be determined numerically—however, the problem of root finding is more easily tractable than that of numerical integration over very sharp resonances. Moreover, once these roots have been found, it is straightforward to adapt numerical integration algorithms for considering small finite losses, for example, by changing variables to effectively "stretch" the integrand in the vicinity of resonances, or by guiding singularity handlers in commercial computer algebra packages.

Bibliography

1. A. Sommerfeld. Propagation of waves in wireless telegraphy. *Ann. Phys.*, 23:665–737, 1909.
2. H. Weyl. Propagation of plane waves waves over a plane conductor. *Ann. Phys.*, 60:481–500, 1919.
3. J. Zenneck. On the propagation of plane electromagnetic waves along a flat conductor and its application to wireless telegraphy. *Ann. Phys.*, 23:846–866, 1907.
4. K. A. Norton. The propagation of radio waves over a plane Earth. *Nature*, 135:954–955, 1909.
5. E. Purcell. Spontaneous emission probabilities at radio frequencies. *Phys. Rev.*, 69:681, 1946.

6. J. McKeever, A. Boca, A. D. Boozer, R. Miller, J. R. Buck, A Kuzmich, and H. J. Kimble. Deterministic generation of single photons from one atom trapped in a cavity. *Science (New York, N.Y.)*, 303:1992–4, 2004.

7. Peijun Yao, C. Van Vlack, A. Reza, M. Patterson, M. M. Dignam, and S. Hughes. Ultrahigh Purcell factors and Lamb shifts in slow-light metamaterial waveguides. *Phys. Rev. B*, 80:1–11, November 2009.

8. Yiyang Gong, Selcuk Yerci, Rui Li, Luca Dal Negro, and Jelena Vuckovic. Enhanced light emission from erbium doped silicon nitride in plasmonic metal-insulator-metal structures. http://arxiv.org/abs/0910.3901v2, 2009.

9. L. Novotny and B. Hecht. *Principles of Nano-Optics*. Cambridge University Press, New York, 2006.

10. Roy J. Glauber and M. Lewenstein. Quantum optics of dielectric media *Physical Review* A 43(1):467–491, 1991.

11. Zubin Jacob, Leonid V. Alekseyev, and Evgenii Narimanov. Optical hyperlens: Far-field imaging beyond the diffraction limit. *Opt. Express*, 14(18):8247–56, September 2006.

12. L. V. Alekseyev, Z. Jacob, and E. Narimanov. *Tutorials in Complex Photonic Media*, chapter 2, pages 33–55. SPIE Press, New York, 2009.

13. G. W. Ford and W. H. Weber. Electromagnetic interactions of molecules with metal surfaces. *Phys. Rep.*, 113:195–287, 1984.

14. W. Lukosz and R. E. Kunz. Light emission by magnetic and electric dipoles close to a plane interface I: Total radiated power. *J. Opt. Soc. Am.*, 67:1607, December 1977.

15. I. Pockrand, A. Brillante, and D. Möbius. Decay of excited molecules near a metal surface. *Chem. Phys. Lett.*, 69:499–504, 1980.

16. W. L. Barnes. Fluorescence near interfaces: The role of photonic mode density. *J. Mod. Opt.*, 45(4):661–699, 1998.

17. Zubin Jacob, Igor Smolyaninov, and Evgenii Narimanov. Broadband Purcell effect: Radiative decay engineering with metamaterials. http://arxiv.org/abs/0910.3981v2, 2009.

18. R. R. Chance, A. Prock, and R. Silbey. *Molecular Fluorescence and Energy Transfer Near Interfaces*, volume 37 of *Advances in Chemical Physics*, pages 1–65. John Wiley & Sons, Hoboken, NJ, January 1978.

8

Bianisotropic and Chiral Metamaterials

Martin Wegener
and Stefan Linden

*Institute of Applied
Physics, Institute of
Nanotechnology, and
DFG-Center for Functional
Nanostructures (CFN)
Karlsruhe Institute of
Technology (KIT)
D-76128 Karlsruhe
Germany*

8.1 Introduction

The vast majority of natural materials used in optics and photonics are dielectrics. Chapter 5 showed that metamaterials additionally allow for obtaining artificial magnetism at optical frequencies, hence leading to effective magnetodielectric materials. For example, combining a negative magnetic permeability, μ, and a negative electric permittivity, ε, leads to backward waves, that is, to a negative phase velocity of light, c, in the sense that the dot product of the phase velocity vector (or, equivalently, the wave vector of light $\vec{k}$) and the Poynting vector $\vec{S}$ is negative, that is, $\vec{k} \cdot \vec{S} < 0$.

The possibilities of metamaterials are much richer than just that. Recall that the electric permittivity ε effectively describes the excitation of electric dipoles in the material as a result of the electric field of the light wave. Similarly, the magnetic permeability μ effectively describes the excitation of magnetic dipoles by the magnetic component of the light field. However, more generally, magnetic dipoles can also be excited by the electric component of the light field. Likewise, electric dipoles can also be excited by the magnetic component of the electromagnetic light wave. In the absence of a static magnetic field, the corresponding coefficients are directly related to each other via reciprocity. If the excited dipole vectors include an arbitrary angle with the exciting field vector, general bianisotropic materials result. As a consequence, phenomena such as asymmetric

225

reflectance and absorbance can occur for opposite directions of light propagation. An interesting, transparent, and especially relevant subclass of bianisotropic optical materials is that of isotropic chiral metamaterials. They are sometimes also referred to as *gyrotropic materials*, as *reciprocal biisotropic materials*, or as *Pasteur media*. Here, the angle between dipole and exciting field is always zero. Examples of resulting phenomena are strong optical activity, negative reflection (also see Section 8.2.1.1)—which is distinct from negative refraction—and giant circular optical dichroism. The latter can, for example, lead to extremely compact (broadband) circular polarizers—a promising candidate for an early real-world application of the far-reaching concepts of metamaterials. We will return to giant circular optical dichroism in Section 8.4.

Furthermore, we will see that the condition of simultaneous negative ε and negative μ is neither sufficient nor necessary (in the usual mathematical sense) for obtaining a negative phase velocity of light, $c = c_0/n$, in general, effective media ($c_0 = 1/\sqrt{\varepsilon_0\mu_0}$ is the vacuum speed of light and n is the refractive index of the medium). In chiral metamaterials, a negative phase velocity of light can, in principle, even result if electric permittivity and magnetic permeability are positive at the same time ("not necessary"). Intuitively, the eigen-polarizations for chiral media are left-handed circular polarization (LCP) and right-handed circular polarization (RCP) of light. The stronger the effects of chirality, the larger the difference between the refractive indices for LCP and RCP, $\Delta n = n_- - n_+$. Their mean value, n, is fixed by the square root of the product of the electric permittivity and the magnetic permeability. Thus, with increasing difference $n_- - n_+$, one of the two refractive indices eventually becomes negative. In natural systems such as a solution of chiral sugar molecules, the difference $|n_- - n_+|$ is on the order of 10^{-4} and, hence, much smaller than their mean value of order one. In contrast, in metamaterials, the index difference can approach or even exceed unity within a certain frequency range.

The situation is quite different for the general bianisotropic case. Here, the phase velocity of light can be positive (or one gets evanescent waves) even if electric permittivity ε and magnetic permeability μ are simultaneously comfortably negative ("not sufficient"). To understand this statement intuitively, let us consider the following example. Suppose that magnetic dipoles are excited by the magnetic field such that $\mu < 0$. Together with $\varepsilon < 0$ this may lead to a negative phase velocity of light. However, further magnetic dipoles can be excited by the electric component of the light field. These magnetic dipoles can, for example, be perpendicular to the exciting electric-field vector; hence, for a plane wave, they can be antiparallel to those magnetic dipoles corresponding to $\mu < 0$. Thus, the two sets of magnetic dipoles can cancel to zero, and the material would exclusively show an electric-dipole response, for which no negative phase velocity of light is expected. In our special example, the waves would become evanescent. This aspect will be discussed mathematically in Section 8.2.2. Altogether, the effects of bianisotropy and chirality are obviously far from being just minor modifications of magnetodielectric materials.

We start with a phenomenological effective-parameter treatment of bianisotropy and chirality in Section 8.2. Several examples as well as important symmetry considerations are given. In Section 8.3, the microscopic origin of the parameters beyond ε and μ is explained in the framework of a simple model, namely, the split-ring resonator (SRR). Chapter 5 discusses this building block or motif in detail. Finally, in Section 8.4, we provide further possibilities for chiral and bianisotropic building blocks of metamaterials and give a review of experiments in this field. Herein, the emphasis is on recently realized structures operating at optical frequencies. Some historic remarks will also be made.

8.2 Phenomenological Effective-Parameter Treatment

The material properties enter into the Maxwell equations via microscopic charge rearrangements and microscopic electric currents. As usual, provided that homogenization is possible and adequate, these can be cast into the form of a macroscopic polarization $\vec{P}$ and a macroscopic magnetization $\vec{M}$. Provided that displacements currents are of minor importance, these two quantities simply describe the electric-dipole density and the magnetic-dipole density, respectively. This leads to

$$\vec{D} = \varepsilon_0 \vec{E} + \vec{P}(\vec{E}, \vec{H})$$

$$\vec{B} = \mu_0 \vec{H} + \mu_0 \vec{M}(\vec{E}, \vec{H}).$$

Within linear optics, for reciprocal materials, the general relations (Lindell et al., 1994; Serdyukov et al., 2001) between the various field components in the frequency domain are given by

$$\begin{pmatrix} \vec{D} \\ \vec{B} \end{pmatrix} = \begin{pmatrix} \varepsilon_0 \ddot{\varepsilon} & -ic_0^{-1}\ddot{\xi} \\ ic_0^{-1}\ddot{\xi}^t & \mu_0 \ddot{\mu} \end{pmatrix} \begin{pmatrix} \vec{E} \\ \vec{H} \end{pmatrix}.$$

Here, the two off-diagonal contributions of the "2×2 matrix" (strictly speaking, this arrangement is not a matrix as its components do not even have the same unit) are identical up to the minus sign and the transposed of $\ddot{\xi}$ in the lower left-hand-side corner. The imaginary unit in the off-diagonal terms of the 2×2 matrix indicates a 90° phase shift. Its microscopic origin will become clear in Section 8.3. These aspects are a direct consequence of reciprocity. Only in the presence of an internal or external *static* magnetic field can this assumption be violated. The Faraday effect is a prominent example of a resulting phenomenon. A more general theoretical treatment has been presented (Lindell et al., 1994; Mackay and Lakhtakia, 2004; Mackay and Lakhtakia, 2008). Here, however, we restrict ourselves to assuming reciprocity (equivalent to zero Tellegen parameter in Reference [Lindell et al., 1994]) because no metamaterial experiments have been published for the more general case so far to our knowledge.

Even under these restrictive conditions, a very large variety of phenomena results. In general (Lindell et al., 1994), the electric permittivity $\ddot{\varepsilon}$, the magnetic permeability $\ddot{\mu}$, and the quantity $\ddot{\xi}$ are all dimensionless tensors ($\ddot{\xi}^t$ is the transposed of $\ddot{\xi}$). In particular, we have

$$\ddot{\xi} = \begin{pmatrix} \xi_{xx} & \xi_{xy} & \xi_{xz} \\ \xi_{yx} & \xi_{yy} & \xi_{yz} \\ \xi_{zx} & \xi_{zy} & \xi_{zz} \end{pmatrix}.$$

This means that the dipole moments excited by the electric $\vec{E}$-field vector and the magnetic $\vec{B}$-field vector can include an angle with the vector that excites them (Serdyukov et al., 2001). Let us consider two obvious limiting cases. First, we assume that the included angle is zero (pure chirality). In this case, ε, μ, and, most importantly, ξ simply become (complex) scalars. Second, we assume that the included angle is 90°. We refer to the latter case as "pure bianisotropy." It is obviously related to the off-diagonal elements of the $\ddot{\xi}$ tensor.

8.2.1 Pure Chirality

Let us consider an incident linearly polarized electromagnetic light wave and the diagonal form

$$\ddot{\xi} = \ddot{\xi}^t = \xi \begin{pmatrix} 1 & 0 & 0 \\ 0 & 1 & 0 \\ 0 & 0 & 1 \end{pmatrix}.$$

In this case, the electric component induces a magnetic dipole parallel to the electric-field vector, and hence perpendicular to the magnetic-field vector. Likewise, the magnetic component of the incident light wave induces an electric dipole parallel to the magnetic-field vector, and hence perpendicular to the electric-field vector. Thus, provided that losses are small, the linear polarization of the wave is just rotated—which is just the phenomenon of optical activity that is well known from solutions of chiral sugar molecules. There, however, the magnitude of $|\xi|$ is typically on the order of 10^{-4} or less. The direction of polarization rotation depends on the sign of ξ (the sign may depend on frequency). This behavior strongly suggests that the eigen-polarizations are not linear but rather left- and right-handed circular polarization of light. Insertion into the Maxwell equations (Lindell et al., 1994) immediately confirms this intuition. In particular, for circularly polarized light, one gets an index of refraction, n, given by

$$n_\pm = \sqrt{\varepsilon\mu} \pm \xi.$$

Here, the subscripts "+" and "−" again refer to right- and left-handed circular polarization of light, respectively. Clearly, for sufficiently large values of the real part of ξ, one index of refraction becomes negative—even if both electric permittivity and magnetic permeability are mainly real and positive.

At this point, it becomes already obvious that the three quantities ε, μ, and ξ cannot be independent. For a passive medium, a necessary condition arising from energy conservation is that exponentially growing propagating waves do not occur. In other words, the imaginary part of the complex refractive index n must be positive; it cannot be negative. If, for example, $\varepsilon = \mu = 1$ and $\xi = \xi_1 + i\xi_2$ with real part ξ_1 and imaginary part ξ_2, one of the two refractive indices $n_\pm$ would have a negative imaginary part.

This intimate connection between the three quantities will also appear in our simple microscopic model to be discussed in the following text (Section 8.3), in which they even share a closely similar resonance behavior. A more detailed discussion regarding the constraints on ε, μ, and ξ based on the Poynting vector can be found in works by Lindell et al. (1994) and Sihvola (2001).

To further illustrate the effects resulting from pure chirality, we consider normal-incidence transmittance and reflectance of light from a slab of such an effective medium. Furthermore, we assume Lorentz oscillator-like resonance behavior of the three complex quantities ε, μ, *and* ξ as illustrated in Figure 8.1. The spectral dependences shown in Figure 8.1a-c will actually result from the simple microscopic model to be outlined in Section 8.3. Generalization of the Fresnel coefficients at the air–medium and medium–air interfaces is straightforward but quite lengthy (Kwon et al., 2008). It leads to the intensity transmittance, *T*, and intensity reflectance, *R*, spectra also shown in Figure 8.1.

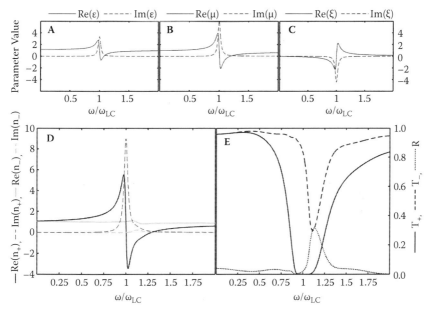

FIGURE 8.1   Illustration of the optical effects of pure chirality. (a) Assumed electric permittivity ε, (b) magnetic permeability μ, and (c) chirality parameter ξ. Real parts are solid, and imaginary parts are dashed. These three dependencies correspond to the simple oscillator model outlined in Section 8.3. (d) Resulting complex refractive indices $n_\pm$ for right-handed circular polarization (RCP or "+") and left-handed circular polarization (LCP or "−") of light. (e) Calculated intensity transmittances $T_\pm$ and reflectance $R_+ = R_+ = R_-$ for normal incidence of light onto a slab on a glass substrate. The model parameters (see Section 8.3) are $F = 0.3$, $\gamma/\omega_{LC} = 0.05$, and $(dc_0/l^2)/\omega_{LC} = 0.75$. The slab thickness is half of an *LC* eigen-wavelength, that is, $c_0\pi/\omega_{LC}$. (Taken from Decker 2010, M., PhD thesis. With permission.)

In transmittance, the circular polarization of light remains unchanged; that is, polarization conversion does not occur. In reflection, right-handed circularly polarized (RCP) light completely turns into left-handed circularly polarized (LCP) light and vice versa (also see Section 8.2.1.1). Clearly, the transmittance spectra for LCP and RCP incident light are different. This phenomenon is known as *circular dichroism*. For the parameters chosen, one circular polarization of light is well transmitted by the slab, whereas the opposite circular polarization is almost completely reflected. This means that the medium simply acts as a circular polarization filter (or as a "circular polarizer"). For single resonances as in Figure 8.1, the effect is restricted to a rather narrow bandwidth. If, however, a distribution (a "band") of chiral resonances is employed, a broadband response can be achieved. We will come back to this broadband aspect in Section 8.4.1 in the context of gold-helix metamaterials.

Optical activity and circular dichroism are two aspects of chirality that are fairly well known from optics textbooks. Metamaterials, however, can boost the magnitude of these effects by orders of magnitude. This increase can lead to qualitatively new and mind-boggling phenomena such as negative reflection—which is not addressed in usual optics textbooks, hence worth some basic thoughts.

8.2.1.1 Negative Reflection

You look into your bathroom mirror, and you see your mirror image. Upon reflection from an (ideal) metal mirror (see Figure 8.2a), RCP light also turns into its mirror image, which is nothing else but LCP light. This statement holds true for normal incidence as well as for oblique incidence. You get the same result when decomposing the circular

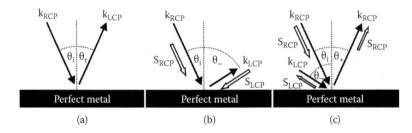

FIGURE 8.2 (a) Scheme of reflection of right-handed circularly polarized (RCP) light from an ideal metal mirror (or perfect conductor). (b) The same translated to a strongly chiral isotropic material with refractive indices $n_- < 0$ for LCP and $n_+ > 0$ for RCP (rather than air as in (a)) leads to an "impossible solution," demonstrating that reflection is not so trivial under these circumstances. (c) The correct solution generally rather shows two reflections (bireflection), one of which includes a negative angle with the surface normal (dashed)—a phenomenon called *negative reflection*. Parameters in (b) and (c) are $\theta_i = 25$ degrees, $\varepsilon\mu = 1$, and $\xi = 3$, hence $n_+ = 4$ and $n_- = -2$, leading to reflection angles $\theta_+ = 25$ degrees and $\theta_- = -58$ degrees, respectively. For clarity, the Poynting vectors of the three partial waves are plotted displaced (schematically) from their corresponding wave vectors.

polarization of light into linear s- and p-polarization. Upon reflection from an ideal metal surface, the sign of the p-polarized electric field remains that of s-polarized light flips sign. This relative change of sign between s- and p-component translates into a change of handedness. Furthermore, the incident and the reflected partial waves must have the same parallel component of the wave vector of light with respect to the metal interface. Without this phase matching, the phase velocities parallel to the interface would be different. Hence, the relative phases of the partial waves would vary along the interface, and the waves could not couple. The length of the wave vector is also conserved. The bottom line is that the reflectance angle equals the incidence angle. Energy conservation is automatically fulfilled.

Let us apply the same procedure to a wave propagating inside of a strongly chiral optically isotropic medium and hitting the metal. The angular frequency of light ω is conserved upon reflection; however, the length (or modulus) of the wave vector of light, $|\vec{k}_\pm| = \omega |n_\pm|/c_0$, is not necessarily conserved because the modulus of the refractive index $|n_\pm|$ is generally different for RCP and LCP. Yet, partial waves with the same handedness must have the same wave vector modulus. If the chirality is sufficiently strong, one refractive index, say, $n_+ > 0$ for RCP, is positive, while the other one, $n_- < 0$ for LCP, is negative. Thus, for the latter, Poynting vector and wave vector of light are antiparallel (see Section 8.1). The net result is that electromagnetic energy would exclusively flow toward the metal. However, for an ideal metal, there is strictly no dissipation. Thus, the scenario depicted in Figure 8.2b just cannot be a proper solution of Maxwell's equations as it violates energy conservation.

The situation can only be resolved by introducing more partial waves (than just the single incident wave and one reflected wave), allowing for satisfying simultaneously all electromagnetic boundary conditions and conservation laws. Indeed, generally two reflected waves appear, a phenomenon that has been named *bi-reflection* (Lakhtakia et al., 1986). For the strongly chiral isotropic case, one of the reflected waves can lie on the same side of the surface normal as the incident wave (Zhang and Cui, 2007), hence the name *negative reflection*. This situation is illustrated in Figure 8.2c. Again, due to phase matching, the tangential components of all partial waves with respect to the metal interface must be the same. The reflected wave must have RCP. If it was LCP, energy conservation could again obviously not be fulfilled. Thus, the angle of reflection, θ_+, of this wave must be identical to the incident angle θ_i with respect to the surface normal, that is, $\theta_+ = \theta_i > 0$. In addition, a second partial wave with LCP is reflected. For negligible losses (i.e., real ε, μ, and ξ, hence real $n_\pm$), the condition of identical tangential components of the wave vectors immediately leads to (Zhang and Cui, 2007)

$$n_- \sin(\theta_-) = n_+ \sin(\theta_i).$$

Due to antiparallel Poynting vector and wave vector of light, this partial wave has a negative reflectance angle in the usual nomenclature. For the special case $n_- = -n_+$ (equivalent to $\varepsilon\mu = 0$), we get $\theta_- = -\theta_i$, that is, the incident beam is back-reflected on its own axis ("perfect retro-reflector" or "perfect cat eye"). With losses, the general expression becomes more complicated (Zhang and Cui, 2007). Neither bireflection nor negative reflection has been observed experimentally on chiral metamaterials at any frequency to date to the best of our knowledge.

8.2.2 Pure Bianisotropy

Let us again consider an incident linearly polarized monochromatic electromagnetic plane light wave, propagating along the z-direction. We consider the form for the transverse components

$$\begin{pmatrix} D_x \\ B_y \end{pmatrix} = \begin{pmatrix} \varepsilon_0 \varepsilon & -ic_0^{-1}\xi \\ ic_0^{-1}\xi & \mu_0\mu \end{pmatrix} \begin{pmatrix} E_x \\ H_y \end{pmatrix},$$

which looks closely similar to what we have seen above for pure chirality. Again, ε, μ, and ξ are scalars, but we refer to the (orthogonal) vector *components* and not to the vectors. Thus, the meaning is quite different. We can alternatively write for the $\overset{\leftrightarrow}{\xi}$ tensor (for more details, see Section 8.2.3)

$$\overset{\leftrightarrow}{\xi} = \xi \begin{pmatrix} 0 & 1 & -1 \\ -1 & 0 & 1 \\ 1 & -1 & 0 \end{pmatrix}, \quad \text{hence} \quad \overset{\leftrightarrow}{\xi}{}^t = \xi \begin{pmatrix} 0 & -1 & 1 \\ 1 & 0 & -1 \\ -1 & 1 & 0 \end{pmatrix}.$$

Now, the electric component induces a magnetic dipole perpendicular to the electric-field vector, and hence parallel to the magnetic-field vector. Likewise, the magnetic component of the incident light wave induces an electric dipole perpendicular to the magnetic-field vector, and hence parallel to the electric-field vector. Thus, the incident linear polarization of the wave is maintained; that is, the linear polarizations are eigenpolarizations. Insertion into the Maxwell equations (Rill et al., 2008; Kriegler et al., 2009) delivers the refractive index n given by

$$n^2 = \varepsilon\mu - \xi^2.$$

From the two mathematical solutions for n, only the one with a positive imaginary part is physically relevant. Clearly, a large (real) value of ξ will eventually lead to a negative n^2, hence, to mainly imaginary n, thus to evanescent waves—even if both permittivity and permeability are both mainly real and negative. This is what we have meant by "not sufficient" in Section 8.1, where an intuitive reasoning has already been given.

Due to reciprocity, the transmittance through a slab of such purely bianisotropic material is still symmetric; that is, if one measures from one direction and from the opposite direction, one gets identical results. However, neither reflectance nor absorbance is generally symmetric, whereas they have both been symmetric for the case of pure chirality (although we have not explicitly emphasized this earlier).

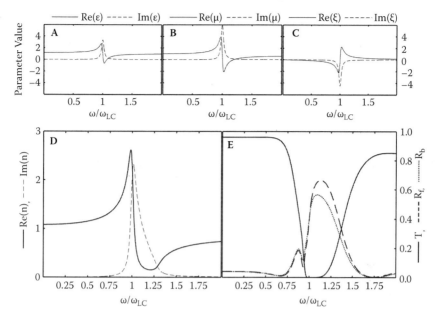

FIGURE 8.3 Illustration of the optical effects of pure bianisotropy. (a) Assumed electric permittivity ε, (b) magnetic permeability μ, and (c) bianisotropy parameter ξ. Real parts are solid, and imaginary parts are dashed. As in Figure 8.1, these three dependencies correspond to the model outlined in Section 8.3. (d) Resulting complex refractive index n. (e) Calculated intensity transmittance $T = T_f = T_b$ and reflectances $R_{f,b}$ for forward ("f") and backward ("b") propagation of linearly polarized light impinging normal to a slab on a glass substrate. All model parameters are identical to those in Figure 8.1. (With permission taken from Decker, M., PhD thesis.)

This behavior is illustrated in Figure 8.3, where we have intentionally chosen the strictly identical resonance parameters for ε, μ, and ξ as for the case of pure chirality in Figure 8.1. Again, working out the generalized versions of the Fresnel coefficients for the interfaces is straightforward in principle but very lengthy in practice. Corresponding details and formulae can be found in References (Rill et al., 2008 and Kriegler et al., 2009). Indeed, the normal-incidence intensity reflectance spectrum for linearly polarized incident light is not symmetric. The two opposite directions are denoted "f" (forward) and "b" (backward). The absorbance, A, is not depicted in Figure 8.3, but it results immediately from $A = (1 - R - T)$. As T is symmetric and R is not symmetric, A will generally be not symmetric either. In principle, one could even design a structure with very low transmittance T (a "wall"). The reflectance could be near unity for propagation from one side and nearly zero for propagation from the opposite side. Thus, the absorbance would be negligibly small for electromagnetic radiation impinging from one side and almost 100% for incidence from the other side. Following Kirchhoff's law derived from the second law of thermodynamics, the emittance equals the absorbance A. Thus, the material would emit thermal Planck radiation (in a certain frequency range) in one direction but not in the opposite direction. However, no such material has been made to date.

This discussion of phenomena arising from effective chiral and bianisotropic (homogeneous) media on the phenomenological level is by no means complete, but sufficient for our purpose. For further details, we refer the reader to corresponding textbooks (Lindell et al., 1994; Serdyukov et al., 2001).

8.2.3 Symmetry Considerations

Both bianisotropy and chirality require three-dimensional structures with broken symmetry. However, the symmetry requirements are different for the two. We will discuss (a) inversion symmetry, (b) mirror-symmetry planes, (c) planar structures (and stacks thereof), and (d) isotropy and rotations.

   a. *Center of inversion.* For space inversion, we have the well-known replacements $\vec{r} \rightarrow -\vec{r}$, $\vec{k} \rightarrow -\vec{k}$ (for all propagation directions), $\vec{E} \rightarrow -\vec{E}$, and $\vec{B} \rightarrow +\vec{B}$. The latter replacement guarantees that the vectors $\vec{k}$, $\vec{E}$, and $\vec{B}$ (in this sequence) continue forming a right-handed orthogonal system according to the second Maxwell equations (induction law). From the preceding relations, we get

$$\vec{B} = ic_0^{-1}\overset{\leftrightarrow}{\xi}{}^t \vec{E} + \mu_0 \mu \vec{H},$$

with the preceding general form for the bianisotropy tensor

$$\overset{\leftrightarrow}{\xi} = \begin{pmatrix} \xi_{xx} & \xi_{xy} & \xi_{xz} \\ \xi_{yx} & \xi_{yy} & \xi_{yz} \\ \xi_{zx} & \xi_{zy} & \xi_{zz} \end{pmatrix}.$$

For simplicity, we have assumed that the magnetic permeability μ is a scalar. It is not important at this point anyway. As $\vec{E}$ changes sign upon space inversion, whereas $\vec{B}$ does not, this equation can generally only be fulfilled provided that the tensor $\overset{\leftrightarrow}{\xi} = 0$. This means that all nine tensor components necessarily need to be zero, that is, neither bianisotropy nor chirality is possible. Lack of inversion symmetry is absolutely crucial!

   b. *Mirror-symmetry planes.* Let us consider a light wave propagating along the z-axis, that is, $\vec{k} = (0,0,k_z)$. (Propagation along the x- and the y-axis is analogous.) Hence, for a transverse plane wave, $\vec{E}$ and $\vec{B}$ lie in the xy-plane, that is,

$$\vec{E} = (E_x, E_y, 0) \quad \text{and} \quad \vec{B} = (B_x, B_y, 0).$$

For a time-harmonic plane wave, the induction law furthermore immediately leads to

$$\vec{B} = \frac{k_z}{\omega}(-E_y, E_x, 0).$$

Now consider a mirror-symmetry plane parallel to the yz-plane. We get the replacements $E_x \rightarrow -E_x$, $E_y \rightarrow E_y$, $B_x \rightarrow B_x$, and $B_y \rightarrow -B_y$. For the transverse components of the magnetic-field vector, we get

$$B_x = ic_0^{-1}\left(\xi_{xx}^t E_x + \xi_{xy}^t E_y\right) + \mu_0\mu H_x$$

and

$$B_y = ic_0^{-1}\left(\xi_{yx}^t E_x + \xi_{yy}^t E_y\right) + \mu_0\mu H_y.$$

As E_x changes sign whereas B_x does not, the diagonal element ξ_{xx}^t must vanish, that is, $\xi_{xx}^t = 0$. As B_y changes sign whereas E_y does not, we also get $\xi_{yy}^t = 0$. In contrast, ξ_{xy}^t can be nonzero as neither B_x nor E_y changes sign. Similarly, ξ_{yx}^t can also be nonzero as both B_y and E_x do change sign. The equation for the $\vec{D}$ field from above can simultaneously be fulfilled for nonzero ξ_{xy}^t and nonzero ξ_{yx}^t. For a mirror-symmetry plane parallel to the xz-plane, we obtain the same result as for the yz-plane. This is not surprising as both the xz-plane and yz-plane include the wave vector of light $\vec{k} = (0,0,k_z)$.

The situation is different for a mirror-symmetry plane parallel to the xy-plane, that is, perpendicular to the wave vector of light. Upon replacing $k_z \rightarrow -k_z$, hence $\vec{k} \rightarrow -\vec{k}$, and $\vec{E} \rightarrow -\vec{E}$, we have to set $\vec{B} \rightarrow +\vec{B}$ such that $\vec{k}$, $\vec{E}$, and $\vec{B}$ continue forming a right-handed orthogonal system according to the induction law. In analogy to our reasoning for inversion symmetry, we conclude that $\xi_{xx}^t = \xi_{yy}^t = \xi_{xy}^t = \xi_{yx}^t = 0$ needs to be fulfilled.

c. *Planar structures.* A planar structure is its own mirror plane. Thus, for excitation under normal incidence of light, a planar structure can neither show transverse bianisotropy nor chiral effects (ξ_{zx}^t and ξ_{zy}^t can be nonzero though). We will come back to corresponding reports on "planar chirality" when discussing early experiments on photonic metamaterials in the following text. Finite structure thickness, the presence of a substrate, and/or oblique incidence of light relax this strict impossibility. We briefly come back to oblique incidence in (d) in this section. One should also remember that our effective-parameter treatment is an approximation in itself. Beyond this approximation, small effects may occur.

Strictly speaking, one cannot assign a $\ddot{\xi}$ tensor to a planar structure as this implies that a magnetic-dipole (volume) density and an electric-dipole (volume) density exist—which is not the case. However, we can envision a sufficiently dense stack of identical planar structures, leading to a three-dimensional material, for which the concept of $\ddot{\xi}$ does become meaningful. In this spirit, our reasoning does apply to planar structures.

d. *Isotropy and rotations.* To avoid confusion, we emphasize that by "isotropy" we refer to isotropic optical properties at this point, that is, to properties that do neither depend on the propagation direction of light inside the material nor on the polarization of light. This does not necessarily mean that the microscopic material structure is isotropic. In this sense, isotropy means that x, y, and z as well as cyclic permutations of the indices xyz need to be equivalent, which brings us to $\xi^t_{xx} = \xi^t_{yy} = \xi^t_{zz}$ and to $\xi^t_{xy} = \xi^t_{yz} = \xi^t_{zx}$. We can also rotate the electric and the magnetic field by an arbitrary unitary rotation matrix $\vec{U}$, which must leave the tensor $\ddot{\xi}$ unaffected. This leads us to

$$\vec{B} = ic_0^{-1}\ddot{\xi}^t\vec{E} + \mu_0\mu\vec{H} = \vec{U}\vec{B} = ic_0^{-1}\ddot{\xi}^t\vec{U}\vec{E} + \mu_0\mu\vec{U}\vec{H},$$

and hence (after multiplying with the inverse of $\vec{U}$ from the left-hand side) to

$$\ddot{\xi}^t = \vec{U}^{-1}\ddot{\xi}^t\vec{U} = \ddot{\xi}^t$$

for all rotations $\vec{U}$. For example, we can choose

$$\vec{U} = \begin{pmatrix} \cos(\varphi) & \sin(\varphi) & 0 \\ -\sin(\varphi) & \cos(\varphi) & 0 \\ 0 & 0 & 1 \end{pmatrix}.$$

From here, it is straightforward to show that holds. If the magnetic dipoles excited by the electric-field vector are oriented along the axis of the electric-field vector, we get

$$\ddot{\xi}^t = \xi \begin{pmatrix} 1 & \cdot 0 & 0 \\ 0 & 1 & 0 \\ 0 & 0 & 1 \end{pmatrix},$$

that is, pure isotropic chirality (pure optical activity), the effects of which we have already discussed earlier. If they are perpendicular to the electric-field vector, we get

$$\ddot{\xi}^t = \xi \begin{pmatrix} 0 & -1 & 1 \\ 1 & 0 & -1 \\ -1 & 1 & 0 \end{pmatrix},$$

that is, pure bianisotropy as already encountered earlier as well.

For less symmetric structures, the clear-cut distinction between diagonal and off-diagonal elements of the $\ddot{\xi}$ tensor vanishes. This is mathematically trivial, however, important to notice. For example, in the following, we will encounter structures (see Figure 8.10) corresponding to the form

$$\ddot{\xi}^t = \xi \begin{pmatrix} 0 & 0 & 0 \\ 1 & 0 & 0 \\ 0 & 0 & 0 \end{pmatrix}.$$

Upon rotation of the sample, hence rotation of the incident linear polarization with respect to the sample around the z-axis, we obtain

$$\tilde{\tilde{\xi}}^t = \ddot{U}^{-1}\tilde{\tilde{\xi}}^t\ddot{U} = \xi \begin{pmatrix} -\sin(\varphi)\cos(\varphi) & -\sin^2(\varphi) & 0 \\ \cos^2(\varphi) & \cos(\varphi)\sin(\varphi) & 0 \\ 0 & 0 & 0 \end{pmatrix},$$

and, for example, for $\varphi = 45$ degree rotation, we get

$$\tilde{\tilde{\xi}}^t = \frac{\xi}{2}\begin{pmatrix} -1 & -1 & 0 \\ 1 & 1 & 0 \\ 0 & 0 & 0 \end{pmatrix}.$$

This means that neither linear nor circular but rather elliptical polarizations are eigen-polarizations. In close analogy, planar structures excited under oblique incidence of light can also lead to nonzero diagonal elements of the $\tilde{\tilde{\xi}}$ tensor ("extrinsic chirality" [Plum et al., 2009b]). The elliptical polarizations may, however, limit the usefulness of this aspect. For example, a planar SRR array (slit along x-axis) under normal incidence can lead to a magnetic-dipole moment along the z-direction, that is,

$$\tilde{\tilde{\xi}}^t = \xi \begin{pmatrix} 0 & 0 & 0 \\ 0 & 0 & 0 \\ 1 & 0 & 0 \end{pmatrix}.$$

For details, see Section 8.4. Upon rotation of the sample around the y-axis (for fixed $\vec{k} = (0,0,k_z)$) via the rotation matrix

$$\ddot{U} = \begin{pmatrix} \cos(\varphi) & 0 & \sin(\varphi) \\ 0 & 1 & 0 \\ -\sin(\varphi) & 0 & \cos(\varphi) \end{pmatrix}$$

we obtain

$$\tilde{\tilde{\xi}}^t = \ddot{U}^{-1}\tilde{\tilde{\xi}}^t\ddot{U} = \xi \begin{pmatrix} -\sin(\varphi)\cos(\varphi) & 0 & -\sin^2(\varphi) \\ 0 & 0 & 0 \\ \cos^2(\varphi) & 0 & \cos(\varphi)\sin(\varphi) \end{pmatrix}.$$

In particular, the diagonal element $\tilde{\tilde{\xi}}^t_{xx} \neq 0$ in general for $\varphi \neq 0$, 90, 180, etc., degrees rotation.

To summarize our symmetry considerations, general bianisotropy (including the special cases of pure chirality and pure bianisotropy) requires absence of a center of inversion of the metamaterial. Chirality requires the absence of mirror-symmetry planes. In contrast, pure bianisotropy can occur in the presence of mirror-symmetry planes. However, one must not have a mirror-symmetry plane perpendicular to the propagation direction of light. Thus, strictly planar metamaterial structures excited under normal incidence of light can neither exhibit pure chirality nor pure transverse bianisotropy. These general and maybe somewhat abstract symmetry requirements will implicitly reappear for the following explicit special model system.

8.3 Split-Ring Resonators as a Model System

The split-ring resonator (SRR) is simply a metallic loop with one or several slits (see Figure 8.4). The SRR is the paradigm building block of electromagnetic metamaterials (Pendry et al., 1999). As an individual resonant circuit, it has already been used in the early days of electromagnetism. In fact, the third sentence of Heinrich Hertz' pioneering paper (Hertz, 1887) on the experimental discovery of electromagnetic waves in 1887 in Karlsruhe describes the SRR as a potential candidate for high-frequency oscillations (in German): "… Schnellere Schwingungen noch als diese lässt die Theorie voraussehen in gutleitenden ungeschlossenen Drähten, deren Enden nicht durch zu grosse Capazitäten

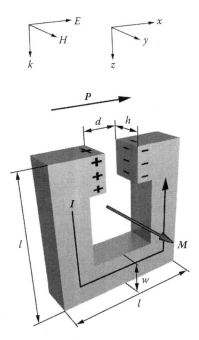

FIGURE 8.4 Illustration of a split-ring resonator and the parameters used. The considered excitation geometry is also depicted. (With permission taken from Kriegler, C. E. et al. Bianisotropic photonic metamaterials, 2010. *IEEE J. Selected Top. Quantum Electron.*: 16:367–375.)

belastet sind, ... " At the time, 100 MHz was a really high frequency. Here, we rather discuss frequencies of 100 THz and beyond. The following simple but instructive treatment very closely follows our recent review (Kriegler et al., 2009); also see (Marques et al., 2002).

The incident light field can induce a circulating and oscillating electric current into the metallic wire of the SRR that gives rise to a local magnetic field (magnetic dipole) normal to the SRR plane. Viewed as an electric circuit, the SRR resonance arises from the inductance of an almost closed loop (with inductance L) and the capacitor formed by the two ends of the wire (with capacitance C). This leads to an LC eigen-frequency $\omega_{LC} = 1/\sqrt{LC}$. For small SRR, the so-called kinetic inductance can add to L (Klein et al., 2006; Busch et al., 2007). A more detailed modeling would also have to account for additional surface contributions to the SRR capacitance C. Furthermore, incident energy can either be dissipated by ohmic losses of the metal or can be reradiated into free space, leading to the so-called radiation resistance (Meyrath et al., 2007). The effect of both aspects can be summarized by an effective resistance R of the circuit.

Using Kirchhoff's voltage law, the differential equation for the electric current I in this simple circuit reads

$$U_C + U_R + U_L = \frac{1}{C} \int I \, dt + RI + L \frac{dI}{dt} = U_{ind}.$$

Here, the source voltage U_{ind} is induced by the incident light field.

Importantly, the SRR has both an electric and a magnetic dipole moment. The electric dipole moment is given by the charges separated on the capacitor plates, $\int I \, dt$, times their distance d. The macroscopic electric polarization, P, is the product of the individual electric-dipole moment times the volume density of the dipoles $\frac{N_{LC}}{V} = \frac{1}{a_{xy}^2 a_z}$. a_{xy} and a_z are the in-plane and out-of-plane lattice constants of the artificial crystal composed of SRR, respectively. This reasoning obviously neglects interaction effects among the SRR in the crystal and leads us to the polarization vector $\vec{P} = (P_x, 0, 0)$ with

$$P_x(t) = \frac{1}{a_{xy}^2 a_z} d \int I \, dt.$$

The magnetic dipole density, $\vec{M} = (0, M_y, 0)$, is the product of the SRR density and the individual magnetic dipole moment. Within the quasi-static limit (i.e., retardation effects neglected), the latter is given by the instantaneous electric current $I(t)$ multiplied by the geometrical area of the loop l^2, that is,

$$M_y(t) = \frac{1}{a_{xy}^2 a_z} I(t) l^2.$$

Note that we have neglected the displacement current at this point. Hence, our reasoning is only strictly valid provided that the ohmic current I dominates over the displacement current. In other words, the slit width d in the SRR must not be too large.

a. We start by discussing an SRR current that is induced by Faraday's induction law. In this case, we have $U_{ind}(t) = -\frac{\partial\phi}{\partial t}$ with the magnetic flux $\phi(t) = \mu_0 H_y(t) l^2$. Assuming a harmonically varying magnetic field $H_y(t) = H_y \exp(-i\omega t) + cc.$, we obtain $M_y(t) = M_y \exp(-i\omega t) + cc.$ with

$$M_y = \frac{F\omega^2}{\omega_{LC}^2 - \omega^2 - i\gamma\omega} H_y .$$

Here, we have used the inductance of a long coil $L = \mu_0 l^2 / h$. The damping is given by $\gamma = R/L$ and the SRR volume filling fraction by

$$0 \leq F = \frac{l^2 h}{a_{xy}^2 a_z} \leq 1 .$$

More importantly in the context of chirality and bianisotropy, induction via Faraday's law also leads to a polarization $P_x(t) = P_x \exp(-i\omega t) + cc.$ given by

$$P_x = \frac{d}{l^2} \frac{iF\omega}{\omega_{LC}^2 - \omega^2 - i\gamma\omega} H_y .$$

Here, we have employed the capacitance $C = \varepsilon_0 wh/d$ for a plate capacitor with large plates. The imaginary unit in the numerator indicates that the polarization P_x is phase delayed by 90° with respect to the exciting H-field. Otherwise, the polarization P_x reveals the same harmonic-oscillator resonance behavior around the LC eigen-frequency as the magnetization M_y. Also note that the induced polarization P_x is proportional to the slit width d of the SRR. Hence, the polarization disappears for decreasing slit width.

b. An electrical current can also be induced by a voltage drop $U_{ind}(t) = E_x(t)d$ over the plate capacitor due to the electric-field component, $E_x(t)$, of the light field. For $E_x(t) = E_x \exp(-i\omega t) + cc.$, we obtain

$$M_y = \frac{1}{\mu_0} \frac{d}{l^2} \frac{-iF\omega}{\omega_{LC}^2 - \omega^2 - i\gamma\omega} E_x$$

and

$$P_x = \frac{1}{\mu_0} \left(\frac{d}{l^2}\right)^2 \frac{F}{\omega_{LC}^2 - \omega^2 - i\gamma\omega} E_x .$$

In this case, the 90° phase delay (see imaginary unit in numerator) occurs for the magnetization M_y.

Provided the displacement current is negligible (Agranovich and Gartstein, 2006)—an assumption that is usually fulfilled in the vicinity of the resonance and for not too small values of the SRR slit width d—we can identify the macroscopic magnetization with the magnetic dipole density M_y discussed in the preceding text. Hence, we get the macroscopic material equations

$$D_x = \varepsilon_0 E_x + P_x \quad \text{and} \quad B_y = \mu_0 H_y + \mu_0 M_y.$$

On this basis, we can summarize our findings from (a) and (b) by

$$\begin{pmatrix} D_x \\ B_y \end{pmatrix} = \begin{pmatrix} \varepsilon_0 \varepsilon & -ic_0^{-1}\xi \\ ic_0^{-1}\xi & \mu_0\mu \end{pmatrix} \begin{pmatrix} E_x \\ H_y \end{pmatrix}$$

—an expression that we have already encountered in Section 8.2.2 in the phenomenological treatment of pure bianisotropy. Now, however, we can provide explicit expressions for the electric permittivity, the magnetic permeability, and the bianisotropy parameter, that is,

$$\varepsilon(\omega) = 1 + \left(\frac{dc_0}{l^2}\right)^2 \frac{F}{\omega_{LC}^2 - \omega^2 - i\gamma\omega},$$

$$\mu(\omega) = 1 + \frac{F\omega^2}{\omega_{LC}^2 - \omega^2 - i\gamma\omega},$$

and

$$\xi(\omega) = -\frac{dc_0}{l^2} \frac{F\omega}{\omega_{LC}^2 - \omega^2 - i\gamma\omega}.$$

These are exactly the expressions that we have used for the model calculations underlying Figures 8.1 and 8.3 in the preceding text. Clearly, all three quantities show resonances around the same *LC* eigen-frequency. Hence, they are closely connected—as briefly anticipated before in our above phenomenological treatment of pure chirality.

The combination of four SRRs shown in Figure 8.5 leads to pure chirality for wave vectors $\vec{k} = (0, 0, k_z)$, that is, neglecting interaction effects among the SRR, to

$$D_{x,y} = \varepsilon_0 \varepsilon E_{x,y} - ic_0^{-1}\xi H_{x,y}$$

$$B_{x,y} = ic_0^{-1}\xi E_{x,y} + \mu_0\mu H_{x,y}.$$

At least for the transverse components of the fields, only diagonal elements of $\ddot{\xi}$ appear. This result can also be obtained mathematically via the rotation matrix described in list (d) in Section 8.2.3. Obviously, this structure is not isotropic but rather uniaxial. Likewise, essentially any $\ddot{\xi}$ tensor can simply be constructed by combinations of individually rotated SRRs (as long as retardation and interaction effects are neglected).

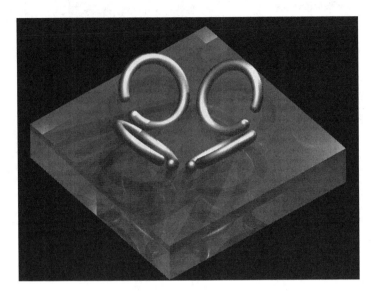

FIGURE 8.5 Scheme of an arrangement of four split-ring resonators that leads to a uniaxial metamaterial unit cell exhibiting pure optical activity for propagation of light along the vertical axis. Likewise, other bianisotropy tensors can also be realized by combinations of split-ring resonators.

Let us finish this section with a remark that is likely confusing at first sight, but potentially enlightening upon reflection. In the preceding list (a), we have discussed a current induced by Faraday's law. This has led to a magnetic-dipole moment induced by the harmonically varying magnetic field, and hence to $\mu \neq 1$. A devil's advocate might argue that it was not really the magnetic field that induced this current but rather the electric field. After all, the magnetic field actually induces a nonzero curl of the electric field, which then leads to an SRR current via Ohm's law. For metals, at the end of the day, it is actually always the electric field that gives rise to an electrical current. Thus, strictly speaking, for a single SRR (Husnik et al., 2008) or other single building blocks, there is no meaning in making this differentiation. It is related to the "chicken-and-egg" problem: in an electromagnetic wave, electric and magnetic components mutually induce each other. However, if a material composed of many SRRs is mapped onto effective parameters, the effects of μ and ξ are quite different and, most importantly, detectable in transmittance and reflectance measurements (see discussion in Section 8.2 and, e.g., Figures 8.1–8.3).

8.4 Experimental Status

8.4.1 Chiral Structures

Chapter 3 shows that metamaterials operating at optical frequencies can actually be fabricated by means of nanotechnology. This also holds true for chiral and bianisotropic photonic metamaterials. In the list (c) in Section 8.2.3, we have discussed that these

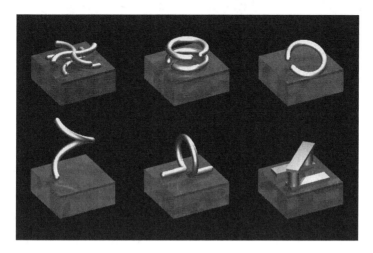

FIGURE 8.6 Scheme of various motifs of three-dimensional chiral and/or bianisotropic metamaterials that have been realized experimentally.

phenomena cannot occur in strictly planar structures but rather require three-dimensional metamaterials, hence truly three-dimensional nanofabrication.

Nevertheless, the possibility of "planar chiral" metamaterials (Papakostas et al., 2003) has been suggested. Works by Kuwata-Gonokami et al. (2005) and Decker et al. (2007) have studied related structures. Indeed, circular dichroism (Decker et al., 2007) and polarization rotation (Kuwata-Gonokami et al., 2005) have been observed experimentally. These effects are not strictly zero but fairly weak (Kuwata-Gonokami et al., 2005; Decker et al., 2007) compared to those of truly chiral three-dimensional metamaterials (Rogacheva et al., 2006). The finite size of the effects results from the finite height of the "planar" structures, from the supporting glass substrate and fabrication imperfections, both of which break the mirror symmetry, from our effective-medium approximation in the list (c) in Section 8.2.3 itself, or from combinations of these aspects.

Figure 8.6 summarizes a variety of motifs of truly chiral and/or bianisotropic metamaterials that have actually been fabricated and characterized experimentally. (Some authors have used the alternative name *stereometamaterials* for general bianisotropic metamaterials [Liu et al., 2009].) The motifs shown in Figure 8.6 are usually arranged into a periodic lattice. They can, however, also be densely packed in random spacing and/or orientation.

8.4.1.1 Helical Metamaterials

Let us start our discussion with the metal helices in Figure 8.6. A metal helix can be thought of as adiabatically emerging from a planar SRR (Figure 8.7). Such helices are used as a paradigm example for explaining optical activity by classical physics in Chapter 8 of Eugene Hecht's famous optics textbook *Optics* (Hecht, 2002). Early microwave experiments on literally hand-made artificial materials composed of metal helices (embedded in cotton) with randomized orientation have already been published in 1920

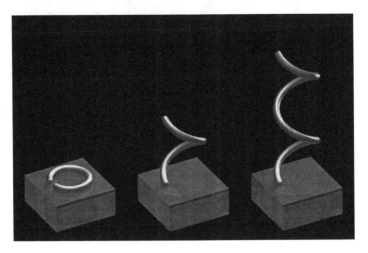

FIGURE 8.7   Scheme of the transition from a planar split-ring resonator to a three-dimensional metallic helix.

(Lindman, 1920). In this reference, Karl Lindman directly measured the rotation of a linearly polarized incident wave (up to a few tens of degrees) arising from such artificial optical activity. Furthermore, metal helices are also well known from electrical engineering in the context of antenna theory. There, they are referred to as so-called helical antennas in the end-fire geometry. Such helical antennas have been invented in 1946 by John Kraus (Kraus and Marhefka, 2003). They are, for example, widely used in today's microwave wireless local-area networks. Let us briefly dive into this instructive analogy, which we have discussed previously (Gansel et al., 2010).

Antenna theory predicts that a circularly polarized wave is emitted along the helix axis upon injection of an electrical current into the metal wire. Optimum performance requires that the pitch p (i.e., the period along the helix axis) is on the order of the helix diameter. Provided that the wire radius r is small compared to the helix radius (i.e., $r \ll R$) and provided that a sufficient number of helix pitches are used, antenna theory states that the free-space operation wavelength lies in the interval

$$\lambda \in 2\pi R \times \left[\frac{3}{4}, \frac{4}{3} \right].$$

This leads to a ratio of maximum to minimum wavelength of 16/9, which is close to one octave. The helical antenna is a subwavelength antenna as the center free-space operation wavelength $\lambda = 2\pi R$ is 2π times larger than the helix radius R. This is an important ingredient for being able to use the helix as a motif of an effective material. The helical antenna operation wavelength neither depends too much on the length of the antenna, nor on the number of helix pitches, nor on the helix pitch p. For an infinite number of helix turns, one obtains perfectly circularly polarized emission.

An electrical current in a left-handed helical "sender" antenna leads to emission of a left-handed circularly polarized electromagnetic wave propagating along the helix axis. Conversely, an incident wave with the same handedness propagating along the helix axis will induce an electrical current in an identical "receiver" antenna. In contrast, the other handedness of light is not expected to interact with the antenna.

Thus, a two-dimensional square array of helical antennas with a certain lattice constant, a, is expected to lead to reflection and/or absorption of light incident onto the array if the handedness of the light is identical to that of the helices. In contrast, light with the opposite circular handedness will be transmitted. We have encountered such "circular polarizer" behavior before in our phenomenological discussion of purely chiral effective materials (e.g., Figure 8.1). More recently, several theoretical studies (Belov et al., 2003; Silveirinha, 2008) have investigated such helical antenna arrays in the metamaterials context and language. Propagation along and perpendicular to the helix axes has been discussed, and effective metamaterial parameters have been retrieved. Antenna theory dealt with such arrays many years ago already—although not for optical frequencies. Indeed, in the "antenna bible" by Kraus and Marhefka (2003)—the first edition of which appeared in 1950—the inventor of the helical antenna, John D. Kraus, states: "... Not only does the helix have a nearly uniform resistive input over a wide bandwidth, but it also operates as a "supergain" end-fire array over the same bandwidth! Furthermore, it is noncritical with respect to conductor size and turn spacing. It is also easy to use in arrays because of almost negligible mutual impedance ...." The latter statement, however, is only really correct in the limit of very long helices. It is an important necessary requirement for being able to consider the arrangement as a truly effective material.

Electron micrographs of chiral metamaterial structures, fabricated using direct laser writing into a positive-tone photoresist and subsequent electroplating of gold (Gansel et al., 2009), are shown in Figure 8.8. Corresponding measured and calculated intensity transmittance spectra for circularly polarized incident light propagating along the helix axis are depicted in Figure 8.9. Obviously, one gets a reasonably sharp resonance for about one pitch of the helices. This resonance is more or less that of a planar SRR for which one end of the wire is pulled out of the plane (see Figure 8.7). For two helix pitches in Figure 8.9, the resonance essentially disappears and a stopband occurs with about one octave bandwidth. In this stopband, light with the same handedness as the helices is reflected (and/or absorbed), hence not transmitted, whereas the other handedness is highly transmitted (Gansel et al., 2009). The conversion of circular polarization is small (not shown). Hence, the structure can be used as a compact and broadband circular polarizer. Intuitively, in the language of solid-state physics, the large bandwidth results from the coupling of the different (here two) pitches within one helix. This behavior is analogous to the transition from sharp atomic levels to broad bands in solids. The one-octave bandwidth is also consistent with the preceding expectations from antenna theory. A systematic numerical study on the influence of the various helix structure parameters can be found in the work by Gansel et al. (2010). For example, reducing the lattice constant for otherwise fixed parameters leads to a blue shift of the resonances. This behavior indicates that the lateral interaction between neighboring gold helices in the array is not negligible either. Thus, these metamaterials can qualitatively be viewed as homogeneous effective materials, but not quantitatively.

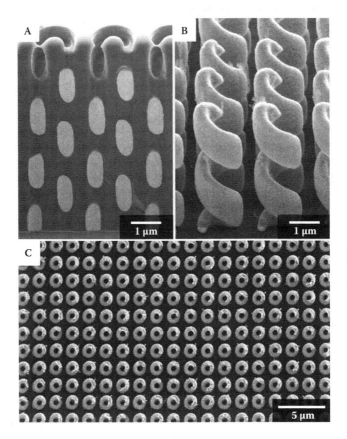

FIGURE 8.8   (a) Focused-ion-beam cut of a polymer structure partially filled with gold via electroplating. (b) Oblique view of a left-handed gold helix structure after removal of the polymer via plasma etching. (c) Top-view image revealing the circular cross section of the helices and the homogeneity on a larger scale. The lattice constant of the square lattice is $a = 2$ μm. (With permission taken from Gansel, J. K. et al. 2009. *Science* 325: 1513–1515.)

8.4.1.2 Layered Structures

The gold helix example has emphasized the huge circular dichroism that can be obtained from chiral metamaterials. Via the Kramers–Kronig relations, circular dichroism is connected to optical activity. Negative refractive indices, precisely negative phase velocities of light due to optical activity (Pendry, 2004; Wegener and Linden, 2009), have been observed experimentally at microwave (Plum et al., 2009b) and at far-infrared frequencies (Zhang et al., 2009). The former structure has been composed of the twisted gammadions shown in Figure 8.6. Both uniaxial structures, however, are one lattice constant or even less in thickness (along the propagation direction of light). Thus, as usual, the retrieved effective "material" parameters have to be taken with a grain of salt. At optical

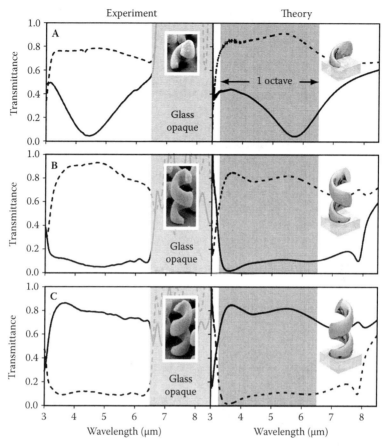

FIGURE 8.9 Normal-incidence measured and calculated transmittance spectra are shown in the left- and right-hand-side column. Left-handed circular polarization of light (LCP) and right-handed circular polarization of light (RCP) are depicted solid and dashed, respectively. (a) Slightly less than one pitch of left-handed helices, (b) two pitches of left-handed helices, and (c) two pitches of right-handed helices (see insets). (Taken from Gansel, J. K. et al. 2009. *Science* 325: 1513–1515. With permission.)

frequencies, the increased damping has hindered corresponding negative phase velocities. The largest difference of refractive indices $|n_- - n_+|$ has been obtained on arrays of pairs of twisted gold crosses (see Figure 8.6) and amounts to $|n_- - n_+| \approx 0.34$ at around 1.36 μm wavelength (Decker et al., 2009). To date, *isotropic* strongly chiral metamaterials operating at optical frequencies have not been reported. However, so-called bi-chiral helix architectures aiming at isotropic optical chirality that have been presented in the context of dielectric photonic crystals (Thiel et al., 2009) can possibly be translated to metallic metamaterials.

8.4.2 Bianisotropic Structures

The essence of bianisotropy and chirality is the cross coupling between the electric/magnetic field of the light and electric/magnetic dipoles in the material. Regarding photonic metamaterials, magnetic-dipole resonances excited by the electric field were first demonstrated by Linden et al. (2004). These samples were fabricated by standard electron-beam lithography (see Chapter 3). Figure 8.10 shows the structures (insets) as well as the measured transmittance spectra revealing the magnetic resonance as dips in transmittance. The excitation of electric dipoles by the magnetic component of the light field on SRR arrays excited under oblique incidence (also see Section 8.2.3) has directly been demonstrated by Enkrich et al. (2005). The gray areas in Figure 8.11 directly show the measured polarization rotation due to the combined effects of optical activity and linear birefringence. This phenomenon has later been named *extrinsic chirality* (Plum et al., 2009a). A discussion can be found in the work by Volkov et al. (2009). For normal incidence of light, both of these experiments (Linden et al., 2004; Enkrich et al. 2005) correspond to

$$\overset{\leftrightarrow}{\xi}^t = \xi \begin{pmatrix} 0 & 0 & 0 \\ 0 & 0 & 0 \\ 1 & 0 & 0 \end{pmatrix},$$

a form already encountered in list (d) in Section 8.2.3, where the resulting form for oblique incidence has been discussed as well.

Figure 8.12 shows different variations of (elongated) SRR that have been fabricated by direct laser writing, silver coating using chemical-vapor deposition (Rill et al., 2008), and subsequent postprocessing by focused-ion-beam milling (Kriegler et al., 2009). In this excitation geometry, the diagonal elements of the bianisotropy tensor play no role, and we get pure bianisotropy, for which linear polarization is the eigen-polarization. This case also exactly corresponds to the geometry that we have considered in Section 8.3. The corresponding bianisotropy tensor is given by

$$\overset{\leftrightarrow}{\xi}^t = \xi \begin{pmatrix} 0 & 0 & 0 \\ 1 & 0 & 0 \\ 0 & 0 & 0 \end{pmatrix},$$

which has also been discussed in list (d) in Section 8.2.3. The retrieved effective parameters (Rill et al., 2008) qualitatively closely resemble those shown for the model in Figure 8.3. In particular, as discussed earlier, the real part of the index of refraction stays positive even though both the real part of the electric permittivity and the real part of the magnetic permeability are comfortably negative. Bianisotropy is not a friend of negative indices. However, negative indices can still be obtained with more elaborate bianisotropic metamaterial structures similar to the one shown in Figure 8.13 (Rill et al., 2009).

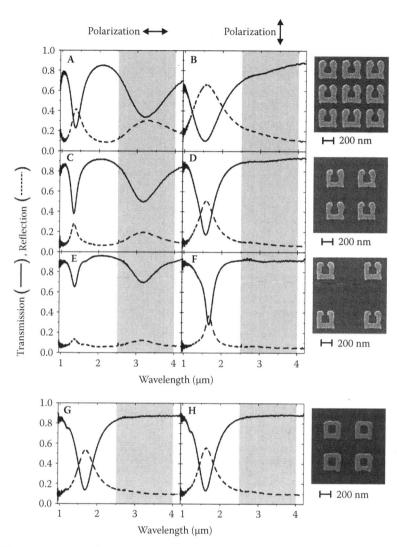

FIGURE 8.10 Measured transmission and reflection spectra. In each row of this "matrix," an electron micrograph of the sample is shown on the RHS. The two polarization configurations are shown on top of the two columns; (a) and (b) lattice constant $a = 450$ nm, (c) and (d) $a = 600$ nm, and (e) and (f) $a = 900$ nm correspond to nominally identical split-ring resonators, (g) and (h) to corresponding *closed* rings, $a = 600$ nm. The resonance at about 3 μm wavelength (highlighted by the gray areas) is the *LC* resonance of the individual split-ring resonators. Their magnetic-dipole moment is excited via the electric component of the incident light field. (With permission taken from Linden, S. et al. 2004. *Science* 306: 1351–1353.)

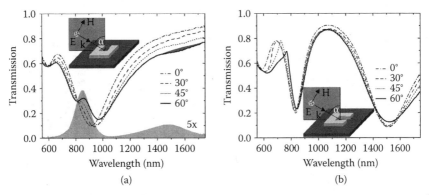

FIGURE 8.11 Measured transmission spectra taken for oblique incidence for the configurations shown as insets. In (a), coupling to the fundamental magnetic mode at 1.5 μm wavelength is only possible via the magnetic-field component of the incident light, and for (b), both electric and magnetic field can couple. Note the small but significant feature in (a) for 60° around 1.5 μm wavelength. The lower gray area in (a) is the transmission into the linear polarization orthogonal to the incident one for α = 60°. This rotation is due to the magnetoelectric coupling. (Taken from Enrich, C. et al. 2005. *Phys. Rev. Lett.* 95: 203901. With permission.)

This structure composed of elongated SRR and intentionally elevated orthogonal metal wires has been fabricated via direct laser writing and shadow evaporation of silver and has revealed a negative phase velocity of light in the 3–4 μm optical wavelength region.

In principle, the effects of bianisotropy also have to be considered for the double-fishnet-type negative-index photonic metamaterials. Chapter 5 discusses the ideal double-fishnet structure. The mirror symmetry of the ideal structure is broken by the presence of the glass substrate as well as by the trapezoidal shape of the edges arising from the lift-off fabrication procedure (Ku and Brueck, 2009). For a few degrees deviation from perpendicular side walls, however, the qualitative behavior remains unaffected.

8.4.3 Perspectives

Bianisotropy and chirality further enhance the possibilities and potential of photonic metamaterials. While several basic phenomena have recently been demonstrated experimentally, all chiral photonic metamaterials presented so far are uniaxial. Gold helix photonic metamaterials acting as compact broadband circular polarizers might be an early application of this class of structures. The experimental realization of isotropic chiral metamaterials operating at optical frequencies poses a significant future challenge, especially regarding three-dimensional nanofabrication. A success might allow for the observation of mind-boggling fundamental optical phenomena such as, for example, negative reflection (not to be confused with negative refraction). Important further possibilities arise if Faraday-active constituent materials could be incorporated into bianisotropic photonic metamaterials (Mackay and Lakhtakia, 2004; Mackay and

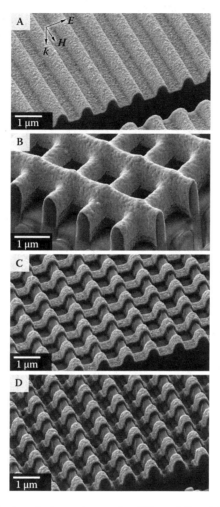

FIGURE 8.12 Oblique-view electron micrographs of three different, though related, recently experimentally fabricated bianisotropic photonic metamaterials composed of different variations of split-ring resonators. The dark parts correspond to the polymeric templates, light gray parts to the silver films. (With permission taken from Kriegler, C. E. et al. 2010. Bianisotropic photonic metamaterials, *IEEE J. Selected Top. Quantum Electron.*: 16:367–375.)

Lakhtakia, 2008). In this case, not only the reflectance but also the transmittance is generally asymmetric, which might, for example, give rise to extremely compact optical isolators. Rather bulky optical isolators are a major cost driver in today's fiber-based telecommunication systems. In addition to all of these linear optical aspects, the nonlinear optics of bianisotropic/chiral metamaterials is quite unexplored as well. Finally, regarding quantum optics, repulsive Casimir forces in chiral metamaterials have recently been predicted theoretically (Zhao et al., 2009). This might lead to quantum levitation, which is one of the most futuristic horizons of the field of metamaterials.

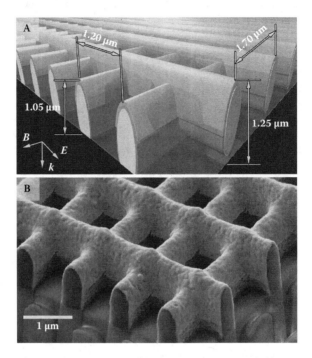

FIGURE 8.13 A bianisotropic negative-index photonic metamaterial. (a) Metamaterial design. The white regions are the polymer located on a glass substrate. The sidewalls of the polymer (encapsulated by silica) are coated with silver. The polarization of the incident electromagnetic field is illustrated on the lower LHS corner. (b) Oblique-view electron micrograph of a structure fabricated by direct laser writing and silver shadow evaporation that has been cut by a focused-ion beam to reveal its interior. (With permission taken from Rill, M. S. et al. 2009. *Opt. Lett.* 34: 19–21.)

Acknowledgments

We thank all members of our Karlsruhe team for their scientific contributions that have been cited in this tutorial chapter; in alphabetical order: Manuel Decker, Georg von Freymann, Andreas Frölich, Justyna K. Gansel, Christine E. Kriegler, Michael S. Rill, Isabelle Staude, and Michael Thiel. Furthermore, we thank our collaborating groups headed by Costas M. Soukoulis (Iowa State), Kurt Busch (Karlsruhe), Sven Burger (Berlin), and Volker Saile (Karlsruhe). We also thank Peter Würfel (Karlsruhe) for discussions regarding Kirchhoff's law. We acknowledge support by the Deutsche Forschungsgemeinschaft (DFG) and the State of Baden-Württemberg through the DFG-Center for Functional Nanostructures (CFN) within subprojects A1.4 and A1.5. The project PHOME acknowledges the financial support of the Future and Emerging Technologies (FET) program within the Seventh Framework Program for Research of the European Commission, under FET-Open grant number 213390. The project METAMAT is supported by the Bundesministerium für Bildung und Forschung (BMBF). The research of S.L. is further supported through a Helmholtz-Hochschul-Nachwuchsgruppe (VH-NG-232).

References

Agranovich, V. M. and Gartstein, Y. N. 2006. Spatial dispersion and negative refraction of light. *Phys. Usp.* 49: 1029–1044.

Belov, P. A., Simovski, C. R., and Tretyakov, S. A. 2003. Example of bianisotropic electromagnetic crystals: The spiral medium. *Phys. Rev. E* 67: 056622.

Busch, K., von Freymann, G., Linden, S., Mingaleev, S., Tkeshelashvili, L., and Wegener, M. 2007. Periodic nanostructures for photonics. *Phys. Rep.* 444: 101–202.

Decker, M., Klein, M. W., Wegener, M., and Linden, S. 2007. Circular dichroism of planar chiral magnetic metamaterials. *Opt. Lett.* 32: 856–858.

Decker, M., Ruther, M., Kriegler, C. E. et al. 2009. Strong optical activity from twisted-cross photonic metamaterials. *Opt. Lett.* 34: 2501–2503.

Decker, M. PhD thesis. In preparation.

Enrich, C., Wegener, M., Linden, S. et al. 2005. Magnetic metamaterials at telecommunication and visible frequencies. *Phys. Rev. Lett.* 95: 203901.

Gansel, J. K., Thiel, M., Rill, M. S. et al. 2009. Gold helix photonic metamaterial as broadband circular polarizer. *Science* 325: 1513–1515.

Gansel, J. K., Wegener, M., Burger, S. et al. 2010. Gold helix photonic metamaterials: A numerical parameter study. *Opt. Express* 18: 1059–1069.

Hecht, E. 2002. *Optics*. Addison-Wesley.

Hertz, H. 1887. Ueber sehr schnelle elektrische Schwingungen. *Ann. der Physik* 31:421.

Husnik, M., Klein, M.W., Feth, N. et al. 2008. Absolute extinction cross section of individual magnetic split-ring resonators. *Nature Photon.* 2: 614.

Klein, M. W., Enrich, C., Wegener, M., Soukoulis, C. M., and Linden, S. 2006. Single-slit split-ring resonators at optical frequencies: Limits of size scaling. *Opt. Lett.* 31: 1259–1261.

Kraus, J. D. and Marhefka, R. J. 2003. *Antennas: For All Applications*. McGraw-Hill.

Kriegler, C. E., Rill, M. S., Linden, S., and Wegener, M. 2010. Bianisotropic photonic metamaterials. *IEEE J. Selected Top. Quantum Electron.* 16:367–375.

Ku, Z. and Brueck, S. R. J. 2009. Experimental demonstration of sidewall-angle-induced bianisotropy in multiple-layer negative-index metamaterials. *Appl. Phys. Lett.* 94: 153107.

Kuwata-Gonokami, M., Saito, N., Ino, Y. et al. 2005. Giant optical activity in quasi-two-dimensional planar nanostructures. *Phys. Rev. Lett.* 95: 227401.

Kwon, D. H., Werner, D. H., Kildishev, A. V., and Shalaev, V. M. 2008. Material parameter retrieval procedure for general bi-isotropic metamaterials and its application to optical chiral negative-index metamaterial design. *Opt. Express* 16: 11822–11829.

Lakhtakia, A., Varadan, and Varadan, V. K. 1986. A parametric study of microwave reflection characteristics of a planar achiral-chiral interface. *IEEE Trans. Electromagn. Compat.* 28: 90–95.

Lindell, I. V., Sihvola, A. H., Tretyakov, S. A., and Viitanen, A. J. 1994. *Electromagnetic Waves in Chiral and Bi-Isotropic Media*. Artech House.

Linden, S., Enrich, C., Wegener, M., Zhou, J., Koschny, T., and Soukoulis, C. M. 2004. Magnetic response of metamaterials at 100 THz. *Science* 306: 1351–1353.

Lindman, K. F. 1920. Über eine durch ein isotropes System von spiralförmigen Resonatoren erzeugte Rotationspolarisation der elektromagnetischen Wellen. *Ann. der Physik* 63: 621–644.

Liu, N., Liu, H., Zhu, S., and Giessen, H. 2009. Stereometamaterials. *Nat. Photon.* 3: 157–162.

Mackay, T. G. and Lakhtakia, A. 2004. Plane waves with negative phase velocity in Faraday chiral mediums. *Phys. Rev. E* 69: 026602.

Mackay, T. G. and Lakhtakia, A. 2008. Negative reflection in a Faraday chiral medium. *Microw. Opt. Tech. Lett.* 50: 1368–1371.

Marques, R., Medina, F., and Rafii-El-Idrissi, R. 2002. Role of bianisotropy in negative permeability and left-handed metamaterials. *Phys. Rev. B* 65: 144440.

Meyrath, T. P., Zentgraf, T., and Giessen, H. 2007. Lorentz model for metamaterials: Optical frequency resonance circuits. *Phys. Rev. B* 75: 205102.

Papakostas, A., Potts, A., Bagnall, D. M., Prosvirnin, S. L., Coles, H. J., and Zheludev, N. I. 2003. Optical manifestations of planar chirality. *Phys. Rev. Lett.* 90: 107404.

Pendry, J. B., Holden, A. J., Robbins, D. J., and Steward, W. J. 1999. Magnetism from conductors and enhanced nonlinear phenomena. *IEEE Trans. Microwave Theory Tech.* 47: 2075–2084.

Pendry, J. B. 2004. A chiral route to negative refraction. *Science* 306: 1353–1355.

Plum, E., Liu, X. X., Fedotov, V. A., Chen, Y., Tsai, D. P., and Zheludev, N. I. 2009a. Metamaterials: Optical activity without chirality. *Phys. Rev. Lett.* 102: 113902.

Plum, E., Zhou, J., Dong, J. V. et al. 2009b. Metamaterial with negative index due to chirality. *Phys. Rev. B* 79: 035407.

Rogacheva, A. V., Fedotov, V. A., Schwanecke, A. S., and Zheludev, N. I. 2006. Giant gyrotropy due to electromagnetic-field coupling in a bilayered chiral structure. *Phys. Rev. Lett.* 97: 177401.

Rill, M. S., Plet, C., Thiel, M. et al. 2008. Photonic metamaterials by direct laser writing and silver chemical vapor deposition. *Nat. Mater.* 7: 543–546.

Rill, M. S., Kriegler, C. E., Thiel, M., von Freymann, G., Linden, S., and Wegener, M. 2009. Negative-index bianisotropic photonic metamaterial fabricated by direct laser writing and silver shadow evaporation. *Opt. Lett.* 34: 19–21.

Serdyukov, A., Semchenko, I. V., Tretyakov, S. A., and Sihvola, A. 2001. *Electromagnetics of Bi-Anisotropic Materials: Theory and Applications*. Gordon and Breach Science Publishers.

Sihvola, A. 2001. Condition for chiral material parameters revisited. *Microwave Opt. Technol. Lett.* 31: 423–426.

Silveirinha, M. G. 2008. Design of linear-to-circular polarization transformers made of long densely packed metallic helices. *IEEE Trans. Antennas Propag.* 56: 390.

Thiel, M., Rill, M. S., von Freymann, G., and Wegener, M. 2009. Three-dimensional bichiral photonic crystals. *Adv. Mater.* 21: 4680–4682.

Volkov, S. N., Dolgaleva, K., Boyd, R. W. et al., 2009. Optical activity in diffraction from a planar array of achiral nanoparticles. *Phys. Rev. A* 79: 043819.

Wegener, M. and Linden, S. 2009. Giving light yet another new twist. *Physics* 2: 3.

Zhang, C. and Cui, T. J. 2007. Negative reflections of electromagnetic waves in a strong chiral medium. *Appl. Phys. Lett.* 91: 194101.

Zhang, S., Park, Y.-S., Li, J., Lu, X., Zhang, W., and Zhang, X. 2009. Negative refractive index in chiral metamaterials. *Phys. Rev. Lett.* 102: 023901.

Zhao, R., Zhou, J., Koschny, Th., Economou, E. N., and Soukoulis, C. M. 2009. Repulsive casimir force in chiral metamaterials. *arXiv:0907.1435v1*.

9

Spatial Dispersion and Effective Constitutive Parameters of Electromagnetic Metamaterials

Chris Fietz and
Gennady Shvets
*The University of Texas
at Austin, Department of
Physics, Austin, Texas*

9.1 Introduction

Optical properties of metamaterials are conceptually different from those of the natural materials. Most of these differences can be traced back to the fact that the unit cell of a natural material is extremely small whereas that of most metamaterials is, generally, not. The relevant spatial scale to compare the unit cell's size a of either natural material or metamaterial is the wavelength of light λ. The unit cell's size of a natural material is only on the order of the Bohr's radius $a_B \approx 0.5 = 5 \times 10^{-9}$ cm. Therefore, it is four orders of magnitude smaller that the wavelength $\lambda = 500$ nm of the visible light in vacuum. One implication of that is the negligible phase shift of a propagating electromagnetic (or optical) wave across the unit cell. Specifically, this phase shift $\phi = 2\pi n a/\lambda \ll 1$ unless the refractive index of the material $n > 10^3$. Such a refractive index would correspond to the dielectric permittivity $\epsilon = n^2 > 10^6$. No such materials exist in nature.

Why is the phase shift per unit cell important? As it turns out, a large ϕ is related to the phenomenon of *spatial dispersion*. Mathematically, spatial dispersion can be expressed as the nonlocal relationship between the current $\mathbf{J}(\mathbf{x})$ (or polarization $\mathbf{P}$) and the electric field $\mathbf{E}(\mathbf{x}')$:

$$\mathbf{J}(x) = \int d^3x' \sigma(x-x') \cdot \mathbf{E}(x'),\qquad(9.1)$$

where σ is the nonlocal conductivity tensor of a homogeneous medium, and all field quantities are assumed to depend harmonically on time, $\propto \exp -i\omega t$. While σ is generally nonlocal, its support has a typical scale a of the material's unit cell. This is not surprising, given that homogenized fields cannot be defined on a spatial scale smaller than the unit cell. Therefore, if the wavelength of light *in the material* λ/n is much longer than the unit cell a, or $\phi \ll 1$, spatial dispersion can be neglected and the conductivity tensor can be assumed to be local. Lack of spatial dispersion has profound implications for the optical properties of any material. An example from Landau and Lifshitz's classic textbook is the prediction of negligible magnetic activity ($\mu = 1$) at optical properties.

Two factors make spatial dispersion in metamaterials much more important than in the natural materials. First, a typical unit cell of a metamaterial is less than one tenth of the wavelength.[1] Therefore, from the standpoint of their optical response, one can view metamaterials as *mesoscale* materials: their unit cell is much larger than that of natural materials yet somewhat smaller than the wavelength of light. Second, because unit cells of many metamaterials contain resonant elements (such as magnetic[2] and electric[3] split ring resonators, etc.), it is not unusual for a metamaterial to have an effective refractive index of order 10 or higher. Therefore, spatial dispersion is very important in most metamaterials. Spatial dispersion can lead to extraordinary phenomena such as, for example, effective optical magnetism (i.e., magnetic permittivity μ different from unity) and bianisotropy (contribution of the electric field to the magnetic inductance vector).

9.1.1 Relationship between Spatial Dispersion and Magnetism

The relationship between spatial dispersion and magnetism has been known[4] for some time, and has become a textbook proof of the impossibility of achieving strong magnetic activity in the optical part of the spectrum. To simplify the argument of Reference 4, assume a homogeneous metamaterial with a spatially dispersive dielectric permittivity satisfying $\bar{\bar{\epsilon}}(\mathbf{k},\omega)=\bar{\bar{\epsilon}}_0(\omega)-\alpha(\omega)a^2\mathbf{k}\times\mathbf{k}\times$, where $\alpha(\omega)$ is the dimensionless function of order unity except in the close proximity of certain resonances that can be[5,6] electrostatic. In the spatial domain, one can express the nonlocal relationship between **D** and **E** as

$$\mathbf{D}=\bar{\bar{\epsilon}}_0(\omega)\cdot\mathbf{E}+\alpha(\omega)a^2\nabla\times\nabla\times\mathbf{E}, \qquad (9.2)$$

and then substitute **D** into Maxwell-Ampere's law. The magnetic field **H** and electric induction **D** can then be redefined as

$$\mathbf{E}'=\mathbf{E}, \quad \mathbf{D}'=\bar{\bar{\epsilon}}_0(\omega)\cdot\mathbf{E}$$

$$\mathbf{H}'=\mathbf{H}\left(1-\frac{\alpha\omega^2}{c^2}\right), \quad \mathbf{B}'=\mathbf{H}, \qquad (9.3)$$

which is equivalent to redefining the magnetic permeability as $\mu=(1-\alpha\omega^2a^2/c^2)^{-1}$. Clearly, away from resonances ($\alpha\sim 1$) we find that $\mu\approx 1$ if $\omega^2a^2/c^2\ll 1$. However, for certain frequencies corresponding to large values of α, and for metamaterials with $\omega a/c\sim 1$, it may be possible to find the conditions for large deviations of μ from unity. In fact, in the limit of $\omega a/c\ll 1$ magnetic activity may still be quite strong in the vicinity of electrostatic resonances. Of course, the specific relationship between **D** and **E** expressed by Equation 9.2 is not general. In fact, even for a strictly isotropic material, the most general (and quadratic in ∇) form of the relationship between **D** and **E** is

$$\mathbf{D}=\bar{\bar{\epsilon}}_0(\omega)\cdot\mathbf{E}+\alpha(\omega)a^2\nabla\times\nabla\times\mathbf{E}+\beta(\omega)a^2\nabla(\nabla\cdot\mathbf{E}), \qquad (9.4)$$

where $\alpha(\omega)$ and $\beta(\omega)$ are dimensionless functions of the frequency. But most metamaterials comprise unit cells belonging to a specific point symmetry group that does not necessarily guarantee an isotropic response. This remains true even for relatively high-symmetry metamaterials such as cubic (in 3-D) and square (in 2-D) lattices of highly symmetric inclusions (e.g., cylinders, spheres, or cubes). In fact, in the vicinity of highly anisotropic resonances of the individual unit cells, the metamaterial does not appear to be isotropic at all. Therefore, it is essential to develop the most general approach to computing $\bar{\bar{\epsilon}}(\mathbf{k},\omega)$, at least in the electrostatic regime where other components ϵ of the constitutive parameters matrix (CPM; see later text) are not important. The procedure for calculating the electrostatic $\bar{\bar{\epsilon}}(\mathbf{k},\omega)$ is illustrated for two-dimensional plasmonic metamaterials in Section 9.2.

In general, an arbitrary material (natural or meta) is characterized by the full CPM, which contains 36 entries. The CPM relates the electric and magnetic induction vectors **D** and **B** to the electric and magnetic fields **E** and **H** according to Equation 9.5:

$$\begin{pmatrix} \mathbf{D} \\ \mathbf{B} \end{pmatrix} = \begin{pmatrix} \epsilon & \xi \\ \zeta & \mu \end{pmatrix} \begin{pmatrix} \mathbf{E} \\ \mathbf{H} \end{pmatrix}, \qquad (9.5)$$

where the field vectors **D**, **B**, **E**, and **H**, are the macroscopic electromagnetic induction and field vectors, respectively. In most natural materials the bianisotropy matrices ξ and ζ, as well as the off-diagonal elements of the ϵ and μ matrices, are negligible because of the high symmetry of the unit cell. Because of the negligible spatial dispersion, it is also reasonable to assume that the CPM of natural materials is a function of the frequency ω but not of the wavenumber **k**. That is no longer the case for metamaterials exhibiting strong spatial dispersion. In fact, the CPM of a typical metamaterial exhibits strong dependence on both ω and **k**, resulting in a wide range of novel optical phenomena such as magnetic activity, bianisotropy, and strong anisotropy of the ϵ and μ tensors. Several examples presented in this chapter illustrate that the aforementioned phenomena are manifested even in metamaterials with a high symmetry of the unit cell.

This raises an important question: how does one properly characterize a metamaterial exhibiting strong spatial dispersion and extract the 36 parameters of the CPM? In Section 9.2, we start with a simple two-dimensional electrostatic example of a metamaterial comprising a square array of plasmonic nanorods and demonstrate how one extracts all four components of the ϵ-tensor by applying inhomogeneous external electric fields of two linearly independent polarizations. This concept is then extended in Section 9.3 to fully electromagnetic cases, where the remaining entries of the CPM need to be extracted. This is done using the technique of current-driven homogenization (CDH), which is described in detail. Examples of application of the CDH technique to two plasmonic metamaterials exhibiting negative refraction are presented in Section 9.4. One of the practical advantages of possessing the full CPM of a metamaterial is the ability to design metamaterial cloaks and shell into which arbitrary antennas are embedded. Examples of directional antennas embedded inside a metamaterial slab are presented in Section 9.5.

9.2 Spatial Dispersion at the Smallest Spatial Scale: The Electrostatic Approach

Spatial dispersion has been extensively discussed in the context of metamaterials exhibiting full electromagnetic response,[7] that is, in the regime where electromagnetic retardation due to the finite speed of light is important for metamaterials' properties. Interestingly, spatial dispersion can fully manifest itself even at the electrostatics level. The electrostatic regime is particularly interesting in the context of subwavelength plasmonic metamaterials.[5,6,8] As was explained earlier, spatial dispersion of the dielectric permittivity tensor must be accounted for in order to describe, for example, one of the most counterintuitive phenomena encountered in metamaterials: magnetic activity of the electromagnetic structures that do not contain any magnetic inclusions. In this section, we develop an approach to computing $\bar{\bar{\epsilon}}\,(\mathbf{k},\omega)$ in the electrostatic regime where other components of the CPM are not important. To illustrate the approach, we consider, for simplicity, two-dimensional metamaterials, that is, unit cells consisting of infinitely extended in the z-direction dielectric or plasmonic cylinders. To further simplify our calculations, it will be assumed that the unit cell is a square of the size $a \times a$ in the xy plane: $-a/2 < x < a/2$ and $-a/2 < y < a/2$. Later, we describe how the four entries $(\epsilon_{xx}, \epsilon_{yy}, \epsilon_{xy}, \epsilon_{yx})$ of the 2×2 dielectric permittivity tensor can be computed numerically as a function of **k** and ω.

The dependence on ω is relatively simple: it is caused by the frequency dependence of the dielectric (or, to a much greater degree, plasmonic) inclusions inside the unit cell. The wavenumber $\mathbf{k}$ is introduced by imposing an external nonuniform electric field $\mathbf{E}_0 = e_1 e^{-i(\omega t - \mathbf{k} \cdot x)}$, where e_1 is the unit vector defining the polarization direction. Note that no relationship between ω and $\mathbf{k}$ needs to be assumed. The total electric field is sought in the form of $\mathbf{E} = \mathbf{E}_0 - \nabla \phi$, where ϕ is the internal potential satisfying $\nabla \cdot (\epsilon(x, \omega) \mathbf{E}) = 4\pi \rho_0$, where $\epsilon = \epsilon(x, \omega)$ describes the frequency-dependent permittivity of a localized plasmonic inclusion inside the unit cell of a metamaterial and $4\pi \rho_0 \, i\mathbf{k} \cdot \mathbf{E}_0$ is the external charge corresponding to the nonuniform external field. The equation for ϕ thus takes on the form of

$$\nabla \cdot (\epsilon \mathbf{E}_0 - \epsilon \nabla \phi) = i\mathbf{k} \cdot \mathbf{E}_0 \tag{9.6}$$

and is solved on a square domain *ABCD* (where *AB* and *CD* are parallel to *y*, *BC* and *AD* are parallel to *x*). The internal potential ϕ satisfies the following phase-periodic boundary conditions: (a) $\phi(x + a_x, y) = e^{ik_x a} \phi(x, y)$, where (x, y) belongs to *AB*, and (b) $\phi(x, y + a_y) = e^{ik_y a} \phi(x, y)$, where (x, y) belongs to *AD*. The cell-averaged electric field $\bar{\mathbf{E}}$ and induction $\bar{\mathbf{D}}$ vectors are defined as

$$\bar{\mathbf{E}} = \int \frac{d^2 x}{A} (\mathbf{E}_0 e^{-ik \cdot x} \nabla \phi)$$
$$\bar{\mathbf{D}} = \int \frac{d^2 x}{A} \epsilon (\mathbf{E}_0 e^{-ik \cdot x} \nabla \phi), \tag{9.7}$$

where $A = a^2$ is the unit cell's area. In practice, the cell-averaged fields are computed by solving the inhomogeneous Poisson's equation for ϕ given by Equation 9.6, and then substituting ϕ into Equation 9.7.

The permittivity tensor can then be computed by solving Equation 9.6 for two linearly independent (noncollinear) polarizations e_1 and e_2 and calculating the two sets of the cell-averaged electric field and induction vectors $\bar{\mathbf{E}}^{(1,2)}$ and $\bar{\mathbf{D}}^{(1,2)}$, respectively. Next, we define the following 2×2 matrices:

$$\mathcal{D} \equiv \begin{pmatrix} \bar{D}_x^{(1)} & \bar{D}_x^{(2)} \\ \bar{D}_y^{(1)} & \bar{D}_y^{(2)} \end{pmatrix}, \mathcal{E} \equiv \begin{pmatrix} \bar{E}_x^{(1)} & \bar{E}_x^{(2)} \\ \bar{E}_y^{(1)} & \bar{E}_y^{(2)} \end{pmatrix}, \tag{9.8}$$

where the relationship between $\mathcal{D}$ and $\mathcal{E}$ is used to define the effective permittivity tensor according to $\mathcal{D} = \bar{\bar{\epsilon}}_{eff} \cdot \mathcal{E}$ or, explicitly:

$$\begin{pmatrix} \epsilon_{xx} & \epsilon_{xy} \\ \epsilon_{yx} & \epsilon_{yy} \end{pmatrix} \begin{pmatrix} \bar{E}_x^{(1)} & \bar{E}_x^{(2)} \\ \bar{E}_y^{(1)} & \bar{E}_y^{(2)} \end{pmatrix} = \begin{pmatrix} \bar{D}_x^{(1)} & \bar{D}_x^{(2)} \\ \bar{D}_y^{(1)} & \bar{D}_y^{(2)} \end{pmatrix}. \tag{9.9}$$

The four equations contained in Equation 9.9 are used to compute the four entries (ϵ_{xx}, ϵ_{yy}, ϵ_{xy}, ϵ_{yx}) of the effective permittivity tensor $\overline{\overline{\epsilon}}_{eff}(\omega,\mathbf{k})$. Clearly, $\overline{\overline{\epsilon}}_{eff}(\omega,\mathbf{k})$ explicitly depends on both $\mathbf{k}$ and ω. The dependence on ω occurs because the dependence of the dielectric/plasmonic inclusion $\epsilon(x,\omega)$ on the frequency affects the solutions of Equation 9.6 and, therefore, the values of $\overline{\mathbf{E}}$ and $\tilde{\mathbf{E}}$. The dependence on the $\mathbf{k}$ arises because of the explicit dependence on $\mathbf{k}$ of the boundary conditions and $\mathbf{E}_0(x)$ in Equation 9.6.

Note that the solutions of the *inhomogeneous* Equation 9.6 can become infinite in the vicinity of the so-called *electrostatic resonances*. These electrostatic resonances correspond to specific values ϵ_r of the dielectric permittivity of the inclusion.[5,9,10] For binary plasmonic metamaterials, that is, metamaterials with a unit cell consisting of a plasmonic inclusion inside a single dielectric host, it is conventional to label these resonant values of the plasmonic permittivity using a variable s defined as [5,9,10] $s \equiv (1-\epsilon_r)^{-1}$. Because of the frequency dependence of $\epsilon(\omega)$, it is possible to identify the frequencies ω_r of such resonances from $\epsilon(\omega_r) = \epsilon_r$. Such electrostatic resonances can be classified as *dipole-active* and *dipole-inactive*. Dipole-active resonances directly contribute to the $\mathbf{k}$-independent dielectric permittivity tensor: it diverges in the vicinities of these resonances. The role of dipole-active resonances has been extensively discussed[9,10] in the literature. But it is the less-known dipole-inactive resonances (which do not directly contribute to the $\mathbf{k}$-independent dielectric permittivity) that can strongly contribute to the spatially dispersive components of the dielectric tensor. Specifically, in the vicinity of the dipole-inactive resonances, coefficients $\alpha(\omega)$ and $\beta(\omega)$ could become very large, even divergent in the absence of resistive losses. As will be demonstrated later, one of the consequences of the enhanced spatial dispersion in the vicinity of an *octupole* (dipole-inactive) resonance is magnetic activity according to Equation 9.3.

As an example, we consider a square array of almost-touching plasmonic cylinders. This structure was originally introduced in References 5 and 6 as the first truly subwavelength ($\lambda/a \approx 10$) negative index plasmonic structure. The metamaterial is assumed to be a square array with period a of cylindrical inclusions (rods of radius R infinitely extended in the z-direction) with dielectric permittivity ϵ imbedded in a host material with dielectric permittivity $\epsilon_h \equiv 1$. A schematic of this plasmonic metamaterial is sketched in the inset to Figure 9.2, where four adjacent unit cells are shown. For a binary plasmonic metamaterial, it is convenient to plot the dispersion relations of various electrostatic resonances in the form of $s \equiv s(\mathbf{k}a)$, where $\mathbf{k}$ is the Bloch wavenumber. Such plot is shown in Figure 9.2,[6] where different resonances are labeled according to their spatial symmetry group.

The two strongest dipole-active resonance are labeled E_y, and their potential distributions are shown in Figure 9.2. Their spectral positions correspond to $s_1 = 0.14$ (or $\omega_1 = 0.38\omega_p$) and $s_1 = 0.40$ (or $\omega_1 = 0.63\omega_p$). These (and other, much weaker) dipole-active resonances directly contribute to the $\mathbf{k}$-independent dielectric permittivity tensor. For example, they directly contribute to the $\overline{\overline{\epsilon}}_0(\omega)$ portion of the dielectric permittivity tensor of an isotropic metamaterial defined by Equation 9.4. They do not contribute to the spatially dispersive portion of the permeability tensor, which is primarily defined by the dipole-inactive electrostatic resonances.

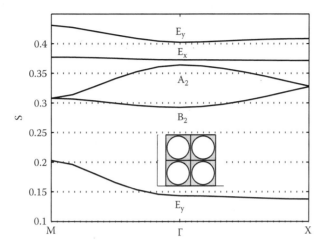

FIGURE 9.1 Electrostatic resonances of the square lattice plasmonic metamaterial consisting of almost-touching plasmonic cylinders with $R/a = 0.45$ (shown as inset), where the period $a = c\omega_p$. Vertical axis: $s = 1/[1 - \epsilon(\omega)] \equiv \omega^2/\omega_p^2$; horizontal axis: normalized Bloch wavenumber. Scanned eigenvalue range: $0 < s < 0.45$. Resonances labeled according to the spatial symmetry at the Γ-point. Dipole-active resonances: E_y; dipole-inactive resonances: B_2 and A_2. (Reprinted from Shvets, G. and Urzhumov, Y. A., *J. Opt. A: Pure Appl. Opt.* 7, S23–S31, 2004.)

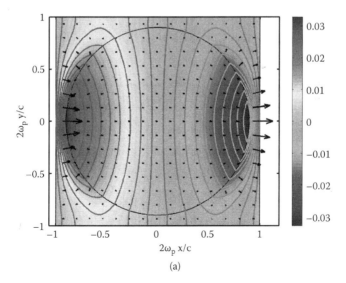

(a)

FIGURE 9.2 *See color insert.* The eigenvalue potentials of the two strongest dipole-active resonances of the plasmonic structure with the parameters from Figure 9.2 (color and contours) and the corresponding electric field (arrows). (a): The strongest dipole-active resonance at $s_1 = 0.14$ (or $\omega_1 = 0.38\omega_p$), and (b) the second-strongest dipole-active resonance at $s_1 = 0.40$ (or $\omega_1 = 0.63\omega_p$). (Reprinted with full permission from Shvets, G. and Urzhumov, Y., *Phys. Rev. Lett.* 93, 243902, 2004. Copyright [2004] by the American Physical Society.)

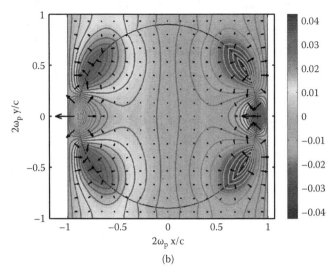

(b)

FIGURE 9.2 (Continued)

Two such resonances, labeled B_2 and A_2, are shown in Figure 9.2. The B_2 and A_2 resonances can be classified as quadrupolar and octupolar, respectively, owing to their angular dependencies inside the plasmonic cylinder:

$$\phi^{(B_2)} = \sum_{n=1}^{\infty} \Phi_n^{(B_2)} \left(\frac{r}{R} \right)^{2n} \sin 2n\theta, \quad \phi^{(A_2)} = \sum_{n=1}^{\infty} \Phi_n^{(A_2)} \left(\frac{r}{R} \right)^{4n} \sin 4n\theta. \qquad (9.10)$$

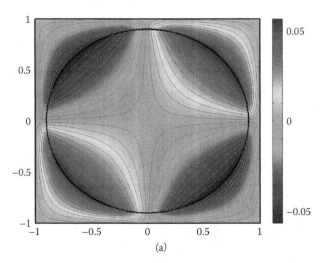

(a)

FIGURE 9.3 *See color insert.* Dipole-inactive electrostatic resonances of the square lattice plasmonic metamaterials with the parameters as in Figure 9.2. Left: quadrupole resonance at $s_Q = 0.29$ ($\omega_Q = 0.54\omega_p$); right: octupole resonance at $s_O = 0.36$ ($\omega_O = 0.6\omega_p$).

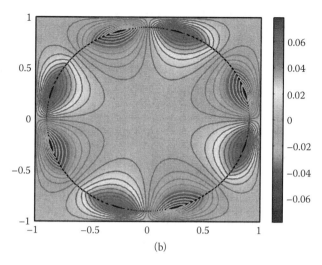

(b)

FIGURE 9.3 (continued)

As shown later, the magnetic activity of this structure can be traced back to the strong spatial dispersion of the $\overline{\overline{\epsilon}}(\mathbf{k},\omega)$ in the vicinity of the octupolar electrostatic resonance A_2.

Despite the high spatial symmetry of the considered square-array plasmonic metamaterial, its dielectric permittivity is not necessarily isotropic. Therefore, even for small $\mathbf{k}$, the dielectric tensor defined by Equation 9.4 cannot be assumed *a priori*. Instead, we assume a more general form of the $\overline{\overline{\epsilon}}(\mathbf{k},\omega)$:

$$\overline{\overline{\epsilon}}(\mathbf{k},\omega) = \begin{pmatrix} \epsilon_0(\omega) & 0 \\ 0 & \epsilon_0(\omega) \end{pmatrix} + \alpha a^2 \begin{pmatrix} k_y^2 & -k_x k_y \\ -k_x k_y & k_x^2 \end{pmatrix} + \beta a^2 \begin{pmatrix} k_x^2 & k_x k_y \\ k_x k_y & k_y^2 \end{pmatrix}$$

$$+ \gamma a^2 \begin{pmatrix} k_y^2 & k_x k_y \\ k_x k_y & k_x^2 \end{pmatrix}, \tag{9.11}$$

where α, β, and γ are frequency-dependent dimensionless functions. Note that the *anisotropic* component of the dielectric tensor is proportional to $\gamma(\omega)$.

Let us examine the spatially dispersive part of $\overline{\overline{\epsilon}}(\mathbf{k},\omega)$ for different values of the metal's dielectric permittivity ϵ. First, we neglect the losses and assume that $\epsilon = 1 - \omega_p^2/\omega^2$ is a real negative number. For very low frequencies, $\epsilon \ll -1$. As an example, assume that $\epsilon = -30$. To determine all numerical coefficients of the expansion given by Equation 9.11, one needs to extract the components of the dielectric tensor for two values of $\mathbf{k}$: (1) $\mathbf{k} = 0$ and (2) $\mathbf{k} = k_x e_x + k_y e_y$, where $k_x \neq k_y$ and $k_x a, k_y a \ll 1$. Simulation (1) yields $\epsilon_0 = 6.24$. Simulation (2) yields the remaining three constants. For example, for $k_x a = 0.4$ and $k_y a = 0.2$, it is found that $\epsilon_{xx} = 6.21$, $\epsilon_{yy} = 6.14$, and $\epsilon_{xy} = \epsilon_{yx} = 0.087$. These three tensor entries are sufficient for extracting $\alpha = -0.88$, $\beta = -0.05$, and $\gamma = 0.26$. As this example indicates, even far away from resonance, the dielectric permittivity tensor is rather anisotropic because of the fairly large value of $|\gamma|$. Therefore, such plasmonic metamaterial cannot be rigorously described using

$\mu_z(\omega)$ and $\bar{\bar{\epsilon}}(\omega)$ even for small $\mathbf{k}$ because it is optically anisotropic. If anisotropy could be neglected, then this metamaterial would be described by $\mu_{zz} \approx (1 - \alpha a^2\omega^2/c^2)^{-1} < 1$, that is, it would be weakly diamagnetic in the limiting case of $\omega^2 a^2/c^2 \ll 1$.

Spatial dispersion intensifies as the frequency approaches one of the dipole-inactive resonances. For example, in the vicinity of the quadrupole resonance ($\epsilon = -2.4$), we find that $\epsilon_0 = -2.25$, $\alpha = 0.03$, $\beta = -0.005$, and $\gamma = 13.4$. The implication of this is that the plasmonic metamaterial is extremely anisotropic in the vicinity of the quadrupole resonance. This conclusion becomes less surprising upon examining the profile of the quadrupole potential shown in Figure 9.2 (left).

As the frequency moves away from the quadrupole resonance, spatial dispersion rapidly decreases. However, it picks up as the octupole resonance is approached. For $\epsilon = -1.76$, it is found that $\epsilon_0 = -0.45$, $\alpha = 16.4$, $\beta = 0.05$, and $\gamma = -0.12$. Here, the implication is that the metamaterial behaves as an almost-isotropic medium ($|\gamma| \ll 1$) with a strong spatial dispersion coefficient α that results in magnetic activity. Indeed, from $\mu_{zz} \approx (1 - \alpha a^2\omega^2/c^2)^{-1}$, it follows that $\mu_{zz} < 0$ can be achieved even for highly subwavelength plasmonic structures with $\omega a/c > 0.25$. Indeed, negative index propagation bands have been theoretically demonstrated[5] for such plasmonic structures. It would be premature, however, to suggest the foregoing structure as a viable path toward plasmonic negative index metamaterial. Even small ohmic losses can dramatically degrade α. For example, for $\epsilon = -1.76 + 0.01i$, it is found that $\alpha = 2 + 5.6i$. Therefore, the effective magnetic permeability becomes imaginary.

In conclusion, strong spatial dispersion can be observed in the simplest plasmonic structures even in the electrostatic limit. The full $\mathbf{k}$-dependent permeability tensor (four entries in the case of the two-dimensional structures considered here) can be extracted by introducing external spatially nonuniform electric fields $\mathbf{E}_0 = e_{1,2}e^{-i(\omega t - k \cdot x)}$, where $e_{1,2}$ are two arbitrary non-collinear polarization directions, and the associated external charge density $4\pi\rho_0 \equiv i\mathbf{k} \cdot \mathbf{E}_0$. The permeability tensor is then obtained by calculating the internal electrostatic potential f from Equation 9.6, computing the averaged electric field and induction from Equation 9.7, and extracting the permittivity tensor from Equation 9.9. While this section dealt with two-dimensional structures, extension to three dimensions is straightforward.

It is found that spatial dispersion is particularly strong in the vicinity of the dipole-inactive electrostatic resonances. Depending on the spatial symmetry of the resonance, highly symmetric plasmonic metamaterials can exhibit either isotropic or strongly anisotropic optical properties. In the former case, it is possible to introduce an effective magnetic permeability originating from the spatial dispersion coefficient a. Because of the inherently narrow spectral width of the dipole-inactive resonances, even a small amount of ohmic loss can strongly reduce spatial dispersion. In the next section, the fully electrodynamic procedure for extracting the full CPM defined by Equation 9.5 is given.

9.3 Electrodynamic Calculation of the Full CPM of Metamaterials

Since the invention of artificial materials with negative magnetic response about a decade ago,[12] considerable work has gone into the study of electromagnetic metamaterials (artificial materials engineered to have exotic optical properties). In that time both analytic[12,13] and numerical[7,14–16] methods have been suggested for determining the constitutive parameters of such metamaterials. Although analytic expressions are useful for approximating a metamaterial's response and providing general intuition, ultimately, numerical methods are needed to accurately calculate constitutive parameters. The full CPM, which relates the electric and magnetic induction vectors **D** and **B** to the electric and magnetic fields **E** and **H** according to Equation 9.5, has 36 components. In this section, we outline a numerical procedure, current-driven homogenization (CDH), for calculating all components of the CPM as a function of ω and **k**. CDH is the full electromagnetic extension of the electrostatic extraction technique described in Section 9.2.

Because electrostatics does not deal directly with propagating waves, the issue of a possible relationship between the frequency ω and the wavenumber **k** did not arise in Section 9.2. Electrodynamics, however, deals with the propagating waves possessing a certain *dispersion relation* between ω and **k**. If ω and **k** satisfy the dispersion relation, they are said to lie on the dispersion curve. An important aspect of the CDH method is that it enables the computing of constitutive parameters for frequencies and wavenumbers both on and off the dispersion curve $\omega = \omega(\mathbf{k})$. Applications such as, for example, novel antennas embedded in metamaterial shells, require the calculation of constitutive parameters both on and off the dispersion curve, that is, for arbitrary and unrelated ω and **k**.

Essential to the CDH approach is the idea of driving a metamaterial crystal with both electric and magnetic charge-current. In Section 9.3.1, we explain why this is necessary and how it enables us to extract all 36 parameters of the CPM. In addition to driving the crystal, a field-averaging procedure that converts the microscopic EM fields in a metamaterial into averaged (macroscopic) EM fields is required. The method of driving the crystal with electric and magnetic charge-current is independent of the averaging procedure used, and different field averaging procedures can be used interchangeably. In Section 9.3.2, a new field averaging procedure is described that is particularly accurate at providing approximately correct boundary conditions.[17] The importance of accurate boundary conditions becomes clear whenever one needs to accurately model reflection/transmission coefficients at the interface between the metamaterials, or when radiation by embedded-into-metamaterials antennas needs to be computed, as explained in Section 9.5.

Finally, in Section 9.4.1, we calculate the constitutive parameters for a 2D plasmonic crystal and then validate the extracted parameters by calculating the transmission of a simple metamaterial antenna driven with a wide spectrum of **k**'s.

9.3.1 Current-Driven Extraction of the Constitutive Parameters

First, let us consider the procedure of extracting the constitutive parameters of a homogeneous medium. The most common definition of the CPM is given by Equation 9.5. The 6×6 CPM is, by convention, separated into four 3×3 matrices known as ϵ, ξ, ζ, and μ.[18] Because the most general CPM contains 36 unknown entries, we need 36 linearly independent equations of constraints to calculate it. A single set of EM fields related through Equation 9.5 provides six equations of constraints. Therefore, we need six sets of linearly independent EM fields obeying Equation 9.5 to compute the CPM. Specifically, we need six linearly independent sets of field vectors providing us with 36 equations of constraints that can be solved for 36 unknowns. Of course, the CPM of a homogeneous medium is known, so what is described in this section is merely an extraction procedure that recovers the *a priori* known CPM.

If we use only EM waves supported by the medium free of source terms (either propagating or evanescent waves), as is conventionally done in the context of metamaterials,[14,15] then for a general material only one set of linearly independent fields is available for a particular set of $[\omega, \mathbf{k}(\omega)]$, where $\omega = \omega(\mathbf{k})$ is the dispersion relation. Three sets of linearly independent fields can be obtained by driving the metamaterial crystal with electric charge-current,[7] but six sets are needed. A solution to our problem becomes apparent when we inspect a modified form[19,20] of the Maxwell equations in a homogeneous medium:

$$\nabla \cdot \mathbf{D} = 4\pi\rho \qquad \nabla \times \mathbf{H} - \frac{1}{c}\frac{\partial \mathbf{D}}{\partial t} = \frac{4\pi}{c}\mathbf{J}$$

$$\nabla \cdot \mathbf{B} = 4\pi\phi \qquad -\nabla \times \mathbf{E} - \frac{1}{c}\frac{\partial \mathbf{B}}{\partial t} = \frac{4\pi}{c}\mathbf{I}, \tag{9.12}$$

where the magnetic charge density ϕ and magnetic current density $\mathbf{I}$ are introduced. In a homogeneous medium, an electric current $\mathbf{J} = \mathbf{J}_0 e^{i(\omega t - \mathbf{k} \cdot \mathbf{x})}$ and magnetic current $\mathbf{I} = \mathbf{I}_0 e^{i(\omega t - \mathbf{k} \cdot \mathbf{x})}$ that are harmonic in time and space will generate a EM waves $\mathbf{E}(t, x) = \mathbf{E}_0 e^{i(\omega t - \mathbf{k} \cdot \mathbf{x})}$, and $\mathbf{H}(t, x) = \mathbf{H}_0 e^{i(\omega t - \mathbf{k} \cdot \mathbf{x})}$ according to Equation 9.12, which can be rearranged in ω and $\mathbf{k}$ space and combined with the constitutive matrices yielding

$$\mathcal{M}\begin{pmatrix} \mathbf{E}_0 \\ \mathbf{H}_0 \end{pmatrix} = \frac{4\pi i}{c}\begin{pmatrix} \epsilon^{-1}\mathbf{J}_0 \\ \mu^{-1}\mathbf{I}_0 \end{pmatrix}, \tag{9.13}$$

where

$$\mathcal{M} = \begin{pmatrix} \omega/c & \epsilon^{-1}(\mathbf{k}\times + \omega\xi/c) \\ -\mu^{-1}(\mathbf{k}\times - \omega\xi/c) & \omega/c \end{pmatrix}. \tag{9.14}$$

Here, the four blocks of the $\mathcal{M}$ matrix are represented by 3×3 matrices, and $(k\times)_{ij} \equiv \epsilon_{ijl}k_l$. All four constitutive matrices are assumed to be functions of ω and $\mathbf{k}$. If ω and $\mathbf{k}$ do not lie on the dispersion curve of the metamaterial crystal (i.e., do not satisfy the $\omega = \omega(\mathbf{k})$ dispersion relation), then $\mathcal{M}$ is invertible, and $\mathbf{E}_0$ and $\mathbf{H}_0$ can be solved for. Therefore, if we limit ourselves to electric current only, then at most three linearly independent field vectors are obtained. Those would be insufficient for solving Equation 9.5 for the constitutive parameters. However, if we allow ourselves to drive the metamaterial crystal with both electric and magnetic current, then six linearly independent field vectors can be obtained. These can be used for extracting all entries of the CPM from

Equation 9.5. Explicitly, the extraction procedure is as follows. First, the following 6×6 EM field matrices are defined:

$$\mathcal{D} \equiv \begin{pmatrix} \mathbf{D}_0^1, \mathbf{D}_0^2 \ \cdots \ \mathbf{D}_0^6 \\ \mathbf{B}_0^1, \mathbf{B}_0^2 \ \cdots \ \mathbf{B}_0^6 \end{pmatrix},$$

$$\mathcal{E} \equiv \begin{pmatrix} \mathbf{E}_0^1, \mathbf{E}_0^2 \ \cdots \mathbf{E}_0^6 \\ \mathbf{H}_0^1, \mathbf{H}_0^2 \ \cdots \mathbf{H}_0^6 \end{pmatrix}, \tag{9.15}$$

$$\mathcal{J} \equiv \begin{pmatrix} \mathbf{J}_0^1, \mathbf{J}_0^2 \ \cdots \ \mathbf{J}_0^6 \\ \mathbf{I}_0^1, \mathbf{I}_0^2 \ \cdots \ \mathbf{I}_0^6 \end{pmatrix}.$$

Each column of the matrices in Equation 9.15 is associated with a single current-driven electromagnetic simulation that involves the solution of Equation 9.12, subject to the medium's constitutive parameters. For example, in the first simulation, the driving current is $\mathcal{J}_{i1} = (1,0,0,0,0,0)$, that is, $\mathbf{J}_0 = \mathbf{e}_x$ and $\mathbf{I}_0 = 0$; in the second simulation, $\mathcal{J}_{i2} = (0,1,0,0,0,0)$, etc. Electromagnetic fields can then be combined into the matrices $\mathcal{E}$ and $\mathcal{D}$, which are, by definition, related through the CPM defined by Equation 9.5. Finally, the CPM is recovered:

$$C \equiv \begin{pmatrix} \epsilon & \xi \\ \zeta & \mu \end{pmatrix} = \mathcal{D}\mathcal{E}^{-1}. \tag{9.16}$$

Note that Equation 9.16 is a fully electrodynamic extension of the electrostatic procedure for extracting the permittivity tensor given by Equation 9.9. Because the six electromagnetic simulations are performed for a particular ω and $\mathbf{k}$, the calculated CPM is a function of ω and $\mathbf{k}$: $C = C(\omega, \mathbf{k})$. For the homogeneous medium, the CDH procedure returns the *a priori* known CPM, so no new information about the medium is gained. The real utility of the CDH is that the same extraction procedure can be applied to an inhomogeneous (periodic) metamaterial if an appropriate field averaging procedure can be introduced. The field averaging procedure resulting in the effective CPM of the metamaterial is now described.

9.3.2 Field Averaging and Homogenization of Periodic Metamaterials

By definition, a periodic metamaterial is highly inhomogeneous. A unit cell of a metamaterial may consist of various arbitrarily shaped material inclusions such as metals or dielectrics. Microscopic EM fields **e, h, d,** and **b** inside each of these inclusions are related through Equation 9.12, subject to the material's local constitutive parameters.

In practice, solving Maxwell's Equation 9.12 for **e, h, d**, and **b** inside a single unit cell for a fixed frequency ω can be accomplished using any commercial finite elements software package, for example, COMSOL Multiphysics, HFSS, or CST Microwave Studio. Microscopic EM fields are subject to phase-shifted periodic boundary conditions determined by the wavenumber **k**. Introducing the effective CPM of such a metamaterial requires averaging strongly inhomogeneous microscopic fields inside the unit cell in order to obtain a matrix of macroscopic fields given by Equation 9.15. The CDH procedure is then applied to obtain the effective CPM according to Equations 9.15 and 9.16.

Field averaging procedures are not unique, and several have been utilized in the past[7,12,15] to predict wave propagation inside a metamaterial. In the following, we describe the field-averaging procedure, which has an important advantage over the earlier ones. Like the earlier procedures,[7,12,15] it predicts the correct dispersion relation for propagating waves. In addition, it provides approximately correct boundary conditions.[21] The general three-dimensional volume-averaging procedure is described later, although only two-dimensional examples are provided in the rest of the section in keeping with the spirit of Section 9.2.

Because of the distinct way different components of the EM field enter Maxwell's equations, the volume-averaging technique also treats these components differently. Specifically, the following averaging formulas are used for the field components of $(\mathbf{D}_0, \mathbf{B}_0)$ parallel to $\mathbf{k}^*$ (the * indicates complex conjugation) and the field components of $(\mathbf{E}_0, \mathbf{H}_0)$ perpendicular to $\mathbf{k}$:

$$-i\mathbf{k} \cdot \begin{Bmatrix} \mathbf{D}_0 \\ \mathbf{B}_0 \end{Bmatrix} S_V = \int_\Omega \frac{d^3x}{V} \nabla \cdot \begin{Bmatrix} \mathbf{d}_0 \\ \mathbf{b}_0 \end{Bmatrix} \qquad (9.17)$$

and

$$-i\mathbf{k} \cdot \begin{Bmatrix} \mathbf{E}_0 \\ \mathbf{H}_0 \end{Bmatrix} S_V = \int_\Omega \frac{d^3x}{V} \nabla \times \begin{Bmatrix} \mathbf{e}_0 \\ \mathbf{h}_0 \end{Bmatrix}, \qquad (9.18)$$

where integration is over the unit cell (Ω), $V = a_x a_y a_z$ is the volume of a tetragonal unit cell with dimensions $a_x \times a_y \times a_z$, and S_V is the three-dimensional generalization of the form factor earlier introduced[22]:

$$S_V = \frac{\sin(k_x a_x/2)}{a_x k_x/2} \cdot \frac{\sin(k_y a_y/2)}{a_y k_y/2} \cdot \frac{\sin(k_z a_z/2)}{a_z k_z/2}. \qquad (9.19)$$

For the remaining field components, a straightforward volume averaging is used:

$$\mathbf{k}^* \times \begin{Bmatrix} \mathbf{D}_0 \\ \mathbf{B}_0 \end{Bmatrix} S_V = \int_\Omega \frac{d^3x}{V} \mathbf{k}^* \times \begin{Bmatrix} \mathbf{d} \\ \mathbf{b} \end{Bmatrix} \qquad (9.20)$$

and

$$\mathbf{k}^* \cdot \begin{Bmatrix} \mathbf{E}_0 \\ \mathbf{H}_0 \end{Bmatrix} S_V = \int_\Omega \frac{d^3 x}{V} \mathbf{k}^* \cdot \begin{Bmatrix} \mathbf{e} \\ \mathbf{h} \end{Bmatrix}.$$

(9.21)

The macroscopic field components $\mathbf{E}_0$, $\mathbf{H}_0$, $\mathbf{D}_0$, and $\mathbf{B}_0$ are the counterparts of the averaged $\overline{\mathbf{E}}$ and $\overline{\mathbf{D}}$ fields introduced in Section 9.2. Note that the propagation wavenumber can, in principle, be complex. That becomes particularly relevant when the CPM needs to be evaluated on the dispersion curve. The propagation wavenumber $\mathbf{k}$ corresponding to real ω is always complex when finite losses are considered. Even in the absence of losses, $\mathbf{k}$ is complex in the spectral range corresponding to evanescent waves.

The approach of applying different averaging recipes to different field components is shared by most averaging schemes.[12,15] For example, without introducing different averaging recipes for $\mathbf{H}$ and $\mathbf{B}$, it would not be possible to predict effective magnetic activity ($\zeta \neq 0$ or $\mu \neq 1$) for metamaterials that do not contain any magnetic inclusions. Also, it should be noted that if $\mathbf{k}$ is parallel to the principal axis of a crystal and if the inclusions in the metamaterial are small, then the present averaging method for the transverse components of $\mathbf{E}_0$ and $\mathbf{H}_0$ is equivalent to the transversely averaged fields used in.[23] Finally, because of the nature of the definitions in Equations 9.17–9.21, it is impossible to rigorously compute the constitutive parameters at the Γ point ($\mathbf{k} = 0$). It is, however, possible to compute the entire CPM for any finite wavenumber. As illustrated below in Section 9.3.3, finite wavenumbers introduce a new feature of electromagnetic metamaterials: finite bianisotropy caused by spatial dispersion. This bianisotropy manifests itself as an angle-dependent (anisotropic) magnetic permeability. This is yet another way in which spatial dispersion gives rise to the phenomena that can be interpreted as magnetic activity.

9.3.3 Bianisotropy due to Spatial Dispersion

It has long been recognized[13] that many recently introduced metamaterials lacking spatial inversion symmetry are bianisotropic, that is, exhibit cross-polarization effects: an electric polarization as a response to an applied magnetic field, and vice versa. It is widely believed that a centrosymmetric crystal cannot be bianisotropic (ξ and ζ must be zero). However, this is only true if the constitutive parameters are functions of the frequency ω only. If the constitutive parameters are functions of both ω and $\mathbf{k}$, then spatial dispersion can cause bianisotropy even in a centrosymmetric crystal, as shown in the following.

Consider the symmetry properties of the ζ pseudotensor under coordinate transformations characterized by the transformation matrix T (limiting ourselves to transformations where $\det(T) = \pm 1$). The pseudotensor ζ, when dependent only on ω, transforms like $\zeta'(\omega) = \det(T)T\zeta'(\omega)T^T$. For a centrosymmetric crystal, the constitutive tensors (pseudotensors) should be unchanged by the inversion $T_{inv} = \text{diag}(-1, -1, -1)$, implying that $\zeta(\omega) = \zeta'(\omega) = -T_{inv}\zeta(\omega)T_{inv}^T = -\zeta(\omega) = 0$. However, if ζ (or ξ) is a function of $\mathbf{k}$ as well as ω, then $\zeta(\omega, \mathbf{k}) = \zeta'(\omega, T_{inv}^T \mathbf{k}) = -T_{inv}\zeta(\omega, T_{inv}^T \mathbf{k})T_{inv}^T = -\zeta(\omega, -\mathbf{k})$. Instead of concluding that ζ (and ξ) vanish for a centrosymmetric crystal, we now conclude that $\zeta'(\omega, \mathbf{k}) = -\zeta'(\omega, -\mathbf{k})$. There are similar symmetry relations for the other constitutive matrices for a centrosymmetric crystal: $\epsilon(\omega, \mathbf{k}) = \epsilon(\omega, -\mathbf{k})$, $\xi(w, \mathbf{k}) = -\xi(\omega, -\mathbf{k})$, and $\mu(\omega, \mathbf{k}) = \mu(\omega, -\mathbf{k})$.

Formally, the $\mathbf{k}$ vector breaks the symmetry of the crystal. This can be seen more clearly if we investigate how the pseudotensors ζ and ξ are restricted under spatial reflections. Consider the case when $\mathbf{k} = k_x \hat{x}$. If we perform a reflection across the x-z plane, our transformation matrix is $T_y = \mathrm{diag}(1, -1, 1)$. Under this transformation, for the zy component of ζ we find $\zeta_{zy} = (\omega, \mathbf{k}) = \zeta'_{zy}(\omega, T_y^T \mathbf{k}) = \zeta_{zy}(\omega, \mathbf{k})$ since $T_y^T \mathbf{k} = \mathbf{k}$ for $\mathbf{k} = k_x \hat{x}$. Therefore, ζ_{zy} can be nonvanishing. Now consider ζ_{zx}, which transforms similar to $\zeta_{zx} = (\omega, \mathbf{k}) = \zeta'_{zx}(\omega, T_y^T \mathbf{k}) = -\zeta_{zx}(\omega, \mathbf{k}) = 0$ for a centrosymmetric crystal. So, constitutive parameters that are functions of $\mathbf{k}$ as well as ω are restricted by symmetry in different ways than constitutive parameters that are only functions of ω.

For simplicity, let us consider a periodic two-dimensional metamaterial with centrosymmetric unit cells arranged in a rectangular array in the x-y plane. If we assume that $\mathbf{k}$ lies in the x-y plane, then the nonzero pattern of the constitutive tensors and pseudotensors due to reflection symmetry of the square array in the x and y directions is given in Equation 9.22:

$$\epsilon(\omega,\mathbf{k}) = \begin{pmatrix} \epsilon_{xx} & \epsilon_{xy} & 0 \\ \epsilon_{yx} & \epsilon_{yy} & 0 \\ 0 & 0 & \epsilon_{zz} \end{pmatrix} \qquad \xi(\omega,\mathbf{k}) = \begin{pmatrix} 0 & 0 & \xi_{xy} \\ 0 & 0 & \xi_{yz} \\ \xi_{zx} & \xi_{zy} & 0 \end{pmatrix}$$

$$\zeta(\omega,\mathbf{k}) = \begin{pmatrix} 0 & 0 & \zeta_{xy} \\ 0 & 0 & \zeta_{yz} \\ \zeta_{zx} & \zeta_{zy} & 0 \end{pmatrix} \qquad \mu(\omega,\mathbf{k}) = \begin{pmatrix} \mu_{xx} & \mu_{xy} & 0 \\ \mu_{yx} & \mu_{yy} & 0 \\ 0 & 0 & \mu_{zz} \end{pmatrix}. \qquad (9.22)$$

Naturally, in the limit of $\mathbf{k} = 0$, the $\mathbf{k}$ vector no longer breaks the symmetry of the centrosymmetric crystal, and ζ and ξ as well as the off-diagonal terms in ϵ and μ vanish.

9.4 Examples of Constitutive Parameters of Various Electromagnetic Structures

In this section, we consider applications of the Current Driven Homogenization technique to a variety of optical metamaterials. We will restrict ourselves to centrosymmetric two-dimensional structures, that is, metamaterials with a unit cell possessing an inversion center. Examples of bianisotropy induced by finite spatial dispersion are given. It will also be shown that $\mathbf{k}$-dependent bianisotropy coefficients can give rise to an electromagnetic response strongly reminiscent of magnetic activity. Two structures are considered: an anisotropic plasmonic structure shown in Figure 9.4 is described in Section 9.4.1, and an isotropic structure consisting of a high-index inclusion immersed in a plasmonic shell shown in Figure 9.7 is described in Section 9.4.2.

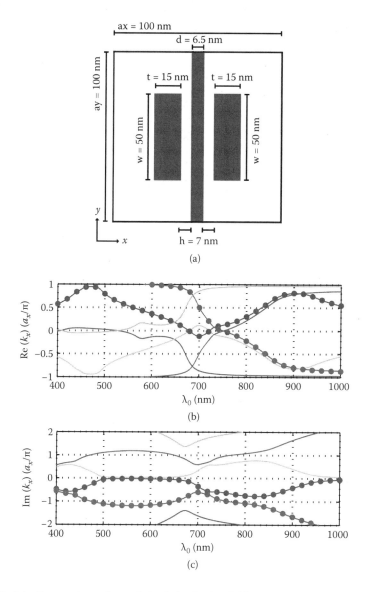

FIGURE 9.4 Propagation of p-polarized (with the field components $\mathbf{H} = H_z(x,y)\hat{\mathbf{z}}, \mathbf{E} = E_x(x,y)\hat{\mathbf{x}} + E_x(x,y)\hat{\mathbf{y}}$) waves through the 2D lattice of SPOF structures. Top: the unit cell (a) of the SPOF (description in the text). Bottom: Real (b) and imaginary (c) parts of k_x versus λ_0 where $\mathbf{k} = k_x\hat{\mathbf{x}}$. Solid lines: dispersion curves $k_x(\omega)$ for a p-polarized wave obtained from an eigenvalue simulation[25] that determines $k_x(\omega)$. Dotted lines: dispersion curves calculated from Equation 9.23 using the CDH procedure, specifically by exciting the SPOF at ω and $\mathbf{k} = k_x(\omega)\hat{\mathbf{x}}$ and using Equations 9.15 and 9.16. (Reprinted from Fietz, C. and Shvets, G., in *Metamaterials: Fundamentals and Applications II*, Noginov, M. A., Zheludev, N. I., Boardman, A. D., and Engheta, N., eds., 739219, SPIE, Bellingham, WA, 2009. With permission from SPIE; Fietz, C. and Shvets, G., *Physica B* 405, 2930–2934, 2010. With permission from Elsevier.)

9.4.1 Example: Constitutive Parameters of an Anisotropic Plasmonic Metamaterial

As the first example, we present the results of applying the CDH approach to extract the effective CPM of a two-dimensional plasmonic metamaterial crystal known as a Strip Pair-One Film or an SPOF.[8,24] A diagram of the SPOF is given in Figure 9.4a; it comprises a square crystal lattice with a thin Au film in the center of the unit cell and two Au strips on both sides of the film. The permittivity of the Au is $\epsilon = 1 - \omega_p^2 / (\omega(\omega - i\Gamma))$ with $\omega_p = 1.32 \cdot 10^{16}/s$ and $\Gamma = 1.2 \cdot 10^{14}/s$. The rest of the SPOF cell encompassing the plasmonic inclusions consists of a dielectric with permittivity $\epsilon = 1.56^2$. Only p-polarized modes (with the field components $\mathbf{H} = H_z(x,y)\hat{\mathbf{z}}$, $\mathbf{E} = E_x(x,y)\hat{\mathbf{x}} + E_y(x,y)\hat{\mathbf{y}}$) are studied in this section. The SPOF structure is clearly anisotropic: it is intuitively clear that $\epsilon_{yy} \neq \epsilon_{xx}$. Mathematically, the point symmetry group of the SPOF structure is C_{2v}.

Recently it has been shown[24] that the SPOF structure exhibits negative index propagation in the x-direction corresponding to $\epsilon_{yy} < 0$ and $\mu_{zz} < 0$. The dispersion curves $k_x(\omega)$ corresponding to wave propagation in the x-direction are obtained using the earlier-described approach[25] and plotted in Figure 9.4b,c as the solid lines. Only one mode (blue line) is ever radiating, and only for certain frequency bands. All other modes are evanescent. Negative index propagation in the close proximity of $\lambda_0 = 700$ nm is apparent from Figure 9.4b. It would be natural to expect that such a structure could be characterized by just three frequency-dependent parameters: ϵ_{xx}, ϵ_{yy}, μ_{zz}. As it turns out, that is not the case because the magnetic permeability μ_{zz} depends on the propagation direction. In the following we demonstrate that the most natural way of describing this anisotropy is by introducing the $\mathbf{k}$-dependent bianisotropy coefficients ζ_{zx} and ζ_{zy}.

As the first demonstration of the power of the CDH approaches, we have extracted the full set of current-driven constitutive parameters as functions of $\omega, \mathbf{k} = k_x(\omega)\hat{x}$ for the eigenmodes of the crystal following the averaging and homogenization techniques described in Sections 9.3.1 and 9.3.2. The unit cell is driven slightly off the dispersion curve so as to prevent the matrix M in Equations 9.13 and 9.14 from being singular. Thus obtained entries of the CPM are used to calculate the real and imaginary parts of the complex wavenumber for each eigenmode according to the dispersion relation of the SPOF (dotted lines), which for a p-polarized wave propagating in the x-direction is given by

$$\left(\frac{k_x - \omega\zeta_{zy}/c}{\mu_{zz}} \right)\left(\frac{k_x + \omega\xi_{zy}/c}{\epsilon_{zz}} \right) - \frac{\omega^2}{c^2} = 0. \tag{9.23}$$

Note from Equation 9.23 that only four entries of the CPM (ϵ_{yy}, ζ_{zy}, μ_{zz}, and ξ_{yz}) affect the mode's propagation in the x-direction. Figure 9.4 indicates that the current-driven constitutive parameters accurately predict correct dispersion for two of the modes, one of which (blue line) is radiative for some frequencies and evanescent for others, while the other one (green line) is always evanescent. The constitutive parameters ϵ_{yy} and ζ_{zy} for the SPOF extracted along the dispersion curve $k_x(\omega)$ of the "radiative" (blue line) eigenmode are shown in Figures 9.5a,b, respectively. The CDH procedure also yields $\xi_{yz} = 0$ and $\mu_{zz} = 1$ for all frequencies. We conjecture that this simplification is due to two

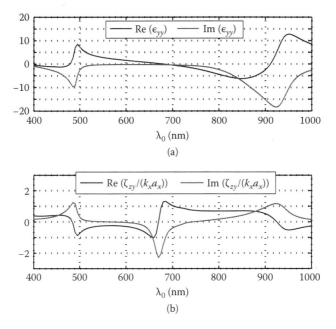

FIGURE 9.5 Constitutive parameters of the 2D lattice of SPOF structures shown in Figure 9.4a: ϵ_{yy} (a) and $\zeta_{zy}/(k_x a_x)$ (b) computed using Equations 9.15 and 9.16 along the "radiative" (least damped) dispersion curve $\mathbf{k} = k_x(\omega)\hat{\mathbf{x}}$ shown in Figure 9.4. Not shown: $\mu = 1$, $\xi = 0$ for all wavelengths. (Reprinted from Fietz, C. and Shvets, G., in *Metamaterials: Fundamentals and Applications II*, Noginov, M. A., Zheludev, N. I., Boardman, A. D., and Engheta, N., eds., 739219, SPIE, Bellingham, WA, 2009. With permission from SPIE; Fietz, C. and Shvets, G., *Physica B* 405, 2930–2934, 2010. With permission from Elsevier.)

factors: that all inclusions of the SPOF possess only a local electric response, and that the SPOF is a centrosymmetric structure.

Extracted $\mu_{zz} = 1$ is in apparent contradiction with the results of the S-parameter retrieval[14] of $\mu_{zz} \neq 1$ for the SPOF [24] or similar negative index structures. In fact, it directly follows from Equation 9.14 and Equation 9.23 that the only relevant quantity for this specific propagation direction and wave polarization is $(k_x - \omega\zeta_{zy}/c)/\mu_{zz}$. Therefore, any set of effective parameters $(\zeta_{zy}^{eff}, \mu_{zz}^{eff})$ satisfying $(k_x - \omega\zeta_{zy}^{eff}/c)/\mu_{zz}^{eff} = (k_x - \omega\zeta_{zy}/c)/\mu_{zz}$, including $\zeta_{zy}^{eff} = 0$ (assumed in S-parameter retrieval) and $\mu_{zz}^{eff} = (1 - \omega\zeta_{zy}/ck_x)^{-1}$, are valid constitutive parameters. Thus defined μ_{zz}^{eff} is indeed very close to the value of μ extracted using the standard S-parameter retrieval.[21] The comparison between the two is shown in Figure 9.6. Figure 9.6 shows a comparison of the extracted ϵ_{yy} and effective μ_{zz}^{eff} versus the ϵ_{yy}^S and μ_{yy}^S determined by S-parameter retrieval.[14] A very good agreement is observed between the CDH-extracted ϵ_{yy} and the S-parameter-retrieved ϵ_{yy}^S. The agreement between μ_{zz}^{eff} given by Equation 9.24 and the S-parameter-retrieved μ_{yy}^S is also very good.

$$\mu_{zz}^{eff} = \frac{1}{1 - \frac{\zeta_{zy}}{ck_x}} \tag{9.24}$$

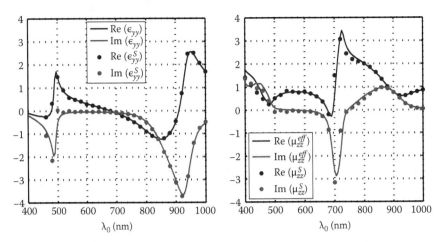

FIGURE 9.6 Left side: real and imaginary parts of ϵ_{yy} (solid lines) extracted from the SPOF using the CDH procedure and ϵ_{yy}^S (dotted lines) calculated using the standard S-parameter retrieval.[14] Right side: real and imaginary parts of μ_{zz}^{eff} (solid lines) calculated from Equation 9.24) using ζ_{zy} and μ_{zz} extracted from the SPOF using the CDH procedure and μ_{zz}^S (dotted lines) calculated using the S-parameter retrieval. Note that between $\lambda_0 = 680$ nm and $\lambda_0 = 695$ nm, both ϵ_{yy} and μ_{zz}^{eff} are negative for this mode, resulting in a negative refractive index. (Reprinted from Fietz, C. and Shvets, G., in *Metamaterials: Fundamentals and Applications II*, Noginov, M. A., Zheludev, N. I., Boardman, A. D., and Engheta, N., eds., 739219, SPIE, Bellingham, WA, 2009. With permission from SPIE.)

Note from Figure 9.6 that the magnetic permeability of the SPOF for the x-direction propagation turns negative around $\lambda_0 = 700$ nm. That is not the case for y-direction propagation: μ_{zz}^{eff} is barely perturbed from unity. The physical reason for this lies in the dependence of ζ_{zx} on k_y. Because the SPOF structure is very anisotropic, the ratios of ζ_{zy}/k_x and ζ_{zyx}/k_y are very different. Therefore, the effective magnetic permeabilities for $\mathbf{k} = e_x k_x$ and $\mathbf{k} = e_y k_y$ are very different.

9.4.2 Example: Constitutive Parameters of a Negative Index Metamaterial Based on a High-Index Dielectric

We now use the CDH procedure to investigate the optical properties of another 2D metamaterial shown in Figure 9.7. It was originally introduced[27] to demonstrate the possibility of effective magnetic response in metamaterials with high-index dielectric inclusions. The metamaterial comprises a 2D array of squares with length $a/2$, where a is the lattice constant. The squares are composed of a high-permittivity material with $\epsilon = 200 - 5i$. They are surrounded by a hypothetical material with the dielectric permittivity $\epsilon = -1$. The high-index inclusion is responsible for the magnetic response while the negative-permittivity shell imparts the electric response. This structure corresponds to a higher spatial symmetry group (C_{4v}) than the SPOF (C_{2v}). Rather than test the extracted constitutive parameters as we did for the SPOF, we investigate the dependence of constitutive parameters on $\mathbf{k}$ near the Γ-point based on the C_{4v} point symmetry of this

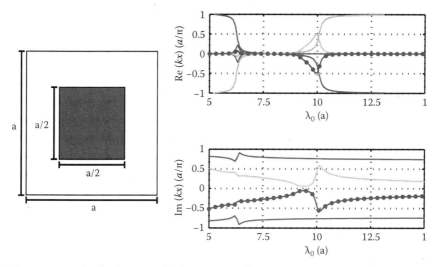

FIGURE 9.7 Left side: the unit cell of a negative index metamaterial[27] with high-index inclusions. The inner square has permittivity $\epsilon = 200 - 5i$ and the surrounding area has permittivity $\epsilon = -1$. Right side: real (upper plot) and imaginary (lower plot) parts of k_x versus λ_0, where $\mathbf{k} = k_x \hat{\mathbf{x}}$. Solid lines are dispersion curves for a p-polarized wave solved using the eigenvalue simulation[25] that computes $k_x(\omega)$). Dotted lines are dispersion curves calculated from Equation 9.27) using the constitutive parameters given by Equations 9.25 and 9.26, which are calculated near the Γ point $k = 0$. (Reprinted from Fietz, C. and Shvets, G., in *Metamaterials: Fundamentals and Applications II*, Noginov, M. A., Zheludev, N. I., Boardman, A. D., and Engheta, N., eds., 739219, SPIE, Bellingham, WA, 2009. With permission from SPIE.)

structure. In addition, it will be shown that, to cubic order in $|\mathbf{k}|$, spatial dispersion of the ζ-tensor can be eliminated by introducing a magnetic permeability $\mu_{zz}^{eff}(\omega)$. Because of the high degree of spatial symmetry of the unit cell, the resulting magnetic permeability $\mu_{zz}^{eff}(\omega)$ is direction independent. This should be contrasted with the lower-symmetry SPOF structure considered in Section 9.4.1, which possessed a direction-dependent magnetic permeability.

Propagation bands of this metamaterial are plotted in Figure 9.7 (solid lines). The blue line corresponds to the optical mode with the least attenuation (smallest imaginary part of the propagation wavenumber). Therefore, we focus on that mode in what follows. Negative index propagation in the vicinity of $\lambda_0 = 10a$ is observed. Next, we extract the constitutive parameters of this subwavelength metamaterial in the vicinity of the Γ-point and reproduce these dispersion curves. Using symmetry arguments similar to those in Section 9.4.1, one can demonstrate that the components ϵ_{xx} and ϵ_{yy} must be even functions of k_x and k_y but components ϵ_{xy} and ϵ_{yx} must be odd functions of k_x and k_y. Specifically, after examining the extracted values of ϵ for small $\mathbf{k}$, it is found that

$$\epsilon_{xx} = \epsilon_0 + \epsilon_1 k_x^2 + \epsilon_2 k_x^2 + \ldots \qquad \epsilon_{xy} = \epsilon_0' k_x k_y + \epsilon_1' k_x^3 k_y + \epsilon_2' k_x k_y^3 + \ldots$$

$$\epsilon_{yx} = \epsilon_0' k_x k_y + \epsilon_1' k_y^3 k_x + \epsilon_2' k_y k_x^3 + \ldots \qquad \epsilon_{yy} = \epsilon_0 + \epsilon_1 k_y^2 + \epsilon_2 k_y^2 + \ldots \qquad (9.25)$$

It is instructive to compare Equation 9.25 to the electrostatically derived Equation 9.11. Because the square array is symmetric under a rotation of 90°, this symmetry manifests as the symmetry relation $\epsilon(\epsilon,\mathbf{k}) = T_{90°}\epsilon(\omega, T_{90°}^T\mathbf{k})T_{90°}^T$, where $T_{90°}$ is a transformation matrix describing a 90° rotation of the x-y plane. One result of this symmetry is that there is a common set of expansion coefficients ($\epsilon_{0,1,2}$) for ϵ_{xx} and ϵ_{yy}, and another common set of expansion coefficients ($\epsilon'_{0,1,2}$) for ϵ_{xy} and ϵ_{yx}. μ has similar symmetry requirements, but it is found that $\mu = 1$ for all w and $\mathbf{k}$.

Likewise, ζ_{zx} must be an even function of k_x but an odd function of k_y. Similarly, ζ_{zy} must be an odd function of k_x and an even function of k_y. After examining extracted values of ζ for small $\mathbf{k}$, it is found that

$$\zeta_{zx} = \zeta_0 k_y + \zeta_1 k_y k_x^2 + \zeta_2 k_y^3 + \dots \qquad \zeta_{zy} = -\zeta_0 k_x - \zeta_1 k_x k_y^2 - \zeta_2 k_x^3 + \dots. \qquad (9.26)$$

Applying to the ζ and ξ tensors the same symmetry arguments that were applied to the ϵ and μ tensors, we can predict the dependence of the former tensors on k in the vicinity of the Γ-point. Owing to the rotation symmetry of the C_{4v} symmetry group, the following symmetry relation is obtained: $\zeta(\omega,\mathbf{k}) = T_{90°}\zeta(\omega, T_{90°}^T\mathbf{k})T_{90°}^T$. The consequence of this symmetry relation is a common set of expansion coefficients ($\zeta_0, \zeta_1, \zeta_2$) for ζ_{zx} and ζ_{zy} given by Equation 9.26. Naturally, ξ should be restricted in a similar way, but we find the extracted values for ξ to identically vanish.

Using the expansions for the ϵ and ζ tensors given by Equations 9.25 and 9.26, one can derive a dispersion relation for the square array when k_x and k_y are small. After lengthy but straightforward algebra, one obtains the following dispersion relation:

$$\epsilon_0 \mu_{zz}^{eff} \frac{\omega^2}{c^2} = (k_x^2 + k_y^2) + \bar{\alpha}(k_x^2 + k_y^2)^2 + \bar{\beta}(k_x^2 + k_y^2)^2 + \dots \qquad (9.27)$$

where the effective magnetic permeability μ_{eff} and the high-order expansion coefficients $\bar{\alpha}$ and $\bar{\beta}$ are defined as

$$\mu_{eff} = \frac{1}{1 + \omega\zeta_0/c}, \quad \bar{\alpha} = \frac{1}{2}\left(\frac{\omega/c(\zeta_2 + \zeta_1)}{1 + \omega\zeta_0/c} - \frac{\epsilon_2 + \epsilon_1 - \epsilon'_0}{\epsilon_0} \right), \quad \bar{\beta} = \frac{1}{2}\left(\frac{\omega/c(\zeta_2 + \zeta_1)}{1 + \omega\zeta_0/c} - \frac{\epsilon_2 + \epsilon_1 - \epsilon'_0}{\epsilon_0} \right).$$

$$(9.28)$$

Note that the dispersion relation given by Equation 9.27 is predicated on the first (quadratic in $\mathbf{k}$) term in the right-hand side of the Equation 9.27 being larger than the remaining two (quadric in $\mathbf{k}$) terms. Under that assumption, it is apparent that for small $\mathbf{k}$'s the dispersion relation is isotropic. For small wavenumbers, wave propagation is determined by $\epsilon_0(\omega) \equiv \epsilon(\mathbf{k} = 0, \omega)$ and $\mu_{eff}(\omega) \equiv \mu_{zz}^{eff}(\mathbf{k} = 0, \omega)$. For larger $\mathbf{k}$'s, the next two terms on the right-hand side of Equation 9.27 become comparable to the leading one to contribute. The first of the two quadrics in $\mathbf{k}$ terms is isotropic in $\mathbf{k}$, while the second is anisotropic. This anisotropy is brought about by spatial dispersion. Direct analogy can be drawn with the electrostatic case described in Section 9.2. Specifically, if one neglects the higher-order spatial dispersion coefficients ζ_2 and ζ_1, then the highly anisotropic condition of $\alpha = \beta = 0$ and $\gamma \neq 0$ (α, β, γ defined in Equation 9.11) would correspond to $\bar{\alpha} = 0$ and $\bar{\beta} \neq 0$. Therefore, the last (anisotropic) term would dominate over the isotropic one, and wave propagation would become strongly anisotropic.

Next, we investigate the effect of spatial dispersion on the effective dielectric permittivity and magnetic permeability, and address the phenomenon of "anti-resonances."[28] In Figure 9.8, we plot the extracted ϵ_{yy} and effective μ_{zz}^{eff} (from Equation 9.24) calculated along the dispersion curve for the "radiative" (least damped) mode in the square array corresponding to the blue dispersion line in Figure 9.7. We also plot ϵ_0 and μ_{eff}, defined in Equations 9.25 and 9.28, respectively. Those are calculated from the constitutive parameters extracted near the Γ point. Therefore, μ_{eff} and ϵ_0 can be thought of as the effective parameters extracted at $\mathbf{k} = 0$.

The earlier discovered phenomenon[28] of antiresonance of ϵ_{yy} is observed in Figure 9.8 (left). Antiresonance of the dielectric permittivity extracted along the dispersion curve manifests itself in two ways: (a) as an unphysical positive imaginary part of the dielectric permittivity $Im(\epsilon_{yy})$ (suggesting gain instead of ohmic loss), and (b) as the unphysical dependence of $Re(\epsilon_{yy})$ on the wave frequency. Both manifestations are seen in Figure 9.8 (left) in the vicinity of the magnetic resonance at $\lambda_0 = 10a$. We see that the antiresonance in ϵ_{yy} does not appear in e_0, indicating the

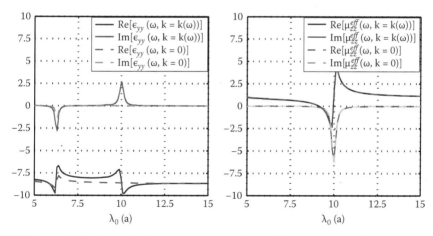

FIGURE 9.8 Left side: real and imaginary parts of ϵ_{yy} (solid lines) extracted from the square array along the dispersion curve of the "radiative" mode (blue line Figure 9.7), and ϵ_0 (dashed line) defined in Equation 9.25) and calculated from constitutive parameters extracted near the Γ point. Note that the antiresonance seen in ϵ_{yy} is not seen in ϵ_0. Right side: real and imaginary parts of μ_{eff} (solid line) extracted from the square array along the dispersion curve of the "radiative" mode using Equation 9.24, and μ_{eff} (dashed line) defined in Equation 9.27 and calculated from constitutive parameters extracted near the Γ point. (Reprinted from Fietz, C. and Shvets, G., in *Metamaterials: Fundamentals and Applications II*, Noginov, M. A., Zheludev, N. I., Boardman, A. D., and Engheta, N., eds., 739219, SPIE, Bellingham, WA, 2009. With permission from SPIE.)

antiresonance is indeed due to spatial dispersion as has been previously suggested by Koschny et al.[28] Note that both ϵ_0 and $\epsilon_{yy}(k_x(\omega),\omega)$ exhibit a weak electric resonance at $\lambda \approx 6a$.

We also observe that μ_{zz}^{eff} and μ_{eff} agree with each other quite well, suggesting that spatial dispersion does not strongly affect magnetic resonance. This holds true even in the proximity of the electric resonance. It can be further observed from Equation 9.26 that, by defining $\mu_{eff} = (1 + \omega\zeta_0/c)^{-1}$, bianisotropy of a metamaterial with C_{4v} symmetry can be eliminated to cubic order in $|\mathbf{k}|$. Moreover, unlike in the lower-symmetry (C_{2v}) SPOF structure, the magnetic permeability μ_{eff} is independent of the propagation direction.

9.5 Modeling Antennas Embedded inside Metamaterial Slabs

One of the exciting applications of metamaterials is the ability to construct shells and cloaks from them. Various emitting devices such as quantum dots for optical emission and antennas for RF applications can then be embedded inside such metamaterials-based structures. It would be highly desirable to be able to describe the properties of such embedding shells in terms of the effective constitutive parameters. Moreover, because emitters (such as, e.g., directional antennas) can have complicated shapes, it is important to be able to extract the full CPM for arbitrary $\mathbf{k}$ and ω because antennas can contain sharp features contributing to a broad spatial spectrum of emitted waves.

An important capability of the CDH approach is that it can calculate the constitutive parameters of a metamaterial for any ω and $\mathbf{k}$ that do not obey the dispersion relation inside the metamaterial crystal. This has important implications for various applications, such as metamaterial-embedded antennas and quantum dots. Because an antenna can have an arbitrary shape and position, detailed knowledge of the CPM on and off the dispersion curve is necessary. In the following we demonstrate a close agreement between radiation patterns of a monochromatic flat (infinitely extended in the y-z plane) directional antenna embedded inside (1) an SPOF array comprised of 5 layers stacked in the x-direction, and (2) a finite homogeneous 500-nm-thick slab with the CPM corresponding to the SPOF metamaterial. Owing to the linear response of the metamaterial, it is sufficient to investigate radiation patterns (transmission from the left and right slab boundaries) for the electric currents in the form of $\mathbf{J_k} = J_0 \hat{y} e^{i(\omega t - \mathbf{k} \cdot \mathbf{x})}$. For example, a flat antenna with a finite size in the x-direction can be represented as a superposition of such currents summed over a spectrum of $\mathbf{k}$.

The results of the simulations for the cases (1) and (2) carried out for a wide spectrum of $\mathbf{k}$'s are shown in Figure 9.9 as solid lines and dots, respectively. All simulations were performed for three antenna frequencies corresponding to $\lambda_0 = 600$ nm, $\lambda_0 = 800$ nm, and $\lambda_0 = 1000$ nm. The wavenumbers $\mathbf{k}$ are varied from $\mathbf{k} = 0$ to $k = \pi/a\hat{x}$ (right half of Figure 9.9) and from $\mathbf{k} = 0$ to $\mathbf{k} = \pi/a\hat{x} + \pi/a\hat{y}$ (left half of Figure 9.9). The finite value of k_y enables steering the beam at an angle with respect to the slab's boundary. For the homogeneous medium, we assumed Maxwell boundary conditions (continuity of

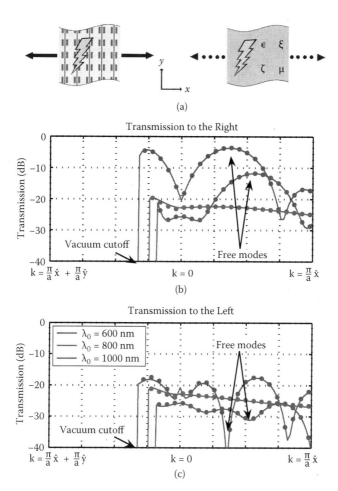

FIGURE 9.9 Radiation by the harmonic current $\mathbf{J_k} = J_0 \hat{y} e^{i(\omega t - \mathbf{k} \cdot \mathbf{x})}$ emulating a directional antenna embedded inside a metamaterial slab. Radiation emerging from the right (a) and left (b) boundaries of the slab is plotted for the current embedded inside a 5-layer thick SPOF metamaterial (solid lines) and an $L = 500$ nm homogeneous slab with the CPM of the SPOF (dotted lines). (Reprinted from Fietz, C. and Shvets, G., *Physica B* 405, 2930–2934, 2010. With permission from Elsevier.)

tangential $\mathbf{E}$ and $\mathbf{H}$ fields). The radiation flux escaping through the right and left slabs' boundaries are shown in Figures 9.9a and 9.9b, respectively.

Strong antenna directionality is observed for $\lambda_0 = 600$ nm and $\lambda_0 = 1000$ nm antennas, which can couple to the low-loss propagating modes: the peaks of the forward flux are matched by the dips of the backward flux. Such phase matching is achieved when $\mathbf{k}$ matches $\mathrm{Re}(\mathbf{k}(\omega))$ on the dispersion curve in Figure 9.4b. The directionality of the $\lambda_0 = 800$ nm antenna is poor because at this frequency all free modes in the SPOF are evanescent according to Figure 9.4c. Very good agreement between the simulations of the cases

(1) and (2) is observed, confirming the accuracy of the homogenization procedure. The small observed discrepancies are believed to be mostly due to the imperfect boundary conditions.

9.6 Conclusions

Spatial dispersion plays an important role in optical metamaterials. While its origin is easy to appreciate—it arises when the unit cell of a metamaterial is not much smaller than the wavelength of light inside the metamaterial—its implications are nontrivial and far-reaching. We have used this tutorial to illustrate how spatial dispersion gives rise to several phenomena that are not usually encountered in naturally occurring media. Those phenomena include bianisotropy (strong coupling between the electric field and magnetic moment of a metamaterial) and optical magnetism (considerable deviation of the effective magnetic permeability of a metamaterial from unity). These phenomena give rise to extraordinary physical effects such as, for example, negative index propagation.

We have demonstrated that spatial dispersion can arise even in very simple nano-structured metamaterials where light retardation is not important, and the system can be viewed as electrostatic. The emergence of strong spatial dispersion in such electro-static systems can be traced back to the presence of high-order (quadrupolar, ocupolar, etc.) electrostatic resonances as has been suggested[17] for the natural materials. A rigorous procedure for calculating all components of the dielectric permittivity tensor as a function of the wavenumber and frequency has been illustrated using a simple two-dimensional plasmonic metamaterial consisting of almost-touching plasmonic rods.

For structures with a nonnegligible light retardation, a fully rigorous procedure of Current Driven Homogenization has been described. The CDH procedure enables extracting all 36 linear electromagnetic constitutive parameters (i.e., its entire CPM, including bianisotropy tensors) for any periodic metamaterial crystal. CDH has been applied to two simple two-dimensional plasmonic metamaterials, which support negative index waves. It revealed that even metamaterials with a centrosymmetric unit cell possess finite bianisotropy due to spatial dispersion. Spatial dispersion-induced bianisotropy manifests itself as optical magnetism, responsible for some of the most exciting phenomena in metamaterials such as negative refraction. The CDH procedure has been demonstrated to be useful for practical applications, such as predicting radiation by arbitrary directional antennas embedded inside a metamaterial cloak.

References

1. Smith, D. R., Padilla, W. J., Vier, D. C., Nemat-Nasser, S. C., and Schultz, S. *Phys. Rev. Lett.* 84, 4184, 2000.
2. Pendry, J. B., Holden, A. J., Robbins, D. J., and Stewart, W. J. *IEEE Trans. Microwave Theory Tech.* 47, 2075, 1999.
3. Padilla, W. J., Aronsson, M. T., Highstrete, C., Lee, M., Taylor, A. J., and Averitt, R. D. *Phys. Rev. B* 75, 041102R, 2007.

4. Landau, L. D. and Lifshitz, E. M., eds., *Electrodynamics of Continuous Media*, Pergamon, Oxford, UK, 1960.

5. Shvets, G. and Urzhumov, Y., Engineering the electromagnetic properties of periodic nanostructures using electrostatic resonances, *Phys. Rev. Lett.* 93, 243902, 2004.

6. Shvets, G. and Urzhumov, Y., Electric and magnetic properties of sub-wavelength plasmonic crystals, *J. Opt. A: Pure Appl. Opt.* 7, S23–S31, 2005.

7. Silveirinha, M. G., Metamaterial homogenization approach with application to the characterization of microstructured composites with negative parameters, *Phys. Rev. B* 75, 115104, 2007.

8. Urzhumov, Y. A. and Shvets, G., Optical magnetism and negative refraction in plasmonic metamaterials, *Solid State Comm.* 146, 208, 2008.

9. Bergman, D. and Stroud, D. *Solid State Phys.* 46, 147, 1992.

10. Stockman, M. I., Faleev, S. V., and Bergman, D. J. *Phys. Rev. Lett.* 87, 167401, 2001.

11. Shvets, G. and Urzhumov, Y. A., Electric and magnetic properties of sub-wavelength plasmonic crystals, *J. Opt. A: Pure Appl. Opt.* 7, S23–S31, 2004.

12. Pendry, J. B., Holden, A. J., Robbins, D. J., and Stewart, W. J., Magnetism from conductors and enhanced nonlinear phenomena, *IEEE Trans. Microwave Theory Tech.*, 1999.

13. Marques, R., Medina, F., and Rafii-El-Idrissi, R., Role of bianisotropy in negative permeability and left-handed metamaterials, *Rhys. Rev. B* 65, 144440, 2002.

14. Smith, D. R., Schultz, S., Markos, P., and Soukoulis, C. M., Determination of effective permittivity and permeability of metamaterials from reflection and transmission coefficients, *Phys. Rev. B* 65, 195104, 2002.

15. Smith, D. R. and Pendry, J. B., Homogenization of metamaterials by field averaging, 2006.

16. Lui, R., Cui, T. J., Huang, D., Zhao, B., and Smith, D. R., Description and explanation of electromagnetic behaviors in artificial metamaterials based on effective medium theory, *Phys. Rev. E* 76, 026606, 2007.

17. Agranovich, V. M. and Ginzburg, V. L., *Crystal Optics with Spatial Dispersion, and Excitons*, No. 42 in Springer Series in Solid-State Sciences, Springer-Verlag, Berlin, second corrected and updated ed., 1984.

18. Kong, J. A., *Electromagnetic Wave Theory*, John Wiley & Sons, New York, 1986.

19. Jackson, J. D., *Classical Electrodynamics*, John Wiley & Sons, New York.

20. Moulin, F., Magnetic monopoles and Lorentz force, *Il Nuovo Cimento B* 116B, 869–877, 2001.

21. Fietz, C. and Shvets, G., Metamaterial homogenization: Extraction of effective constitutive parameters, in *Metamaterials: Fundamentals and Applications II*, Noginov, M. A., Zheludev, N. I., Boardman, A. D., and Engheta, N., eds., 739219, SPIE, Bellingham, WA, 2009.

22. Smith, D. R., Vier, D. C., Kroll, N., and Schultz, S., Direct calculation of permeability and permittivity for a left-handed metamaterial, *Appl. Phys. Lett.* 77, 2246–2248, 2000.

23. Simovski, C. R., Bloch material parameters of magneto-dielectric metamaterials and the concept of bloch lattices, *Metamaterials* 1, 62–80, 2007.

24. Lomakin, V., Fainman, Y., Urzhumov, Y., and Shvets, G., Doubly negative metamaterials in the near infrared and visible regimes based on thin film nanocomposites, *Opt. Express* 14, 11164–11177, 2006.

25. Davanco, M., Urzhumov, Y., and Shvets, G., The complex bloch bands of a 2d plasmonic crystal displaying isotropic negative refraction, opt. Express 15, 1981, (2007). 2007.
26. Fietz, C. and Shvets, G., Current-driven metamaterial homogenization, *Physica B* 405, 2930–2934, 2010.
27. Felbacq, D. and Bouchitte, G., Left-handed media and homogenization of photonic crystals, *Opt. Lett.* 30, 1189–1191, 2005.
28. Koschny, T., Markos, P., Smith, D. R., and Soukoulis, C. M., Resonant and antiresonant frequency dependence of the effective parameters of metamaterials, *Phys. Rev. E* 68, 065602(R), 2003.

Index